本书获中国科学院重点部署项目"地质学在中国的本土化研究"
（项目编号：KZZD-EW-TZ-01）资助

葛利普
与中国古生物学

孙承晟　编著

Amadeus William Grabau and
Paleontology in China

科学出版社

北　京

内 容 简 介

葛利普是 20 世纪世界著名的古生物学家。1920 年他应丁文江之邀来华，担任地质调查所古生物室主任和北京大学地质学系教授。葛利普以其卓越的才华和过人的勤奋，不仅撰写了大量关于中国古生物学和地层学的著作，而且培养了一大批年轻的古生物学家，使中国古生物学从无到有，并为世所瞩目，堪称"中国古生物学之父"。

本书共分三部分：上编为中外学者不同时期为葛利普所撰写的传记或纪念性文字，下编为葛利普撰写的纪念中外学者的文章及相关科普作品，附编则是作者近年来发表的关于葛利普的三篇学术论文。本书勾勒出一个丰富、立体的葛利普形象，并呈现了民国时期活跃的中西方科学交流和学术生态。

本书可供科学技术史、地质学、古生物学等领域的学者和师生阅读参考。

图书在版编目（CIP）数据

葛利普与中国古生物学 / 孙承晟编著. -- 北京 ：科学出版社，2024.11. -- ISBN 978-7-03-079441-3

Ⅰ. Q91-53

中国国家版本馆 CIP 数据核字第 2024130XK7 号

责任编辑：邹　聪 / 责任校对：宁辉彩
责任印制：师艳茹 / 封面设计：有道文化

科学出版社出版
北京东黄城根北街 16 号
邮政编码：100717
http://www.sciencep.com

北京九州迅驰传媒文化有限公司印刷
科学出版社发行　各地新华书店经销

*

2024 年 11 月第 一 版　　开本：720×1000　1/16
2024 年 11 月第一次印刷　印张：26 1/2
字数：450 000

定价：198.00 元

（如有印装质量问题，我社负责调换）

序

　　明清之际和晚清时期，西方科学两度较大规模传入中国，而近代科学在中国的真正发展则始于20世纪初。从科学史的角度看，中国本土科学家的成长、科研机构的建立，以及科学精神的养成，乃是近代科学在中国得以扎根的主要因素。民国初年，在逐渐形成的各门学科中，地质学无疑是发展得最早和最快的学科之一。1913年中国地质调查所和地质研究所成立之初，面临政治动荡、战争频仍、经费紧缺等诸多困难，但中国的地质学仍得以迅速发展。此后，地方地质调查所和高校地质系相继建立，年轻地质学家不断成长，大量研究成果涌现，20世纪上半叶中国在古生物学、地质学等方面均取得了举世瞩目的成就。1948年，中央研究院首批81位院士中，地质学家即占据6席，可见一斑。

　　地质学在中国何以发展得如此迅速？这得益于章鸿钊、丁文江、翁文灏、李四光等学者的领导有方，也离不开当时一些著名西方学者，如安特生、葛利普、德日进、步达生、巴尔博等的鼎力合作。可以说，20世纪上半叶中国地质学的发展与国际合作密切相关。以丁文江为代表的中国地质学领导者，通过国际合作，培养本土研究人才，产出重要的研究成果，同时为外国学者提供友好的研究环境和条件，反映了他们长远的战略眼光。

　　关于中国地质学的早期历史，学界已取得不少重要的研究成果。但对于中外地质学者的合作，中国地质学如何实现本土化等问题，因原始材料发掘不足和研究视角的单一，还有很大的研究空间。2012年，在时任中国科学院副院长丁仲礼院士的鼓励和支持下，"地质学在中国的本土化研究"作为中国科学院重点部署项目得以立项，由我来主持项目的研究工作。我邀请了中国科学院自然科学史研究所的张九辰研究员、孙承晟副研究员作为核心成员。我主要负责跨国的竞争、合作与交流，研究欧美地质学家与中国地质学本土化的关系。张九辰从多元体制角度讨论民国时期地质学的本土化进程，着力刻画科学传播的地方印记。鉴于地质学和古生物学的紧密联系，以及葛利普在古生物学发展过程中的重要地位，加之孙承晟有 3 年在北京大学的求学经历，我自然想到他是研究葛利普和中国古生物学这一课题的最佳人选。同时，我也让一些硕士、博士研究生和博士后参与课题，从事晚清地质学的传入、晚清民国地质学教科书、民国地质调查所与地质事业史料汇编、章鸿钊著述整理、美国中亚考察团，以及外国在华地质学家（如维理士、德日进、巴尔博）的研究。在项目进行过程中，课题组成员系统搜集并整理了中国、瑞典、美国、法国等相关机构所藏的丰富档案资料，对相关人物、机构、事件展开了深入研究，完成书稿 4 部，发表重要研究论文 20 余篇，并依托此项目培养了 3 名硕士、3 名博士和 1 名博士后。在 2017 年的结项评审会上，成果获得专家组的高度肯定。项目成果中，杨丽娟的博士论文《地质学在中国的传播与发展——以地质学教科书为中心（1853～1937）》，2022 年已由浙江古籍出版社率先出版。很高兴看到孙承晟编著的《葛利普与中国古生物学》将在科学出版社付梓，其他项目成果则将由北京生活·读书·新知三联书店陆续推出。

葛利普是 20 世纪上半叶享誉世界的古生物学家，1920 年应丁文江之邀来华，担任地质调查所古生物室主任和北京大学地质学系教授。他不仅撰写了大量有关中国古生物学的著作，如《中国地质史》（*Stratigraphy of China*）、《蒙古之二叠纪》（*The Permian of Mongolia*）、多卷本《脉动理论下的古生代地层》（*Palaeozoic Formations in the Light of the Pulsation Theory*）、《年代的节律：从脉动理论和极控理论看地球的历史》（*The Rhythm of the Ages: Earth History in the Light of the Pulsation and Polar Control Theories*）等；1922 年协助丁文江创办的《中国古生物志》（*Palaeontologia Sinica*），成为当时世界著名的学术期刊。葛利普还培养了一大批年轻的古生物学家，如孙云铸、赵亚曾、杨钟健、张席禔、尹赞勋、斯行健、黄汲清等，他们均成为 20 世纪中国地质学和古生物学研究的中坚力量。中国古生物学从无到有，并为世瞩目，很大程度上是葛利普的功劳，他当之无愧是"中国古生物学之父"。1943 年，葛利普不幸被日军禁于北京东交民巷的英国大使馆，抗战胜利后获释，但不久即于 1946 年 3 月20 日因病去世。遵照他的遗愿，他的骨灰被葬于北京大学地质馆。1982 年，值中国地质学会成立 60 年之际，葛利普之墓被迁至北京大学（燕园）。

孙承晟 2002 年随我攻读博士学位，致力于明清之际西学在中国的传播的研究，发表了多篇有创见的论文，并在此基础上完成了《观念的交织：明清之际西方自然哲学在中国的传播》一书，深得国内外学术界的广泛赞誉。他加入"地质学在中国的本土化研究"课题之后，通过发掘国内外档案文献，就葛利普的地质学理论、在北京大学的教学活动，以及对中国学术社团的建设，进行了深入研究，发表了系列学术论文，受到学界的好评。同时，他通过国内外期刊报纸，广泛搜求中外学者为葛利普写的传记或纪念性文字，以及葛利普为中外学者撰写的纪念性文章或相关科普文章。现在，他将这两个有机部分汇为一编，并附有自己的专论，很好地反映了葛利普对中国地质学所做出的不可磨灭的贡献，以及他作为一名著名学者在其他领域的影响。通过这些文章，读者将能感受到一个丰富、立体的葛利普形象，以及当时中外学者相互之间交往和互动的生动情形，相信此书的出版将大大推动中国近现代科学史的研究。

我很欣喜地看到，孙承晟近年来还在晚清民国地质学史领域继续耕耘，不断有论著发表。他为人诚恳，治学严谨，思路敏锐，厚积薄发。此书的问世不仅是他学术上的新突破，也反映了"地质学在中国的本土化研究"项目的研究潜力，以跨国科学史和国际合作的视角深入探讨中国早期地质学史仍是我们有待努力的研究方向。

借此机会，我谨向长期关心和支持此项目的丁仲礼院士、邱占祥院士、邓涛研究员、张鸿翔先生等致以诚挚的谢意。

韩 琦

香港理工大学

2024 年 7 月

前　言

在 20 世纪上半叶来华的外国科学家中，正如翁文灏所说，"在中国服务最久而贡献最多的，要算一位地质学大师葛利普（Amadeus W. Grabau）先生了。"[1] 葛利普自 1920 年来到中国，以其卓越的见识和融通中西的勇气在中国开展古生物学和地质学研究，撰写了大量有重要影响的论著，并怀着极高的热忱在北京大学教书育人，培养了中国第一批古生物学家，奠定了中国古生物学的坚实基础，堪称"中国古生物学之父"。

葛利普 1870 年 1 月 9 日生于美国威斯康星锡达堡（Cedarburg）一个德国血统的新教家庭。少年时即对自然有一种天然的爱好，对植物学、地质学和古生物学怀有特别的兴趣。因受克罗斯比（William O. Crosby，1850—1925）赏识，1890 年作为特招生进入麻省理工学院地质学系，1896 年获学士学位。次年入哈佛大学地质学系，1900 年以《纺锤螺及其同源动物之系统演化》（*Phylogeny of Fusus and Its Allies*）一文获博士学位。1901 年前往哥伦比亚大学任古生物学讲师，1903 年任副教授，1905 年升任教授，1918 年因故被哥伦比亚大学解聘。

1918 年底，丁文江随梁启超等前往欧洲考察，并出席巴黎和会。丁文江此行还受蔡元培委托，物色地质学家到北京大学任教。他分别将当时在英国留学的李四光和刚被哥伦比亚大学解聘的葛利普引荐到北京大学。葛利普 1920 年来到中国，任地质调查所古生物室主任和北京大学地质学系教授。

来华后，葛利普不仅致力于古生物学和地质学研究，在北京大学地质学系的教学亦极为热心，深受学生爱戴。中国第一代古生物学家和地质学家，大部分都受到他的直接影响，如孙云铸、杨钟健、赵亚曾、张席禔、俞建章、

1　翁文灏：《悼地质学大师葛利普先生》，《大公报》（重庆），1946 年 3 月 28 日。

陈旭、田奇瑻、乐森璕、丁道衡、斯行健、黄汲清、尹赞勋、许杰、朱森、计荣森、赵金科、卢衍豪、王鸿祯等。此外，秉志、李四光等也在他的鼓励下开展古生物学研究。葛利普还积极参与创建中国早期的地质学、生物学学术团体，如北京大学地质研究会（1920 年）、中国地质学会（1922 年）、北京博物学会（1925 年）、中国古生物学会（1929 年）等。

1937 年，全面抗战开始，北京大学和地质调查所南迁，葛利普因腿脚风湿滞留北平。他痛恨日本人的侵略行径，拒绝与伪北京大学合作。太平洋战争爆发后，他的经济来源断绝，生活极为艰苦，且不久被日本人送到英国大使馆集中，身心受到极大摧残。抗战胜利后，中华教育文化基金会恢复葛利普薪俸，虽有政府慰问和地质学界同人悉心照料，但终因胃部出血，于 1946 年 3 月 20 日在北平逝世。遵其遗愿，在其逝世一周年之际，他被安葬于北京大学地质馆（沙滩旧址）前，并由汤用彤代胡适主祭。1982 年 8 月 13 日，时值中国地质学会成立六十周年，葛利普墓被迁至北京大学（燕园）西门附近。

葛利普在来华前，已是著名的地质学家，发表了很多有影响的著作，重要者如《北美标准化石》（*North American Index Fossils*，Vol. I—II, 1909—1910）、《地层学原理》（*Principles of Stratigraphy*, 1913）、《硅酸盐以外的非金属矿床地质》（*Geology of the Non-metallic Mineral Deposits Other than Silicates*, 1920）、《地质学教程》（两卷）（*Text Book of Geology*, 1920—1922）。来华后，则以中国地质材料致力于古生物学和地质学的研究，著述不断，并协助丁文江创办《中国古生物志》（*Palaeontologica Sinica*），卷帙浩繁（他自己撰著的就有 7 种），深受国内外学者重视。他所撰写的《中国地质史》（两卷）（*Stratigraphy of China*，1924—1928）、《蒙古之二叠纪》（*The Permian of Mongolia*, 1931）、《脉动理论下的古生代地层》（四卷）（*Palaeozoic Formations in the Light of the Pulsation Theory*，1934—1938）、《年代的节律：从脉动理论和极控理论看地球的历史》（*The Rhythm of the Ages: Earth History in the Light of the Pulsation and Polar Control Theories*，1940）、《我们居住的世界：地球历史新论》（*The World We Live In: A New Interpretation of Earth History*，1961）等，更是为他赢得了世界性的声誉。终其一生，他发表著作 300 余种，累计 20 000 多页，这在现代科学史上亦不多见。葛利普的科学贡献主要在于古生物学和地质学，进而归于全球性的脉动和极控理论，乃 20 世纪最重要的古生物学家和地质学家之一。

由于在古生物学和地质学上的重要贡献，葛利普于 1925 年获得中国地质学会首届以其名字命名的葛利普奖章（王宠佑捐资设立），1934 年获北京博物学会的金氏奖章（金绍基捐资设立），1936 年获美国国家科学院颁发的汤普森奖章（The Mary Clark Thompson Medal）[1]。为纪念他在科学上的卓越成就，1976 年月球上的一个山脊以葛利普命名。

葛利普对中国极为友好，常呼吁中国人要自强，尤其要在科学上取得进步。1930 年，在其六十寿辰时，以丁文江和美国驻华大使为证人立下遗嘱，将其全部图书赠予中国地质学会，现部分仍存于中国地质图书馆、中国科学院地质与地球物理研究所、中国科学院南京地质古生物研究所等机构。

本书分为上编、下编、附编三部分。上编"葛利普其人"，收录葛利普来华后中外学者于不同时期从不同角度为他所撰写的纪念性文字，包括传记、回忆、介绍、讣告等。下编"葛利普论中国和科学"，主要为葛利普撰写的一些科普、回忆文章。这些文章散见于 20 世纪 20—40 年代的中外期刊，有些颇不易见，现汇为一编，基本以出版时间为序排列，以期勾勒出一个丰富、立体的葛利普形象，并可感受当时的中西科学交流和学术生态。附编"葛利普研究"，为笔者近几年发表的关于葛利普的三篇学术论文，较为深入地探讨了葛利普为建立北京博物学会所作的独特贡献，他一生最为看重的脉动和极控理论，以及他在北京大学的教学活动，展示了葛利普在华 26 年科学活动中最重要的三个方面（正好与三篇论文的顺序相反）：教书育人、科学研究和社团建设。

以下对上、下两编所收文章略作说明。

1926 年，斯行健和丁道衡为葛利普撰写了一篇传记（英文），是为中国第一篇对葛利普较为全面的介绍。该传记原拟刊于北京大学丙寅（1926 年）毕业同学录上，后因故转投于《北京大学日刊》（1926 年 6 月 11 日和 12 日）。斯行健和丁道衡在文中表达了对葛利普的感激之情，同时鼓励北京大学同学以葛利普为榜样，奋发有为。

1930 年，适逢葛利普六十寿辰，北京大学为其举行了隆重的聚会，蒋梦麟、胡适、丁文江、翁文灏、斯文·赫定（Sven Hedin，1865—1952）等均前往祝贺。《中国地质学会志》为此出版专辑以作纪念。其中，有斯文·赫定为葛利普手绘的肖像一幅，颇受葛氏喜爱，后多赠予他人。章鸿钊赋诗一首，盛赞葛利普在中国及世界地质学的重要地位及其对中国地质学教育的卓越贡

---

1 以美国著名慈善家玛丽·克拉克·汤普森（Mary Clark Thompson，1835—1923）命名，自 1921 年开始颁发。早期每年授奖一次，后频次多变，现与丹尼尔·吉罗·艾略特奖章（Daniel Giraud Elliot Medal）每三年轮流颁发。

献。章鸿钊、丁文江、翁文灏、李四光等 8 位中国地质学家则代表中国地质学会撰写了一封致葛利普的公开信，陈述葛氏对中国地质学所作的不可替代的贡献，表达了中国地质学界对他的感激之情。接着是丁文江为葛利普撰写的传记，对他 60 年的生涯作了详细介绍，并列举了他所获得的学术荣誉及其论著目录。曾在哥伦比亚大学受教于葛利普，回国后成为中国著名实业家的王宠佑追忆了他在哥伦比亚大学受教于葛氏的珍贵片段。除上述纪念性的文字外，该专辑还刊登了丁文江、翁文灏、秉志、李四光、德日进（Pierre Teilhard de Chardin，1881—1955）、步达生（Davidson Black，1884—1934）、巴尔博（George B. Barbour，1890—1977）、谢家荣、孙云铸、田奇瑈、黄汲清、计荣森、杨钟健、裴文中、张席禔、俞建章等的学术论文 16 篇。

1936 年，葛利普荣获美国国家科学院颁发的汤普森奖章。该奖章颁给美国在地质学或古生物学有突出贡献的科学家。1921 年，首枚奖章颁给沃尔科特（Charles Doolittle Walcott，1850—1927），后葛利普在美国的同事乌尔里克（Edward Oscar Ulrich，1857—1944）、怀特（David L. White，1862—1935）、舒切特（Charles Schuchert，1858—1942）等分别于 1930 年、1931 年、1934 年获得该奖章。1936 年葛利普获奖，但因腿疾未能前往领奖，由原在北京担任过其秘书的伍德兰（Alice Woodland）女士代读其谢词。刘咸以《葛利普博士之荣誉》一文对该奖项作了详细的报道，记述了美国国家科学院的颁奖词和葛利普的答谢词，并提及葛利普 1920 年前来中国时与秉志同船，期间还讨论了中国生物学的发展。

1946 年 3 月 20 日，葛利普从日军集中营（北京东交民巷英国大使馆）释放不久，即因胃出血不幸逝世。3 月 28 日，翁文灏撰文《悼地质学大师葛利普先生》刊于重庆《大公报》，后 4 月 11 日天津《大公报》全文转载。翁文灏还撰有英文纪念文章，发表于《中国杂志》（The China Magazine）。英国地质学家托马斯（H. Dighton Thomas，1900—1966）亦撰有纪念文章，发表于《自然》（Nature）杂志。

葛利普逝世后一个月，中国地质学会与中央研究院地质研究所、中央地质调查所于 1946 年 4 月 20 日在重庆北碚联合举行了隆重的追悼会，由俞建章（国立重庆大学地质学系主任）代表时任中央研究院地质研究所所长李四光主祭，李春昱报告生平事略，尹赞勋报告学术贡献，均对其学术成就及在中国的地质教育工作作出了高度评价。朱夏为此撰有《葛利普教授悼会记》一文，刊于 1946 年第 1—2 期《地质论评》（另载 1946 年第 6 期的《科学》），其后为

《葛利普先生追悼会挽联》。

葛利普的逝世在学界之外也产生了较大的反响，当时不少媒体多有报道。陶钝和陈光还分别结合时局，发表《悼葛利普教授》和《为中国地质学打下根基的外国人——葛利普》两文，赞誉葛利普的不朽贡献，并借此针砭时弊。

1946 年 9 月 16 日，在教育部部长朱家骅和经济部部长翁文灏的呈请下，国民政府颁发褒扬令，高度赞扬葛利普对中国地质学所作出的杰出贡献。褒扬令刊于 9 月 30 日出版之《教育部公报》（第 18 卷第 9 期）。本书收录了国史馆收藏的《中央地质调查所古生物研究室主任葛利普褒扬案》文档两份，一为褒扬令，一为朱家骅、翁文灏所撰的褒扬事由及葛利普传略。

葛利普逝世一周年之际，《新生报》开设"葛利普教授逝世周年纪念特刊"，发表了孙云铸、王竹泉、袁复礼、王烈、张席褆等撰写的纪念文章。寿振黄和当时著名的记者徐盈则分别撰写《葛利普周年祭》和《地质大师葛利普入土了》，发表于《华北日报》和上海《大公报》，颂扬葛利普人格之伟大及其对中国地质学的杰出贡献，并对当时的社会局势和葛利普去世后中国地质人才的成长产生担忧。如寿振黄称葛利普学问好比一座埃及的金字塔，且"不希望后辈做狭小的专门家，越钻越小，钻进牛角尖去"，徐盈则说"葛利普完成了播种的工作，但是种子的萌苗，却赖于自己的环境"等，对现在中国的科研工作者仍发人深省。

因葛利普的逝世，《中国地质学会志》继 1930 年六十寿辰为其出版专刊，1948 年再次为他出版纪念专辑。首先是章鸿钊赋词一首，深切缅怀葛利普："君去已多时！梁坏！山颓！门墙桃李尽含悲！留得神州新地史，星日同辉！才把凯旋卮，一笑长辞！名山事业后人思！廿载他乡成故国，魂也依依！"其次是孙云铸仿照 1930 年丁文江的葛利普传，为葛利普撰写了更为详细的传记。斯文·赫定再次以老友的身份，深切悼念葛利普。美国著名的动物学家和古生物学家格里高利（William King Gregory，1876—1970）回忆了葛利普生平和科学活动中的一些事迹。葛利普早年在哥伦比亚大学的同事夏默（Hervey W. Shimer，1872—1965）于 1946 年在《美国科学杂志》（*American Journal of Science*）上发表了葛利普的讣告，并于次年在《美国地质学会 1946 年度报告》（*Proceedings Volume of the Geological Society of America Annual Report for 1946*）上撰写了纪念葛利普的文章。

为纪念葛利普，《科学》1948 年第 3 期亦出版专栏，分别发表了杨钟健、孙云铸和寿振黄等的纪念文章。此外，杨钟健并于 1947 年在《人物杂志》上发表《地质学家葛利普》一文，指出葛利普是对他影响最大的人。《北平博物杂志》（*Peking Natural History Bulletin*）1948 年第 3—4 期上则综合先前的一些相关文章，刊发北京博物学会创始人葛利普的简传。

最后为葛利普自编论著目录。其中收录葛利普 1890—1943 年共发表的论著 299 种，总计 20862 页，加上 1943 年后不少正在付印或撰写的手稿[1]，数量更为可观。纵观葛利普的著作，数量在科学史上罕有其匹，其影响也十分深远。值得指出的是，此论著目录（除 1943 年后的部分）后亦附于孙云铸 1947年为他写的英文小传之后；1931 年丁文江在葛利普六十寿辰时为他撰写的英文小传，则附录了葛氏 1890—1931 年的论著目录，共 103 种，12183 页。为节省篇幅，本书所收丁文江和孙云铸的葛利普小传之后的论著目录，均从略。

自 1920 年来华后，除大量学术著作之外，葛利普也撰写了不少科普性、介绍性及纪念性的文章，发表于中外杂志上，为中国读者介绍地质学、古生物学、进化论、人类起源等新知，极力向西方宣传中国科学（尤其是地质学和古生物学）的进展，缅怀赵亚曾、步达生、丁文江、祁天锡（Nathaniel Gist Gee，1876—1937）、奥斯朋（Henry Fairfield Osborn，1857—1935）等学人朋友。

葛利普 1920 年刚到中国，即于 1920—1921 年在北京大学开设了"地球与其生物之进化"的 16 次系列演讲，从地球历史的角度提供了古生物学的概貌，而且以古生物发展演化为背景，系统介绍了当时最新的生物进化论和遗传学理论。本书所收《生物进化的误解》一文当为斯行健据葛利普"地球与其生物之进化"系列演讲笔记所译。

《六十年的达尔文学说》为江锦梁据葛利普该来华前发表于美国自然史博物馆的《自然史》上的"Sixty Years of Darwinism"一文翻译，对当时生物进化论在中国的传播提供了一个权威而普及的材料。然因当时白话文刚行不久，且一些中文术语尚不明确，此文读起来颇为拗口，这可作为研究当时科学翻译的一个样本。葛利普重视在中国进行生物进化论的普及。1922 年 2 月 12 日，适值达尔文诞辰 113 周年暨"南北统一纪念日"，北京高等师范学校博物学会举行纪念大会，下午即有葛利普关于"达尔文的天然淘汰说"的演讲。

1925 年，葛利普应邀在北京扶轮社（Peking Rotary Club）发表题为《科

---

1 1943 年葛利普被日军囚于英国大使馆后，仍笔耕不辍，总结自己的脉动理论，修订出版《脉动理论下的古生代地层》（*Palaeozoic Formations in the Light of the Pulsation Theory*）各卷，并完成其最后一部著作《我们居住的世界：地球历史新论》（*The World We Live In: A New Interpretation of Earth History*）手稿。该手稿几经辗转，后由胡适购得，转交葛利普当时在北京大学的学生阮维周整理，于 1961 年在台湾出版。

学与迷信》的演讲，阐述科学的逐渐兴起和宗教的逐渐衰落，从博物学的角度号召人人起而研究自然这本大书，推动科学发展和人类进步。其中涉及不少科学史和科学方法论问题，反映了他科学素养之深厚和涉猎之广泛。

葛利普不仅教书育人、执着研究，而且积极参加中国科学社团的建设，这也构成了他在中国科学活动最为主要的三个方面。在他的倡导及金绍基的资助下，北京博物研究所及北京博物学会于1925年成立，旨在推动中国的博物学（生物学）研究，后来确实产生了较为广泛的影响。作为这两个组织的真正奠基者，北京博物研究所甫一成立，葛利普即撰写《北京博物研究所》（"The Peking Laboratory of Natural History"）一文，1926年发表于《中国科学与美术杂志》（*China Journal of Science and Arts*），对该研究所的宗旨、机构及所要开展的研究作了详细的介绍。

20世纪30—40年代因战乱、疾病、意外等因素，中国地质学界损失了不少杰出人才，诚为可痛之事。赵亚曾1929年在云南考察时被土匪杀害，深受当时科学界的痛悼，丁文江为此痛哭好几次，并专门为他安排后事，收留遗孤。葛利普对这位"畏友"尤其惋惜，专门撰写了纪念赵亚曾的文章，对其才华赞不绝口，刊于《中国地质学会志》。后胡伯素将之翻译为中文，发表于《国立北京大学地质学会会刊》。此外，葛利普并在其他多篇文章中特别提及赵亚曾的学术成就。

1934年3月15日，步达生在研究"北京人"头盖骨化石的过程中，因心脏病猝逝于北京协和医院的办公室中，令科学界惊愕不已。5月11日，在北京举行了步达生追思会，翁文灏发来追悼信（谢家荣代读），许文生（Paul H. Stevenson，1890—1971）介绍了步达生的生平，丁文江、葛利普、德日进、杨钟健、巴尔博、裴文中、顾临（Roger S. Greene，1881—1947）等先后发言。本书中收录了葛利普的发言。

1936年丁文江在湖南考察途中因煤气中毒不幸去世，引起学界震动。葛利普极为伤心[1]，撰写了纪念文章，发表于丁文江等创办的《独立评论》，介绍丁文江在地质学和古生物学上的贡献，称丁为"先锋"和"热心工人"。此外，本书还收录了葛利普在奥斯朋和祁天锡逝世之后为他们撰写的纪念文章。

葛利普亦有不少介绍中国地质学、古生物学研究进展的文章。1929年，葛利普在北京大学成立31周年纪念会上，发表了《北大毕业生对于地质学之

[1]　1936年1月8日葛利普在与胡适等聚会时，谈起丁文江之死，大家都十分伤感，潸然泪下。见《胡适全集》，第32卷，安徽教育出版社，556页。

贡献》的长篇演说。此文后刊于 1930 年《国立北京大学自然科学季刊》(*The Science Quarterly of the National University of Peking*),后被胡伯素翻译为中文发表于 1931 年的《北大学生》。其中,对北京大学地质学系学生对中国地质学事业的贡献进行了细致的梳理,从一个侧面反映了他自 1920 年在北大地质学系执教以来中国地质学教育的发展。

1929 年 8 月 21—25 日,中国科学社第 14 次年会在燕京大学举行,到会者有任鸿隽、竺可桢、葛利普、秉志等 74 人。葛利普作了《中国科学的前途》的演讲,从欧美科学(尤其是地质学)的历史及他个人的求学和研究经验谈起,高屋建瓴地指出中国科学未来发展的方向,并建议中国科学社应将科学普及、科学教育和科学研究三者并行,将来成为中国科学促进协会(Chinese Association for the Advancement of Science)。该演讲立意深远,且不乏一些实际的经验与建议,对当时的学界有很大的影响。此文后由任鸿隽翻译,发表于 1930 年的《科学》和《东方杂志》上。1933 年,此文与姚万年的《化学与中国建设》和秉志的《国内生物科学近年来之进展》一起,被收录于《中国与科学》(《东方文库续编》,东方杂志社三十周年纪念刊)一书中。

1929 年,葛利普在《岭南科学杂志》(*Lingnan Science Journal*)上发表《中国的古生物学》("Palaeontology in China")一文,力陈中国 10 年来古生物学的迅速发展,已远超此前 60 年的古生物学成就,这正好反映了他 1920 年来华以后对中国古生物学的卓越贡献。他在文中不仅对中国古生物学如何从化石鉴定与描述到全球地层的对比进行展望,还对青年研究人才培养提出了不少切中肯綮的建议,如不能让青年人才过早从事管理职务,若没打好研究基础,不宜出去留学等。这些建议至今仍具有警醒的意义。

1929 年太平洋国际学会(Institute of Pacific Relations)京都年会之后,中国学者为给 1931 年将在中国召开的年会提供深入讨论的材料,决定由陈衡哲(1890—1976)负责编写一本关于中国文化的论集。1931 年,《中国文化论集》(*Symposium on Chinese Culture*)[1]由中国太平洋国际学会出版,并提交给在上海举行的第五次年会。该论文集意在反映中国文化中的一些新变化,约请丁文江、胡适、蔡元培、赵元任、朱启钤、任鸿隽、葛利普、翁文灏、李济等文化名人共撰写了 17 章,另有陈衡哲撰写的结论一章,其中葛利普撰写了《古生物学》("Palaeontology"),翁文灏撰写了《中国地质学》("Chinese Geology")。

1　此书中译本参见陈衡哲主编,王宪民、高继美译:《中国文化论集:1930 年代中国知识分子对中国文化的认识与想象》,福州:福建教育出版社,2009 年。

葛利普是参与撰写的唯一外国学者。他在《古生物学》一章中，对中国 20 世纪 20 年代的古生物学的发展作了详细的论述，反映了中国文化中变的方面——古生物学的迅速发展。此文后被张鸣韶节译为中文，刊于 1931 年第 8 期的《科学》。

1930 年在《自然界》上的《亚细亚和人类的进化》一文，乃是译自葛利普 1930 年发表于《中国杂志》（*The China Journal*）上的 "Asia and the Evolution of Man"，阐述了亚洲在人类进化史上的重要位置。

1931 年，葛利普发表《我们为什么学习地质学》（"Why We Study Geology"）一文，指出地质学在现代科学中的重要性，并强调学者不能仅仅拘泥于某一狭小的领域，而应对其他相关学科都要有深入的了解，这才能做出更杰出的成就。与此类似，他在第十二届中国地质学会年会上的专题发言《现代地层学者应具的勇气》，论述了现代地层学者不能拘泥于局部的研究和旧的理论，而应以新的化石证据和全球地层学的眼光寻求新的理论。他提出的脉动理论便是基于对东西方地层学的深入理解所作的一个新的尝试，同时冀望中国学者基于中国的地层学发展能提出新的理论。

如上文所述，葛利普 1936 年荣获美国国家科学院颁发的汤普森奖章，颁奖典礼上由原在北京担任过其秘书的伍德兰（Alice Woodland）女士代读其谢词。颁奖词和葛利普答谢词还被译为中文发表于《葛利普博士之荣誉》一文。事实上，该谢词的原文后以 "The Development of the Natural Sciences in China" 为题发表于 1937 年的《科学》（*Science*）杂志。本书亦收录此文，以便读者参考。

总之，回首 20 世纪上半叶他人为葛利普所写的纪念文章和葛氏撰写的系列科普著作，不仅能使我们了解这位伟大学者的一些珍贵片段和对中国地质学所作出的不可替代的贡献，而且能使我们感受他在科学上的宽广视野和深厚的人文情怀。他的科学精神和对中国科学的卓越贡献值得我们永远铭记。

在本书编校过程中，原文一些明显的书写或编排错误径予改正，不一一注明，与现代行文习惯不一致但不影响文意的地方则基本保留。文中脚注除"编者注"外，均为原文附注。因所收文章作者不同，发表时间不一，其中一些人名、地名、机构名称多有不统一，且常与现在通行用法或译法不相一致，除少数影响理解之处有所改动之外，其他多保留原初用法，敬希读者明鉴。编校中的错误或不当之处，恳请读者不吝批评指正。

本书的编著和出版得益于业师韩琦教授主持的中国科学院重点部署项目"地质学在中国的本土化研究"，在编撰过程中经常得到韩师的悉心指导。张九辰研究员、胡宗刚研究员一直关心本书的进展并提出很多宝贵的建议。胡大年教授、王光旭研究员、阮学勤博士、杨丽娟博士、张井飞博士、李融冰女士、秦硕荟女士、胡阳东先生在资料收集和编校中给予大力支持和帮助。责任编辑邹聪女士在编辑过程中付出大量辛劳，避免了许多疏漏。对于以上师友的好意与帮助，谨此一并志谢。

孙承晟

2024 年 5 月

目　录

下编　葛利普论中国和科学

附编 葛利普研究

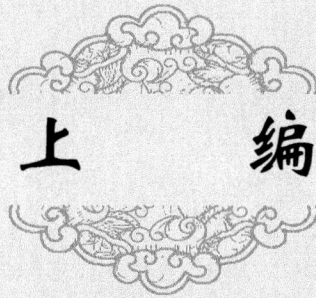

上　编

葛利普其人

葛利普博士传

斯行健　丁道衡

Introduction

Professor A. W. Grabau is one of the most famous geologists and palaeontologists of the world. He was invited to come to China in 1920 by the Peking National University and Geological Survey[1] of China.

During these six years, he has organized the work in palaeontology and stratigraphic geology in China where previously only a mass of disorganized facts and unrelated data had enlisted.

He has gathered around himself a group of Chinese students and has inspired them by his example and encouragement, so that through them his influence is spreading itself to remote corners of the land.

Now, we have the honour to present a sketch of his life to our schoolmates.

T. H. Ting, H. C. Sze.
May, 30th, 1926.

Sketch of the life of Professor Amadeus William Grabau

Professor Grabau was born on January 9, 1870 in the small village of Cedarburg in the middle western state of Wisconsin. His father was a Clergyman and so were both of his grandfathers, as well as other male members of the family. The environment of the village of his birth was well adapted to develop an interest in nature in the young boy, for it is located up on a broad surface of Silurian limestone rich in fossils. Among his earliest recollections are memories of

---

1　原文为 Surveying。——编者注

fossil corals brought by members of his father's parish from the quarries in the vicinity and treasured by the boy as stone horns. In later years on revisiting the scenes of his childhood and examining the quarries he found that many of them were opened in or near old Silurian coral reefs. Another recollection is the interest aroused by the fragments of a richly micaceous gneiss which were obtained by the blasting of a huge boulder brought by the Pleistocene ice from the mountains far to the north, and deposited in what later became a neighboring farm land.

Normal schooling ended when the day 11 years old after which he tried his hand at several occupations. When fifteen years of age the family removed to Buffalo, New York State, where the boy's father had accepted a position as senior professor in a theological school founded by his grandfather. For five years young Grabau was engaged in a business establishment in Buffalo, attending night school for part of the time. He founded a Natural History Society among the students of the seminary and frequent excursions in the vicinity of the city were made, first for the study and collection of plants and later of fossils in which this region abounds. Within the city limits are extensive exposures of Upper Silurian limestones rich in Eurypterids, those remarkable fossils which were probably the ancestors of the scorpions. Here also are exposures of Devonian coral reefs abounding in fossils corals and other animals on the shore of Lake Frie, south of Buffalo, are extensive cliffs of Devonian shales and in the gorge of Eighteen Miles Creek these shales are especially rich in fossils. It was this region which first decided the young man to devote himself to Palaeontology, and his first scientific monograph (*Faunas of the Hamilton Group of Eighteen Miles Creek*, etc.) and his first book (*Guide to the Geology and Palaeontology of Eighteen-Mile Creek and the Lake Shore Sections of Eire County, New York*) were the result of his early studies. Twenty miles north of Buffalo are the famous Niagara Falls and here the opportunity to study the Silurian sections is exceptionally good. Professor Grabau's second book, *Geology and Palaeontology of Niagara Falls*, with a new geological map, embodies the results of the studies in this field. At the age of 21 Grabau went to Boston and entered the Massachusetts Institute of Technology in the new department of Geology, to study under the famous geologist Professor

Crosby. At the same time he became a pupil of Professor Hyatt, one of the former students of Louis Agassiz, who himself was a student of Cuvier. Hyatt was the leading Palaeontologist in America and Grabau studied with him for six years. While a student Grabau gave regular instruction in palaeontology to his classmates and the younger students at the Institute. At the same time he lectured on Natural History in the Museum of the Boston Society of Natural History on Wednesdays and Saturdays. In 1896 he graduated with the degree of B. S. and after acting a year as instructor at the Institute he went to Harvard University to study advanced Palaeontology under Prof. Jackson, Physiography under Prof. Wm. Davis, Geology under Prof. Parker, Zoology under Professors Rarker[1] and Marhs[2], and Mineralogy and Petrography under Professors Palache and Wolff, all men of foremost rank in their sciences. In 1898 he received the degree of Master of Science and in 1900 that of Doctor of Science. He was for a time Instructor in Geology and Mineralogy at Tufts College and Professor of Geology and Mineralogy at the Rensselaer Polytechnic Institute, the oldest geological school in America. For a time he also was connected with the Geological Survey of New York and the Geological Survey of Michigan. In 1901 he was called to New York where for eighteen years he was Professor of Palaeontology in Columbia University. In 1910 he spent eight months travelling in Europe studying most of the important sections and meeting many of the European men of science.

In 1920 Professor Grabau came to where he has remained since as Professor of Palaeontology of the National University and Chief Palaeontologist of the Chinese Geological Survey. In 1925 he founded with others the Peking Society of Natural History of which he is a councilor and the Peking Laboratory of Natural History, of which he is dean. His publications number about 200 titles, aggregating over 10,000 pages. Those on China are *Stratigraphy of China* Vol. I, Ibid Vol. II now in press. Monographs on the Ordovician fossils of North China, the Silurian Fossils of Yunnan, Palaeozoic corals of China, and others still unpublished besides many other smaller papers. He is a fellow of the Geological Society of America, the Palaeontological Society of America, the American

1 应为 George H. Parker（1864—1955）。——编者注
2 应为 Edward L. Mark（1847—1946）。——编者注

Association for the Advancement of Science, the Science Society of China, the Geological Society of China, the Peking Society of Natural History, Honorary member of the Chinese Institute of Mining and Metallurgy, and Foreign member of the Kaiserlich Deutsche Akademie der Naturforscher zu Halle (Germany). In 1925 the Geological Society of China founded the Grabau Gold Medal for research in Geology and Palaeontology, the first award of which was made to Professor Grabau at the annual meeting of the Society in May 1926.

此传本拟登于丙寅毕业同学录上，嗣以未蒙编辑许可，故宣布于此，用作介绍吾辈受博士殷勤训诲。于兹四载，获益良多，一旦赋别，黯然久之。吾同学读此传而有所奋起，则此传之作，为不虚矣！

斯行健附识

——原载《北京大学日刊》1926 年 6 月 11 日和 12 日

葛利普教授六秩之庆

老眼看从开辟时，小周花甲似婴儿。
藏山事业书千卷，望古情怀酒一卮。
故国莼鲈添晚思，他乡桃李发新枝。
东西地史因君重，灿烂勋名奕叶期。

章鸿钊敬祝

——原载 *Bulletin of the Geological Society of China*, 1931, vol. 10

Letter from the Council

Dear Dr. Grabau:

On behalf of all the members of the Geological Society of China, we, your fellow-councillors, wish to express to you our best wishes on the occasion of your 60th birthday, which happily coincides with your 10th year of service in this country. We hope that you will live long and in good health to continue the important work in helping us to lay the foundation of Chinese stratigraphy and paleontology.

Nine of the papers in this volume are written by your own pupils including the youngest, who graduated last year. The rest is the work of your colleagues, Chinese and foreign, who have been in close association with you and owe you a great debt in inspiration and encouragement.

We realise what your influence has been on the development of geological science in China. Not only that enormous volume of work accomplished by you during the last ten years will serve always as the indispensable source of information and guidance, but your example of sustained effort in the face of physical suffering, inflexible devotion to duty under the most difficult conditions and unselfish readiness to help the young at all times inspires us to redouble our endeavour to carry on our work which is so often overshadowed by economic distress and political disorder.

We want particularly to tell you that ever since your arrival in China we have felt that you are one of us. We have long since forgotten that you are a foreigner, because we realise that your heart is here, and that your devotion to science is strong enough to transcend race and nationality. We do not forget, for example, that on several occasion when, owing to political disturbance, the University temporarily closed its doors and your salary was in arear-when, as a foreigner, you had every right to complain-instead, you struggled to carry on your classes at your own house. In return for such loyalty, our present effort can only express very inadequately our deep gratitude.

H. T. Chang　C. H. Chu

V. K. Ting　L. F. Yih

W. H. Wong　C. Y. Hsieh

J. S. Lee　Y. C. Sun

——原载 *Bulletin of the Geological Society of China*, 1931, vol. 10

Biographical Note

By V. K. Ting

The parents of Amadeus W. Grabau were of German origin. His grandfather, a pastor of St. Andrew's church in Erfurt, Germany, migrated to America in 1839, after suffering repeated imprisonment for refusing to conform to the services of the reformed Lutheran church. One thousand of his followers, amongst whom was captain von Rohr, the father of A. W. Grabau's mother, accompanied him to America.

On January 9th 1870 he was born in the village of Cedarburg, Wisconsin, where his father was pastor of the Lutheran church. He was the second son of a family of 7, two of whom died in their infancy.

He attended first his father's parochial school, and later the Cedarburg Public High School. At the age of eleven he lost his mother, and 4 years later, when his father moved to Buffalo, N. Y., to take charge of the Seminary founded by his grandfather, he entered the book-binding business in Buffalo as an apprentice, attending the evening school at the same time. He was soon attracted to natural history and spent all his spare time to educate himself. At first he was mainly interested in botany, but his botanical excursions to Eighteen-Mile Creek, south of Buffalo, and to the Genesee Valley at Portage turned his attention to geology and palaeontology. He then joined the correspondence course in mineralogy conducted by Prof. W. O. Crosby of the Massachusetts Institute of Technology, who, impressed by his brilliant promise, offered him a post as assistant in the mineral supply establishment at the Boston Society of Natural History which enabled him to enter the Massachusetts Institute of Technology in 1890 as a special student.

At the Institute of Technology he studied geology, mineralogy and physical geography under Crosby, Barton and Niles. At the Society of Natural History he came under the influence of S. Henshaw, J. W. Fewkes, R. T. Jackson and, above all, Alpheus Hyatt, then Chief Director of the Museum of the Society and Curator of Palaeontology in the Museum of Comparative Zoology at Cambridge. Henshaw and Fewkes interested him in marine biology, and Hyatt and later on, Jackson, became his teachers in palaeontology. Hyatt also appointed him as public lecturer at the Society's Museum. In addition to all these duties and classes, he had to attend the evening high school courses at the Boston Latin School in order to pass the entrance examination at the Institute.

Upon matriculation in 1891 he was appointed student assistant in the Geological Department under Prof. Niles. Three years later he published his first scientific paper on *The Pre-Glacial Channel of the Genesse River* in which he had been interested even before he went to Boston. This is the beginning of a series of papers on pre-glacial drainage in which he developed the hypothesis of a southward drainage system in the Tertiary. In 1896 he graduated, and his Bachelor's thesis, *The Faunas of the Hamilton Group of Eighteen Mile Creek*, was subsequently published by the Geological Survey of New York. From 1896-1897 he served as assistant in the Department of Geology in the Institute of Technology giving courses in palaeontology.

Even in his student days he became acquainted with many of the leaders in American geology, including A. Agassiz, Major J. W. Powell, C. D. Walcott, J. Hall, J. le Conte and T. C. Chamberlin all of whom directly or indirectly influenced his career. In 1897 he obtained a scholarship in the Department of Geology at Harvard University which was later followed by fellowships. Although chiefly working with Jackson, then Professor of Palaeontology at Harvard, he came under the influence of two master minds; N. S. Shaler, the last of the old time geologists, and W. M. Davis, the creator of physiography. In 1900 he received his degree of Doctor of Science from Harvard, and his dissertation, *Phylogeny of Fusus and Its Allies*, was subsequently published by the Smithsonian Institution. While completing his work for the doctor's degree he held the lecturership of geology in Tufts' College, and during one semester, that of mineralogy and geology at the Rensselaer Polytechnic Institute where he was subsequently appointed professor, but resigned to accept the lecturership in paleontology at

Columbia University in 1901. After one year he was appointed adjunct professor, and three years later, Professor of Palaeontology which position he held until 1919.

In analysing his contributions to science during these years one is struck by his versatility. The accident of living in a fossiliferous region during his formative years determined his career as a palaeontologist. Western New York is not only rich in fossiliferous formations, but is also abundantly provided with good sections. Eurypterids from the Upper Silurian of the quarries in North Buffalo, corals from the Middle Devonian limestones of Williamsville, Silurian brachiopods, cystoids, trilobites etc. from the Niagara Gorge and fossils from the Hamilton group of Eighteen Mile Creek provided him with rich material for study and identification. Thus his first book was a palaeontological and geological guide to Eighteen Mile Creek in which all the fossils were briefly described and figured. The Niagara and the Schoharie regions were treated in the same way, each accompanied by a coloured geological map.

His palaeontological work is however, by no means confined to descriptive monographs. Intimate contact with Hyatt and Jackson naturally resulted in developing his interest in the phylogenetic problems furnished by fossils. His contribution in this line is mainly on the Gastropoda, especially the Fusidae and their allies. He was the first to point out the importance of the protoconch in classification. Similar studies, though of less extent, were made on corals, crinoids and brachiopods.

Speaking German from his childhood, he is equally at home with German scientific literature. Thus he was the first American geologist to follow J. Walther whose book, *Einleitung in die Geologie als historische Wissenschaft* led him into bionomy and lithogenesis. His interest in the bionomic relations resulted in the detailed study of Palaeozoic coral reefs, including those of the Silurian on which his native village was built. His papers on the habitat of the eurypterids, and of the Devonian fishes, and on the origin, distribution and preservation of the graptolites were inspired by the same principles.

In lithogenesis the formation of clastic rocks and desiccation deposits especially claimed his attention. In this he was again influenced by Walther's *Gesetz der Wüstenbildung.* His main ideas were first crystallised in a new classification of sedimentary rocks based on genetic principles, which was followed by studies of individual types. As an extension of the principles of Walther he formulated the criteria for recognising the various types of desiccation deposits in the older series in his *Principles of Salt Deposition*, forming the first volume of his *Geology of Non-Metallic Mineral Deposits other than Silicates*, published in 1920.

Even more original is his introduction of physiography into stratigraphy. As a student of Crosby he early interested himself in the glacial geology in the neighbourhood of Boston. As far back as 1897 he interpreted the sand plains of Cape Cod as formed in successive stages of a glacial lake embayed in the ice front and progressively lowered by the uncovering of outlets across the terminal moraine. Coming under the influence of W. M. Davis, he mapped the successive shore-lines and delta plains of one of the ice-border lakes near Boston which he named Lake Bouvé. In all his works on stratigraphy the development of landforms in older geological periods serves as the key to the distribution of sediments with references to their sources, the formation of larger overlaps and disconformities. The last term was first proposed by him in a paper, *Physical Characters and History of Some New York Formations*, published in *Science* in 1905.

Physiography, lithogenesis and bionomy were finally combined and systematised in his great book, *The Principles of Stratigraphy*, published in 1913. Of this the late Professor Barrell said: "Many chapters could be used without change in a work on physiography".

In 1910 he for the first lime visited Europe. In addition to attending the International Geological Congress at Stockholm, he spent half a year studying European geology in the field in England, Scotland, France, Germany, Austria and Russia. His paper, *Comparison of American and European Lower Ordovicic Formations*, published in 1916, was the direct result of his visit to Europe, but his first-hand knowledge of European geology has left its mark on all his subsequent works. His *Text Book of Geology*, for example, is the only American text-book in which geological formations outside the American continent have received any attention.

In 1920 he was offered by the writer, through Dr. David White of the U. S. Geological Survey, the post of Chief Palaeontologist to the National Geological

Survey of China to act concurrently as Professor of Palaeontology at the National University of Peking, which posts he accepted with enthusiasm. The Geological Survey then had been in existence for 5 years and considerable palaeontological material had been accumulated. It was felt that not only was it urgently necessary to work out the collections in order to lay the foundation of Chinese geology, but also that Chinese students must be trained to undertake independent researches in palaeontology so as to continue the work in the future. To him therefore was entrusted the double task.

As a teacher his success was immediate and complete. Although his lectures were delivered in English to Chinese students who had never been outside their own country, he soon attracted many brilliant young men to his classes. His inspiring enthusiasm, hard work and clear exposition won for him not only the respect but also the love of his pupils. The fact that out of the 25 monographs written for the *Palaeontologia Sinica* by Chinese palaeontologists, 19 were written by men who owed their training to him, speaks more than anything else for his contribution to the education of the younger generation of Chinese geologists.

With regard to descriptive palaeontology, these years have been very productive: 9 papers have appeared in the Bulletin of the Geological Survey and that of the Geological Society; 5 monographs have been published in the *Palaeontologia Sinica*, 5 more are ready for printing. In the meantime he also worked out the Permian Fauna of the Jisu Honguer limestone of Mongolia collected by the Central Asiatic Expedition of the American Museum of Natural History. The total number of published pages amounts to more than 1,800, and the subjects range from Cretaceous ammonites to Ordovician corals.

Characteristically he was not satisfied with descriptive palaeontology as such. He inevitably systematises his palaeontological work into scientific stratigraphy. In his monograph on the Permian of Mongolia the whole Permian question of the world was discussed in all its aspects with characteristic insight and originality. Within four years of his arrival in China the first volume of *Stratigraphy of China* appeared. The second volume followed 4 years later. With German thoroughness he summarised all existing data, stated problems and formulated working hypotheses. It serves as a compendium for all Chinese geologists as well as a text-book for his students. When one of his own pupils succeeded in disproving some of his hypotheses, he was more delighted than any one else, and said rightly that the greatest pleasure of a teacher was to have his mistakes corrected by his own pupils.

Nor did he neglect his favorite researches begun in America. The migration of geosynclines, first outlined in a short paper in 1919, was further elaborated in full in 1924. Part V of *Studies of Gastropoda* appeared in 1928. Only some of the problems bear more local colour: he now constructs palaeogeographical maps of China instead of America and speculates on Asia and the evolution of man.

During the ten years he has been in China he spent only one short holiday at the seaside of Peitaiho, and, as a by-product, published jointly with S. G. King a booklet on the *Shells of Peitaiho*.

He took an active part in the organisation of the Geological Society of China, and in 1925 was awarded by that Society the first gold medal founded in his honour.

He married Mary Antin in 1901 and has one daughter Josephine Esther.

Scientific career and honours

1890.　Appointed assistant in the Mineral Supply Establishment at the Boston Society of Natural History.

1891.　Matriculation at the Massachusetts Institute of Technology. Appointed student assistant in the Geological Department.

1896.　Took the degree of Bachelor of Science. Appointed assistant in the Geological Department.

1897.　Entered Harvard University.

1898.　Took the degree of Master of Science at Harvard. Elected fellow of the Geological Society of America.

1899.　Appointed lecturer on geology in Tufts' College.

1900.　Took the degree of Doctor of Science at Harvard. Appointed Professor of Geology at the Rensselaer Polytechnic Institute.
　　　Appointed member of the Geological Survey of Michigan.

1901. Appointed lecturer in palaeontology at Columbia University.
Elected Fellow of the New York Academy of Science.
Appointed member of the Geological Survey of New York.

1902. Promoted to adjunct Professor at Columbia.

1905. Promoted to Professor of Palaeontology at Columbia.

1910. Travelling in Europe.

1920. Appointed Chief Palaeontologist of the National Geological Survey of China and Professor of Palaeontology of the National University of Peking.

1922. Elected Fellow and Councillor of the Geological Society of China.

1923. Appointed Research Associate of the Central Asiatic Expedition.

1924. Elected Honourary member of the Science Society of China.

1925. Elected member of the Kaiserlich Deutsche Akademie der Naturforscher zu Halle.
Awarded the Grabau Medal by the Geological Society of China.

1927. Elected Honorary member and Life Councillor of Peking Society of Natural History.

1928. Elected Correspondent of the Philadelphia Academy of Natural Science.

1929. Elected Research Associate of Academia Sinica.

葛利普著作目录（从略）。

——原载 *Bulletin of the Geological Society of China*, 1931, vol. 10

Dr. Amadeus W. Grabau: A Reminiscence

By Chung-Yu Wang[1]

It is a great privilege that I, as Dr. Grabau's first Chinese student, should be allowed to write a few words in this volume. I have always treasured the memory of the years when, as a student, I sat at his feet in Columbia University. Little did I dream then that in years to come, he, whom I revered as my master, would come to China to become the mentor and inspirer of so many young Chinese in the science of geology.

I remember distinctly that, when I first attended Prof. Grabau's courses in the autumn of 1904, almost the first thing I was asked to do was to make drawings of a set of fossils, and to draw any conclusions I could from such a study. I did not understand then, as I do now, that such a method of teaching was designed to draw from the student his own mental reaction towards a particular problem, and to show that painstaking and detailed work was a necessary prelude to any broad generalisation. I was greatly impressed with the attractive and lucid delivery of his lectures, as well as with the earnestness and enthusiasm of his manner. I was one of the few graduate students who were admitted to his lectures on Stratigraphy which he later developed into the now standard work, *The Principles of Stratigraphy*. In reading this book I lived over again the scenes when it was delivered to us in its embryonic form.

In connection with these lectures, I remember an incident which may serve to illustrate his spirit of tolerance toward some of the innocent pranks of students. There were about eight of us in this class; and we sometimes used to pass peanuts beneath the desk. On one occasion we suspected detection, and one of us courageously passed some up to Dr. Grabau, who, with a kind smile, joined us in this diversion while continuing his lecture.

1　王宠佑。——编者注

The occassional soiree for graduate students given at his home in New York was an event I always looked forward to. Away from the conventional constraint of the classroom, we were privileged to see the man in Prof. Grabau, revealed in anecdote and jest. On one occasion, to the mystification of all present, I acted the part of a magician in passing a coin through the table. Somehow, unseen by us, the serious professor hid under the table and detected the manner of my performing the trick.

One spring day in 1922, after a long period of seventeen years, I met again in Peking the professor of my student days. I was struck then more than ever before by "his brow bent like a cliff o'er his thoughts", which reminds one of the countenance of Quenstedt of Jurassic fame. It may be recalled that, on the heights of the Swabian Alps, a monument has been erected to the memory of Quenstedt for his achievement in stratigraphy. In like manner, perhaps, in years to come the future geologists of China will commemorate the work of one who has laboured so untiringly for the advancements of the science of geology in China.

The evaluation of Prof. Grabau's work on geology in general, and on Chinese geology in particular, I shall leave to more competent hands. While the visible aspects of his work are known to all, its invisible counterpart as discerned from the virility of the work of his former students and associates, will surely live on in the work of the future Chinese geologists.

Now in trying to express in a few words our tribute to Dr. Grabau, I cannot do better than to quote the words of Le Conte, written in his autobiography, concerning his teacher, the great Agassiz: "There are two types of great men: those of one class who are great by the quantity and importance of their work, but when one comes in contact with them and measures them intellectually, they seem of ordinary stature... those of the other class, the nearer they are approached, the greater they grow—they are themselves greater than all their visible results. These are the great teachers; their spirit and enthusiasm are contagious; their personality is magnetic". Dr. Grabau is pre-eminently of this latter class.

——原载 *Bulletin of the Geological Society of China*, 1931, vol. 10

葛利普博士之荣誉[1]

重 熙[2]

美国华盛顿国立科学院（National Academy of Sciences）于本年 4 月 27 日晚举行年宴时，以四种奖章分别赠予成绩卓越之科学家，就中特以汤穆逊[3]奖章（The Mary Clark Thompson Medal）授予我国国立北京大学古生物教授兼中国地质调查所古生物学总技师，本社特社员葛利普博士（Dr. Amadeus William Grabau），嘉其在地质学与古生物学上之不朽贡献，弥足纪也。兹将举行赠授典礼时汤穆逊奖金委员会主席林格仁氏（Waldemar Lindgren）颂词及葛教授答词分志于下，藉飨读者，兼留史料。

"汤穆逊基金委员会于 1936 年集会时，曾一致议决将本年奖章赠予中国国立北京大学教授兼中国地质调查所古生物学总技师葛利普氏，以符奖赠'地质学及古生物学上最有贡献者'之规定。自从 1924 年以来，曾获本奖章者大多数为古生物学家，计有华尔阁（C. D. Walcott）、戴玛哲理（E. de Margerie）、克拉格（J. M. Clarke）、斯美士（J. Perrin Smith）、斯各德（W. B. Scott）、乌尔立希（E. O. Ulrich）、怀德（David White）、巴士（F. A. Bather）及 1934 年之舒霍尔德（Charles Schuchert）。

本委员会对于葛教授之一般地及地层地质学、非金属矿物学，尤其在古生物学上之优异贡献，曾加以深切注意。彼在古生物学上之研究，除纽约州之古生代（Palaeozoic）外，过去十七年来致力于中国古生物学之工作，尤足称道，其结果均发表于富丽之《中国古生代及中生代化石》之专集中，最后

---

1 本篇材料及照片系由孙洪芬、林伯遵两先生供给，特此志谢。

2 刘咸（1901—1987），字仲熙、重熙，江西都昌人。1921 年入国立东南大学学习，师从秉志、陈桢、陈焕镛等，1925 年毕业。先后任职于东南大学、清华大学生物系。1928 年考取公费留学，入英国牛津大学学习人类学，1932 年获硕士学位。回国后担任山东大学生物系主任。1935 年，经秉志介绍，出任《科学》杂志主编，同时主编《大公报》科学周刊、《申报》科学副刊等。1947 年后一直担任复旦大学生物系、社会学系教授，直至 1983 年退休。1956 年与秉志、朱洗等一起创办《中国动物学》杂志，1981 年发起成立中国人类学会。在生物学、人类学上有重大贡献，为中国人类学研究的开拓者之一。——编者注

3 现一般译为汤普森，下同。——编者注

一卷，亦如以前各卷，均由北平中国地质调查所刊布。

葛教授以 1870 年生于威士康新州（Wisconsin），初肄业于麻省理工大学[1]（Massachusetts Institute of Technology），继自 1892—1897 年，任该校教员，1900 年获哈佛（Harvard）大学科学博士学位后，自 1902 至 1905 年任哥伦比亚（Columbia）大学副教授，1905 至 1919 年任正教授，1920 年被聘来华任前述之两要职，直至今日。统观十七年来致力于中国古生物学及地质学之研究，无间寒暑，成绩斐然，所有贡献除翔实之外，皆为具基本性质之要图。晚年虽身体欠佳，但并不因此而减少其工作之勇气与毅力，葛氏赋性强毅，热心奖掖，善于擘划，勇于实行，以毕生之力，尽忠事业。

葛利普博士画像（斯文哈定画）

葛教授著作等身，择其重要者言之，1910 年与著名古生物学家麻省理工大学席默教授（Prof. H. W. Shimer）合著《北美之指引化石》（*North American Index Fossils*），学者采用至广。稍后有《美国中泥盆纪动物志序论》（*Succession of Faunas in the Middle Devonian in U.S.A.*）；《密歇根之哈弥尔顿动物群》（*Hamilton Fauna of Michigan*）；《无脊椎动物，尤其斧足类之群体发生论》（*The Phylogeny of Invertebrates, Chiefly Gastropods*）；《地层学原理》（*Principles of Stratigraphy*）（1921 年二版）；《地质学教本》（*Textbook of Geology*）2 卷，1921；

1　现一般称为麻省理工学院。——编者注

《非金属矿产》(*Non-metalic Deposits*) 2 卷，1922；《华北奥陶纪之化石》(*Ordovician Fossils of North China*)，1921；《中国地层学》(*Stratigraphy of China*)，1925；其他专门论文多由地质调查所之《中国古生物志》(*Palaeontologica Sinica*)发表，此外短篇论著，不知凡几，诚可谓竭毕生之力获大成功者。

此次本会以本奖章献与葛教授，正所以表示吾人对于此最卓越之古生物学家及地质学家以一适当之钦敬，彼在美国及世界地质学上，均博得无上光荣，在坐同仁，谅有同感。

今晚所感美中不足者，即葛教授未能亲自出席，但可欣幸者，有葛夫人在座[1]，代受奖章，彼将向葛教授代为转达吾人景仰之诚意。

麻省理工大学

林格仁（Waldemar Lindgren）。"[2]

葛教授答词，系在北平作就，于 3 月 17 日寄往美国，译文如下：

"国立科学院院长暨诸位院员公鉴：

我国最高科学团体此次以殊荣惠加鄙人，至深感谢！

予之所以能于中国自然科学发达过程中参加积极工作者，实由于予至北平时，正当中国知识界对于西方科学教育有深厚兴趣，力加提倡，是时中国地质调查所已成立有年，但直至近来方完成其第一步工作，即造就多数之中坚地质学人材是也。

当该所设立之时，所屋陈旧，设备简陋，图书馆仅有书数百册，中国古生代化石亦仅数抽屉而已，但不久之后新建筑即在兴工，在章鸿钊、丁文江、翁文灏三博士热心主持之下，增加设备，罗致材料及所作中国地质之调查进步极速，所中外籍顾问安特生博士（Dr. J. G. Andersson）曾组织科学探险队开发重要之脊椎动物化石堆，同时并由外籍古生物学家担任研究。

予之职务为担任中国无脊椎动物化石之研究，同时在大学兼授功课，训练中国之青年古生物学家及地层学家。

由于安特生博士之斡旋，《中国古生物志》之最初出版费，遂有着落，

---

1 按系葛教授之侄媳，曾任葛氏书记者。

2 原文参见 Waldemar Lindgren, "Presentation of the Mary Clark Thompson Medal to Amadeus William Grabau," *Science*, New Series, 1937, 85(2210): 436. ——编者注

此志计分四组：A.《古植物志》，B.《古无脊椎动物志》，C.《古脊椎动物志》，及 D.《古人类志》。以后此项费用完全由地质调查所担任。

《古生物志》之最初两号，系由予编制，于 1922 年四月及九月先后出版。自兹以往，继续不断，现已出至 95 号，都 8760 页（四开页），及 844 图版，若将正在印刷中，及已作就者一并计入，已超过 100 号，约 9000 余页，就中论文最短者 14 页，1 图版，最长者竟达 441 页，31 图版。

第一篇古生物学专门论文由中国古生物学家写成者，为孙云铸博士，出版于 1924 年。全志中 B 类论文已出版 43 号，就中有 32 篇为中国人之著作，执笔者几全为北大毕业生。

起初毕业生多派送出洋研习脊椎动物，但近年工作人员之训练即在北平就地养成之。此乃由于调查所中创设新生代研究室，专门从事脊椎动物之研究，在魏登瑞（Weidenreich）、杨钟健、德日进（Père Teilhard de Chardin）诸博士指导之下。"北京人"（Sinanthropus pekinensis）之研究，亦在其内。此外因美国第三次亚洲探险队之来华，由安德鲁士[1]（Andrews）领导，国际著名科学家如格兰恪（Granger）、马修（Matthew）、奈尔逊（Nelson）、钱乃[2]（Chaney）、柏奇（Berkey），及由斯文哈定[3]（Sven Hedin）领导之瑞典科学家多人，以及科学赞助人瑞典皇太子殿下等之先后莅平，于调查所事业之推动，不无关系。

自兹以往，调查所觉得周口店为发掘场所，设置各种必需仪器，作大规模之发掘，该场现已成为著名之古人类考古场所。

土壤调查室、西山地震观测室，以及其他多方面之事业，均由所中分出，此外关于地质构造之研究，中国经济矿产之调查，及地质图之制绘，均在积极进行中。

1920 年北大地质系在李四光、谢家荣及故丁文江诸博士主持之下，加以整顿，此后进步甚速，现在该学系独占屋一座，设备完善，有讲演厅、实验室、标本室及图书馆。地质学系共有教授 7 人，讲师 2 人，助教 4 人。

该系数百毕业生中，大多数仍积极从事地质工作，或服务于国立及

---

1 即安得思（Roy Chapman Andrews，1884—1960），曾于 20 世纪 20 年代在中国组织 5 次大规模的中亚考察。——编者注

2 现一般译为钱耐。——编者注

3 即斯文·赫定。——编者注

省立各地质调查所，或在各大学任教职。该系现发行刊物，已出版 12 种，其他在印刷中。

1922 年中国地质学会成立，当时仅有发起会员 26 人，第一次年会时（3 月 2 日）加入新会员 36 人，学生会员 9 人，是年发行会志，第一卷共刊论文 15 篇，都 99 页，就 1936 年出版之会志，为 XV 卷，都 574 页，每年分四期出版。

1937 年 2 月举行年会时，有论文 60 篇当场宣读。会员人数分配如下：

正会员	320
学生会员	66
外国通讯会员	36
现在荣誉会员	1
机关会员	6
总计	429

当予来华时，有一中国同船旅客，秉志博士，于康乃尔（Cornell）大学学成之后，亦正返国，乃得有机会，相与讨论发展研究中国博物学之计划，盖秉博士正回国主持中国科学社生物研究所事业[1]。该社成立于数年之前，以提倡及传布科学为宗旨，出版中文《科学》月刊，现已出至第 XXI 卷。研究所成立后，在秉博士领导之下，首先积极从事于华中动植物之研究，结果用英文刊布，陆续出版，数量甚富[2]。

在北平吾人更得金叔初先生之赞助，有北京博物学会之组织，自此以后，金氏为中国科学赞助最力之一人。彼曾在北戴河设立海滨实验室，经两时季之采集，吾人以所得材料，作成《北戴河贝壳指南》一书，书中描述贝壳 120 种，并有插图，金先生此后被举为许多外国软体动物学及贝壳学会会员，并在中国设立一最完善之贝壳学图书馆[3]，并不时与秉博士合作研究华南贝壳动物，论文时有发表。

当静生生物调查所组织时，将北平博物学研究所之原拟计划，几全部加以采用，并包括发行《中国动物学志》（Zoologica Sinica）及其他发展计划。该所自从迁入设备完善之新所屋后，现已成为中国主要生物学

---

1 按秉博士初应国立东南大学之聘，继主办本社生物研究所。
2 按本社生物研究所发刊生物论文分动植两组，动物已出至 XII 卷，植物 X 卷。
3 金先生之贝壳图书于民国 25 年全部捐赠于本社明复图书馆，以专室藏之，颜曰"金叔初贝壳学图书室"以纪念之。

研究机关之一。

北平博物学研究所之前期活动之另一直接产物，即为 1925 年北平博物学会之成立，当时由鸟类学家祁天锡博士（Dr. N. Gist Gee）任组织干事，予为召集人，于 9 月 21 日举行第一次集会，到发起会员 38 人。此后每月举行讲演会一次，讨论中国博物学。在第一年度中，会员增加至 101 人，其中有外国通讯会员 26 人，荣誉会员 4 人，会志现正出第 XI 卷，除此以外该会并刊行手册，已出版者有四种：（1）韦克斯（R. D. Wickes）著《北戴河之花卉》；（2）葛利普与金叔初合著《北戴河之贝壳》（1928年第二版）；（3）波灵（Boring）[1]、刘及周[2]合著《华北两栖类爬行类手册》；（4）周汉藩著《河北习见树木志》。

此外该会又出版专集 5 种，为伊博恩（Bernard Read）著之中国药用植物与动物药[3]，及伊、朴[4]两博士合著之医用矿物与石头，及尼登（J. G. Needham）著之中国蜻蜓手册[5]，插图甚多。

自然历史之科学的研究，在中国现已成为一种知识的钻研。方其开始发达之际，凡吾人躬逢其盛，略有献助者，皆深信将来在地质学、古生物学、生物学及考古学各方面之贡献，中国科学家渐著成绩，不但对乎本国，而且对于世界科学，将有重要之贡献。

中国博物学家亦如予所感觉，今晚贵会所授于予之荣誉，实不啻承认中国科学事业之进步，彼等将因此种鼓励，而继续努力。

兹因贵会以汤穆逊奖章授予鄙人，故作以上之陈述，谨以私衷感谢之忱，接受此荣赉！

葛利普。"[6]

1　即长期在中国从事动物学研究和教育的美国学者博爱理（Alice M. Boring, 1883—1955）。——编者注

2　此处刘、周分别指刘承钊、周淑纯。——编者注

3　伊博恩撰有：Chinese Materia Medica: Animal Drugs, Avian Drugs, Dragon and Snake Drugs, 分别刊于《北平博物杂志》（Bulletin of the Peking Society of Natural History）1931 年第 5 卷第 4 期，1931 年第 6 卷第 1 期，1932 年第 6 卷第 4 期，1934 年第 8 卷第 4 期。——编者注

4　伊、朴分别指伊博恩和朝鲜学者朴柱秉（Chubyung Pak），他们合著 A Compendium of Minerals and Stones used in Chinese Medicine from Pen T'sao Kang Mu, Li Shih Chen, 1597 A. D., 发表于《北平博物杂志》（Bulletin of the Peking Society of Natural History）1928 年第 3 卷第 3 期。——编者注

5　尼登 1930 年出版 A Manual of the Dragonflies of China: A Monographic Study of the Chinese Odonata 一书。——编者注

6　原文见 Amadeus William Grabau, "The Development of the Natural Sciences in China," Science, New Series, 1937, 85(2215): 551-553，亦收于本书 329—333 页。——编者注

汤穆逊奖章（正面）　　　　　　　　　　汤穆逊奖章（反面）

吾人于读毕双方颂辞与答辞之后，有不能已于言者，即葛利普教授以高年硕学，侨居吾国，研究科学，培植人材，历十七年如一日，热心毅力，至堪钦佩！中国古生物学之所以有今日之发达者，葛氏提倡之功为不可没，且其他于吾国科学发达之一般影响，亦至重大。葛氏道德高尚，蔼然学者，凡曾与之晋接者，莫不肃然起敬，此次美国科学院以氏在地质学及古生物学上有特殊贡献，荣以汤穆逊奖章，是乃葛氏实至名归，应有之荣誉，乃教授逊谢不遑，于答辞中历数中国科学事业之进步，而不及自身之成绩，谦谦君子，尤可敬佩！如葛氏者诚可以风矣。

——原载《科学》1937 年第 9—10 期

悼地质学大师葛利普先生

翁文灏

　　近数十年来世界知名的外籍科学家，在中国服务最久而贡献最多的，要算一位地质学大师葛利普（Amadeus W. Grabau）先生了。但是不幸得很，这位大师于本年三月二十日下午五时半在北平逝世，这不仅使中国地质界人士痛悼，亦是世界上科学家所认为不幸的消息。他是一位学问渊博、忠于职务、学不厌、诲不倦的学者。在中国工作达二十六年之久，未曾稍闲，对于奠定中国地质学术的基础，实有莫大之功绩。兹略述葛先生之生平与事迹，藉以悼念这位科学界的伟人。

　　葛先生的祖父原属德籍，一八三九年迁至美国，一八七〇年葛先生生于美国威斯康新州之一乡村，童年时代即对自然科学感具兴趣，选习植物、地质等科，在 MIT[1]研究矿物学，在自然科学会研究古生物学，旋任 MIT 的助教。一八九七年入哈佛大学，一九〇〇年得该校理学博士学位，翌年即任教于哥伦比亚大学地质系，以迄来华之前，在这个学校任教授将及二十年。至一九二〇（即民国九年）应中国之聘，任中央地质调查所的古生物研究室主任，兼北京大学地质系教授。二十多年来，作育的生徒，为数甚多。现在国内从事于地质工作者，总有四分之三以上是直接或间接受过他的讲解或指导的。"七七事变"之后，中央地质调查所和北京大学分别南迁，这位七十老翁以患腿疾不良于行，因而羁留北平，精神上则非常痛苦，生活尤感窘迫。及珍珠港事件发生，初则被集中于德国医院，继则迁移到英国使馆旧址，与外边益形隔绝，营养自甚缺乏，因之身体日渐消瘦，憔悴不堪。光复后，他的友人生徒相继到平，地质调查所和中华教育文化基金董事会并恢复了他的薪俸，给予他以经济上的援助，这使他得到很大的安慰，精神日渐好转。他自己亦希望康复后重作研究工作，但是实际上他已是骨瘦如柴，远非昔比，并且他的钢琴、文学书籍以及衣物用具已在这几年中卖的光光了。今年一月他为得要脱离曾经被集中的地方，为得表示又回到地质调查所里工作，而且为得使调

1　即麻省理工学院（Massachusetts Institute of Technology）。——编者注

查所同仁照料方便起见，他一再的商量要住到地质调查所里，可是所里的宿舍被某机关占用，坚不腾出，没奈何，在丰盛胡同三号陈列馆的后院给他修理了三间房子，不料这便成了他寿终之所了。

葛利普先生的著作非常宏富，统计其已出版的约二百多种，不下二万余页，其中最重要的如《地层原理》、《非金属矿床》、《北美之标准化石》、《普通地质学及地史学》，皆是来华以前之闻名巨著；到中国以后著的有《中国地质史》两巨册，《蒙古之二叠纪》一巨册及《中国古生物志》十一册[1]。晚年致力于地球脉动说之研究，他就地史的观察，认为某一个地史时期海水侵入大陆，全球皆然，海水退时亦是同样现象，他以为不是海水流到某一处去了，而是地球表面略有涨缩，宛如人体之脉搏，故为是说。在他被集中以前所完成的五大册均已印出，就古生代的地史旁征博引，资料丰富，尚有一些稿子等待整理，在集中以后，还写了几篇文章，据他自己口头告诉，有一本叫着《我们生存的世界》（*The World Where We Are Living*）[2]，听说已经付印了。

葛利普先生教书极其认真，非有疾病从不缺课，他本以腿疾不大能走路，但是坐着人力车到北京大学，扶杖步入课堂，数十年如一日，或因疾扶杖亦不能行时，则召学生到他的寓所，照常讲解，往往讲到兴趣浓处，每不知道已经下课，且已侵占了下一点钟的时间。在民国十五六年，北京政府往往欠薪甚久，或折扣过半，而葛先生则从未因此而影响了学生的功课。相反的，或有其他先生因故不来授课时，葛先生辄忿曰"凡是没有人教时，我都可以替他教"，因此他的学生都非常爱戴他，钦佩他，每见到他扶杖到校时辄上前挽扶着他。就不是他的学生，凡相问者，亦必殷殷指导，因亦多敬以师礼。

葛利普先生研究从不懈怠，初到中国的几年勉能行走，常按时到地质调查所工作，后来因为行动实在不便，乃在家里研究，他把一天的时间充分的利用了，他的夫人和女儿都没有跟他来中国，家里只用一位女秘书替他打字照管书籍。几时到他家里去，都看见他坐在书桌前面，手不停挥，或是疲累了，躺在靠椅上，亦不断的抚弄他所研究的化石，四壁图书便是他随身的伴侣。

葛利普先生不计薪酬，而专以研究服务为目的，他在中国二十六年，每月的收入都是中国国币，亦从来没有希望取得比中国人较优厚的待遇，因此他没有积蓄，亦没有拿钱赡养他的眷属，偶有余资，则以购置书籍，而且这些书籍他早已立下遗嘱，等他死后，全部捐赠给中国地质学会了。他对中国

1　实际应为 7 卷。——编者注
2　应为 *The World We Live In*。——编者注

非常同情，在中国终身服务，已早具决心，在他到中国的二十六年中，仅于民国二十二年参加世界地质学会回过美国一次，这亦就是他最后的去美国之一次。

葛利普先生待人和蔼，视年长者如弟兄，年幼者如子侄，从不与人龃龉，因此每闻人有蛮横之行为者，常不相信，盖以个人的情形来判断，不了解人类何以有暴行也。初到中国，曾到唐山研究地质，看见多人，嬉笑追随，葛先生从不瞋怒，濒晚散归的时候，犹颔其首说"再见"，"再见"。民国十八年他的高足赵亚曾先生在云南遇匪被戕，他非常伤悼，赵先生的长子松岩那时才十三四岁，到了北平，他抚如自己的儿子，时常教和他同桌吃饭。我还记得他给松岩买了一顶空军皮帽子，现在松岩之入空军，或未始不是前缘已定吧。

本月二十日上午裴文中先生由北平来信，说他当日咯出血很多，我们看了很替他的健康担心，那知到夜里高振西先生和裴先生再发来电报和航空信，说这位大师已于五时四十五分与世长辞了，享年七十六岁。虽说葛利普先生已经死了，而由他的赞助奠定了中国的地质学术基础，则是永久不会磨灭的。

——原载重庆《大公报》1946 年 3 月 28 日，
亦载天津《大公报》1946 年 4 月 11 日

In Memoriam: Amadeus W. Grabau[1]

By Wong Wen-Hao[2]

Not only Chinese geological circles but also world scientists will grieve at the loss of Dr. Amadeus W. Grabau, the famous American geologist who died in Peiping on March 20 after offering the longest service and the utmost contribution to China of any foreign scientist. His twenty-six years' strenuous work laid a solid foundation for Chinese geology.

The exhaustive, devoted and studious Dr. Grabau was born in the State of Wisconsin in the year 1870. His grandfather was a German who settled in the New World after 1838. His profound interest in natural science during his boyhood impelled him first to take up botany and geology and later to do research in mineralogy at M. I. T. as well as archaeology in the Institute of Natural Science. He became assistant professor at M. I. T. before entering Harvard University, where he received a D. Sc. degree. He had been teaching geology at Columbia University when he was invited by the Chinese Government to become director of the National Geological Service and simultaneously professor in Peking University in 1920.

Dr. Grabau's more than twenty years in China resulted in the fact that more than three-fourths of all Chinese geologists are either directly or indirectly his disciples. With the moving of the National Geological Survey and Peking University southward after the outbreak of the war, the old teacher who had been suffering from podagra, had to remain in Peiping where he later endured great hardships, both materially and spiritually. As a result of the Pearl Harbor attack, the sick geologist was interned by the Japanese first in the German Hospital in the city and then at the former site of the British Embassy. Being ill-fed and

---

1　Reprinted with special permission from *The China Magazine* for September, 1946.

2　Vice President, Executive Yuan, The National Government.

completely cut off from the outside world, he was so much changed that not even his friends and students who came to the old capital at the end of the war could recognize him.

A War Casualty

The liberated geologist found himself a pauper, but the C.G. I D. and the Chinese Cultural Endowment Committee soon resumed his salary and rendered him every possible financial help. He was eager to return to his original work, but his newly recovered health did not allow him to do so. Neverthe-less, he insisted on staying at the department. The authorities were unable to house him since all the dormitories had been occupied by another organization. However, they managed to repair three rooms behind the museum of the department in Fung Shing Hu Tung as his quarters where not long afterwards he took his last breath.

Dr. Grabau wrote more than two hundred books, totaling more than twenty thousand pages. Such famous works as *Non-metallic Mineral Ore Deposits, Standard Fossils of North America, General Geology, and Geological History*, were all written before he came to China. During his stay in China he published three books, namely, *Mongolia; China's Geological Outlook*, two volumes; and *China's Archaeolog*, eleven volumes. In his later years he devoted himself to the study of the theory of global forces in which he envisaged, according to geological history, a singular phenomenon will happen throughout the world if the ocean at a certain geological period overlaps the continent.

He held that water does not come from one place and ten flows somewhere else; it is entirely the effect resulting from the expansion and contraction of the earth just like human forces. Five volumes of this theory had been published and the rest were ready for the press at the time when he was interned. Even in the concentration camp he did not forget his book writing.

Taught Even When Ill

Dr. Grabau was a very responsible professor and was never absent one day from his class unless he was confined to bed. Unable to walk, he usually came to the university by ricksha and went to the classrooms with the help of a stick. Day after day and year after year the professor never changed his habits. Even when ill, he would give lectures at home and many a time unknowingly trespassed on other professors' hours when talking on an interesting subject. The treatment he received from the Peking government did not in the least interfere with the professor's arduous work. When other faculty members did not come to attend classes because of their dissatisfaction over the delayed payment and the deduction of their salaries, he would take over their places furiously crying: "I am ready to take any vacant hours." That explained why the students liked him so much and would surge up to help him upon his arrival at the university. Since his advice and teaching was open to all, he never failed to satisfy any inquirer.

Nor did he relax once in his research work. During the first few years when he could still walk with great difficulty, he went to his office at regular hours but was later compelled to do his research at home. As his wife and daughter did not come to China, his secretary who typed his notes and managed all his books was the only person to stay with him, and so he was free to devote all his time to his work. Fossils and books served as his intimate companions.

Worked for China

With conducting research and rendering service as his goal, the famous geologist never thought much about his salary. He was paid in national currency as any other Chinese; therefore, he had no savings nor any money to support his family in America. All he had was spent on books which he put down in his will as donations to the Chinese Geological Society. His deep sympathy toward China urged him to sacrifice his life working for her. He left China only once when he went back to the States to participate in the World Geology Society Meeting.

This great friend of China was kind to everyone, looking upon the elders as his brothers and the young ones as his children. Not only that, he never had contentions with others and could never be convinced of the occurrence of riots. Being a stranger to this country, he was once mocked by town boys while on a geological study in Pangshan. Nevertheless, he was not the least bit angry and bade them goodbye on his departure.

His love for his students was fathomless. In 1929 when one of his beloved students, Chao Ya-tseng, was murdered by bandits in Yunnan province, he voluntarily took up the responsibility of caring for his son, and with his assistance the young man was able to join the Chinese Air Force afterwards.

Although this 76-year-old man is now dead, his selfless efforts in laying the cornerstone of geology for China will always be remembered by the Chinese people and his working spirit will no doubt forever shine in the pages of human history.

——原载 *The China Magazine*, 1946, vol. 16, no. 5
本文录自 *Quarterly Bulletin of Chinese Bibliography*, 1946, vol. 6, nos. 1–4

Prof. Amadeus W. Grabau

By H. Dighton Thomas

Amadeus W. Grabau was born of German stock at Cedarburg, Wisconsin, on January 9, 1870, his father and paternal grandfather being Lutheran Church pastors. His grandparents had left Germany in the middle of last century to seek refuge in the United States from persecution for refusal to conform to the practices of the reformed Lutheran Church. Perhaps it was this ancestry which bred in Grabau that stubbornness and refusal to accept current geological dogma without challenge which characterized his career. His radical opinions, expressed with a forthrightness not always to the liking of more conservative minds, touched not only American and Asian geology, but also impinged forcibly on the fundamentals of world geology.

At the age of fifteen, Grabau was apprenticed to a bookbinder in Buffalo, N.Y., but he continued his education at evening classes, discovering the delights of botany and geology. His ability in a correspondence course in mineralogy led Prof. W. O. Crosby, of the Massachusetts Institute of Technology, to offer him in 1890 a post at the Boston Society of Natural History and a special studentship at the Institute. The young man now came into contact with a brilliant gathering of teachers and he responded readily to their influence. His interest in physiography was sharpened, but to it was added a lively knowledge of marine bionomics and, through Alpheus Hyatt and R. T. Jackson, of palaeontology. He graduated in 1896, and entered Harvard University in 1897, where he took his master's (1898) and doctor's (1900) degrees. By way of Tufts College, the Rensselaer Polytechnic Institute and the Geological Survey of Michigan, Grabau passed to a lectureship in palaeontology at Columbia University in 1901, becoming full professor in 1905.

Grabau had already published a number of studies on Pleistocene geology, such as "The Pre-Glacial Channel of the Genesee River" (his first paper) and on glacial phenomena of Cape Cod and of Glacial Lake Bouvé. But the richly fossiliferous Lower Palaeozoic and Devonian rocks of New York State were an equal attraction, and he published papers on their faunas. In addition, he speculated (sometimes from unsound premises, as with the Fusidae) on the phylogeny and bionomics of the fossil groups-Devonian fishes, Palaeozoic corals and coral reefs, graptolites, gastropods, etc., all passed beneath his scrutiny. These essays were related to numerous others on the classification, nature and formation of sedimentary rocks and of salt deposits; for example, his book *Geology of the Non-Metallic Mineral Deposits other than Silicates. Vol. I. Principles of Salt Deposition* (1920). Grabau's widely ranging interests in stratigraphy, paleogeography, palaeontology and sedimentation were, however, inter-related in his mind, and were synthesized into a whole in his "Principles of Stratigraphy" (1913). He stoutly advocated his views at meetings of the Geological Society of America, where his clashes with E. O. Ulrich and A. F.[1] Foerste, who held other opinions equally strongly, became legendary. But the storms of the meeting-room were always followed by peace-making discussions afterwards.

The War of 1914–18 brought a crisis in Grabau's affairs. In an America where the teaching of German at public schools was banned and where streets and places with German names were re-christened, any defence of German literature, arts and science could not be tolerated. But Grabau commended these contributions to world culture, stubbornly refused to explain his views more fully, and ultimately left Columbia. China seized the opportunity and offered him the post of chief palaeontologist to the National Geological Survey of China and professor of palaeontology at the National University of Peking, which he accepted in 1920.

Grabau plunged with vigour into his new tasks, and quickly built up a flourishing school at the University of Peking. The mass of palaeontological material in the collections of the Survey was a mine he explored eagerly. The faunas of China were revealed, with adequate descriptions and figures, in a flood of monographs and papers by Grabau and by students he had trained. His own palaeontological contributions were more particularly on the Palaeozoic corals and brachiopods, though they embraced other groups from almost every age, and inevitably included discussions of their bionomics. The implications of this work on Chinese stratigraphy were quickly grasped by Grabau, and within four years of his arrival in that vast country he issued the first volume of his *Stratigraphy of China* (volume 2 came out in 1928), wherein he put forth new hypotheses to solve the difficult questions, while he later published a series of papers on "Problems in Chinese Stratigraphy". His *Permian of Mongolia*, volume 4 of "Natural History of Central Asia", was the vehicle for a discussion of the Permian of the world, where his original ideas once more coloured a long-drawn-out debate. He held strong views on the migration of geosynclines, and in a number of papers on the pulsation theory strenuously advocated universal transgression and regression for a given geological period. His *Rhythm of the Ages*, published in 1940, is characteristically stimulating and full of ideas.

The Geological Society of China honoured Grabau by founding the Grabau Gold Medal, of which he was the first recipient in 1925, while later it celebrated his sixtieth birthday by dedicating volume 10 of its Bulletin to him. Its preface, signed by the eight leading Chinese geologists, is a moving testimony to the regard which he had won in his adopted country by the same inspiring enthusiasm, kindly understanding and homely hospitality that are not forgotten by his old American students.

In 1933 Grabau re-visited America and was pleased at his welcome and at renewing personal contact with his old friends and antagonists, Foerste and Ulrich. Though some of the old narrow prejudices persisted even later in some quarters, this visit did much to relieve the mental suffering he had so long endured. By then he was already painfully crippled by rheumatism which progressively worsened. He kept bravely on, however, and it is remarkable that such a mass of research work and of ideas could have been produced by one who suffered so severely. The Japanese invasion of China brought increasing difficulties to Grabau. When the Geological Survey and the National University of Peking moved to Kunming, Yunnan, in 1937, his illness forced him to be left behind; but he struggled on, formulating and publishing his ideas, while his

Chinese friends got food and money to him whenever possible. After the Pearl Harbour incident, he was housed by the Japanese in the old British Embassy in Peking, but the lack of food and attention and his utter hatred of the Japanese aggression told heavily on the old man. He was very ill bodily and mentally when he was liberated in September, 1945, and despite the care of the authorities of the Geological Survey, he died on March 20, 1946, after internal haemorrhage. He was a widower with one daughter.

——原载 Nature, 1946, 158 (4003)

葛利普教授悼会记

朱 夏

地质学大师葛利普教授，于民国三十五年三月二十日在北平城内逝世。噩耗传来，学术界人士莫不同声哀悼。本会与中央研究院地质研究所及中央地质调查所二机关特于四月二十日在北碚联合举行追悼大会，以寄哀思。是日上午十时在中央地质调查所礼堂开会，到学术界人士、研究所及调查所同人暨本会会员凡百数十人。会场布置简单肃穆，中悬葛先生遗像，白发皤然，神采奕奕，其下罗列翁文灏、朱家骅诸氏致送之花圈，四壁遍悬挽联及诔词。葛氏遗著数十种，亦一并陈列。遗教常存，哲人竟萎！与会者追怀硕德，莫不怆痛弥深。大会由

俞建章氏代表李四光氏主祭

全体肃立致哀后，俞氏致词，略谓：英美德法各国研究地质科学之历史，均在百年以上，我国则开始在民国纪元以后，在此短短三十余年之中，虽未必遽能与欧美并驾齐驱，然亦卓然可以自立。究其原因，实由于诸先进之领导有方，而葛先生之丰功伟绩，尤不可没。葛先生学识渊博，著作宏富，在来华以前，即已名满世界，来华以后，对于中国尤具深厚之同情，早以终老斯土自许，执教北大二十余年如一日，恺悌慈祥，诲人不倦，回忆本人（俞氏自称）当初受业葛先生之门，尝以菊石化石之某一问题，一再请益，先生谆谆解释，不以为忤，盖先生虽生长美国，而视中国为故乡，视友生如家人，形貌虽殊，精神实与我人相融合。今先生之形骸虽逝，而其伟大精神，犹永铭我人之心目。赓继前规，发扬光大，其责端在我人，愿以此告慰于先生在天之灵。词毕并宣读浙江大学，北京师范大学地质系及翁文灏等氏之唁电。继由

李春昱氏报告葛先生生平事略

略谓：葛氏生于 1870 年，诞生于美国威斯康辛州，行二，有弟 John 现居美国，三十一岁与安亭女士（Mary Antin）结婚，生女一，现安亭夫人及其女

公子亦均在美。先生十一岁丧父，十五岁习订书业，同时在夜校攻读，好习植物学；后入麻省理工大学，习地质矿物，自然地理等学科，并在波士顿博物馆研究，二十六岁毕业，得学士学位；二十七岁入哈佛大学，三十岁得博士学位；嗣即任哥伦比亚大学讲师，三十二岁升任副教授，三十五岁复升任教授，自此执教哥大直至四十九岁，其间曾一度赴欧游历，遍经德法奥瑞（典）诸国，在执教期间，著有《非金属矿床》、《地层学原理》等巨著多种。民国九年，先生年五十岁，应丁文江先生之邀来华任北京大学教授及地质调查所古生物研究室主任技师。先生患风湿症，不良于行，但来华初年，仍时赴野外工作，曾详细研究河北唐山附近寒武奥陶纪地层，其后因病势日增，策杖而行，仅能致全力于室内研究，先后完成《震旦纪之研究》《中国地史》《蒙古之二叠纪》等巨著，及《中国古生物志》多种。抗战初起，北平沦陷，先生因病滞留，初居豆芽菜胡同继续著述，以阐扬其所倡之《脉动说》。珍珠港事变后，为日寇集中于德国医院，继迁至英大使馆，身心疲苦，饮食菲薄，初仍不废著述，继续完成《脉动说》七卷，及《我们生存的世界》一书（共二十三章）。嗣以环境日劣，体力遂致不支，幸赖其秘书吴兰芝女士（Volange）为之扶持照拂。北平光复后本人（李氏自称）抵平趋访，见先生面容消瘦，神智亦不甚清明。当时生活仍极艰苦，中央地质调查所及中华文化教育基金董事会均设法予以接济，北京大学及教育部朱部长亦均有所馈赠。先生神智虽衰，而每与人言，犹辄以未能早日恢复工作为憾事。去年十二月二十四日，先生迁入地质调查所北平分所，继续养病，惟病势终无起色，今年三月循环系统亦起病态，至三月二十日，呕血甚多，是日下午五时四十五分，遂与世长辞！遗体遵遗嘱，于三月二十二日举行火葬，经北京大学教授会之一致通过，骨灰将下葬于北京大学之地质馆前，并将由教育经济二部联合呈请国府予以褒扬，以彰先生对中国学术界之伟大功绩。李氏于述及先生逝世情形时，潸然泪下，悲不自胜，与会者亦莫不动容，最后李氏并列举事实，以称颂葛先生之学识广博，工作热忱，以及其待人接物之虚怀若谷及一贯对中国之深切了解与同情。继由

尹赞勋氏报告葛氏生平工作

略谓：先生于十九世纪之末叶，是时理化科学已有良好基础。自然科学正当发扬光大之际，硕学鸿儒，声华辉映，生物学有达尔文（Darwin），地质学有许士（Suess），地形学有彭克（Penck），古生物有季特尔（Zittel），构造

地质学有汉谟（A. Heim），均为一代宗师。先生生兹盛世，早年即已获得丰富之学识，复从名师益友游，所得更多。如哈叶脱（Hyatt）之生物进化理论（Recapitulation, Acceleration, Retardation），台维斯（Davis）之地形，克罗斯贝（Crosby）之冰川学，华尔脱（Walther）之岩石生因论（Lithogenesis），以及沙勒（Shaler）、彭克（Penck）、许士（Suess）之著作，对于先生尤有深厚之影响。同时契友复有乌利计（Ulrich）及巴士勒（Bassler）、福斯脱（Foerste）诸人，益收切磋之功。故先生之成为一大儒，其时代背景，良亦有以促成之。先生三十岁得博士学位，其论文为 Physiology[1] of Fusus and its allies（一九四四出版），为研究腹足类化石及腹足类生物进化之一重要著作，材料丰富，研讨精详，叙述与解释并重，不愧为一名著。一九〇九至一九一〇年，先生完成《北美标准化石》一书，集北美古生物研究之大成。不久（一九一三）又著《地层学原理》，都一千一百余页，分为七篇三十二章，特别注意于 Lithogenesis 及 Bionomy，此书受 J. Walther 之影响颇多，继往开来，足与许士及季特尔之巨著比美。一九一七年起，先生研究笔石之生活状态，持论与路特曼（Ruedemann）相反，曾引起不少论争。一九二〇年先生完成《非金属矿产》（*Non-metallic Mineral Deposits other than Silicates*）之第一部：《盐矿沉积之原理》（*Principles of Salt Deposition*）为研究盐矿地质之权威著作。一九二〇至二一年，完成地质学教科书（*Textbook of Geology*）二巨册，共一千九百余页，附图亦多至一千九百余。第一册述 Physical Geology，先叙岩石，后论风化，其方式属于一般教科书而有其特殊风格与优点；第二册述历史地质，以地层发育与生物进化分别论列，并采纳许多欧洲材料，为一般美国教科书所不及。一九二二至一九三六年，先生在中国工作，著古生物志多种，自奥陶纪至二叠纪，自珊瑚至腕足类等，均有所记述与发明。同时并著《中国地史》（*Stratigraphy of China*）完成二册，分述古生代与中生代，此外有《中国地层学上之问题》（*Problems of Chinese Stratigraphy*）及《新生代与灵生代》（*Cenozoic and Psychozoic*）二文，论中生代以后地层特详，可视为《中国地史》之续编。一九三一年著《蒙古之二叠纪》，中论并及苏、印、对世界二叠纪之研究，贡献甚大。一九三四年起先生致力于阐扬其"脉动学说"（Pulsation Theory），分古生代为十四个脉动期（1. Sinian, 2. Taconic, 3. Cambrian, 4. Cambrovisian,

---

1　应为 Phylogeny。——编者注

5. Skiddawian，6. Ordovician，7. Siberian，8. Siluronian，9. Devonian，10. Fengninean，11. Visemurian，12. Doubassian，13. Uraliuskian，14. Permian），其间之间脉动期，亦略赋以名称，实如地史学上之一革命性的理论。预计阐述全部须十四巨册，先后完成者，已得七册，已出版者五册。关于脉动说之辅助理论如 Paugaea 及 Polar Control 之研讨，亦各有专文。一九三五年以后，先生复主人类起源于西藏高原之说，谓由于喜马拉雅山之隆起，原在喜马拉雅地向斜北岸之 Dryopithecus，因环境变异，逐渐进化而具人形，北迁至塔里木盆地复四向散布，向东如北京人，向南如爪哇原人，向西如 Eoanthropus 及 Palaeanthropus，此一崭新理论，自有其伟大的价值。纵观先生毕生，治学至广，而其成就光辉灿烂如日中天。致其成功之原因乃时代环境之启迪，良师益友之敦促，固为重要之原素，而先生足迹所至，于美欧亚三洲之地质均能有丰富的智识，乃非常人所能及，且为学至勤，孜孜矻矻，耄而不倦，更足供我人之效法。其为我地质学上之一代巨人原非幸致也。尹氏词毕，来宾

唐佩璜、卢于道二氏相继致辞

均对葛氏之成就，备致钦崇，对葛氏之殂谢，备致哀悼，而于葛氏晚年困顿敌手之不幸遭遇，尤为扼腕，而深表感慨于良好研究环境之难得。二氏词毕，时已逾午，大会遂隆重闭幕。无限哀思，迄仍弥漫于与会者之脑际。

祭文

时维中华民国三十五年四月二十六日，国立中央研究院地质研究所所长李四光谨以时馐清酌，致祭于故葛利普教授之灵前，曰：公之降生，原在美国，矢志地学，一心一德，学识精微，器宇宏阔，早着令名，寰球是则。应聘来华，讲学北大，循循导诱，谆谆启发，桃李盈门，春滋吾夏，丰功伟绩，不在禹下。我国地质，初具雏型，提之携之，赖公有成，缅怀贤哲，景仰群伦。忽传噩耗，病逝北平，燕云遥望，涕泗沾襟，一樽设祭，魂其来歆，呜呼哀哉。尚飨。

祭文

时维中华民国三十五年四月二十六日，国立重庆大学地质系系主任俞建章暨全体师生谨以时馐之奠，致祭于故教授葛利普之灵前，曰：呜呼我公，

士林所宗，精研地质，贯澈始终。三十而立，七十心从，学既渊博，器亦宏通。公之执教，循循善诱，公之持身，卫道自守，公之著述，名山不巧。缅怀哲人，旷代难有。毕生讲席，栽培后进，坐拥皋比，春风明镜，悠游杖履，乐天知命。一介不染，高旷宁净，乐育英才，咸被手泽。遽闻溘逝，痛失贤哲。甘棠遗爱，永留吾国。去矣导师，中心悱恻。呜呼哀哉。尚飨。

葛利普遗像

——原载《地质论评》1946 年第 1—2 期，亦见《科学》1946 年第 6 期

葛利普先生追悼会挽联

法度谨严，万夫辟易；

焦桐绝响，多士兴悲。

——中央研究院地质研究所挽

毕生地质精研，著述等身师不朽；

老去诲人弗倦，薪传弈世我兴悲。

——常隆庆挽

异域殊方，同申痛悼；

他乡桃李，竞发新枝。

——新疆地质调查所挽

痛失导师。

——中国科学社

师事逾廿年，仰钻高坚，望尘莫及嗟驽骀；

仙游际三月，江山巴蜀，闻耗追思痛杜鹃。

——王恒升挽

奠定华夏大地史；

忆望春风十年前。

——杨敬之、萧安源、杨登华挽

学无止境欤，年逾古稀犹手卷；

师真不朽矣，他乡弟子续薪传。

——刘祖彝挽

大名共泰岱齐高，有楷模型留万卷文章垂不朽；

遗爱与神州并寿，任沧桑脉动千秋俎豆荐常新。

——中央大学地质系全体同人拜挽

廿载育生徒，瘁毕生精神贡献地质学术；

一朝成永诀，痛国际导师中外同声举哀。

——矿冶研究所挽

快意时能念及少辈之流，于今无几；

大人者不失其赤子之心，惟公有诸。

——南延宗挽

皓首穷经，教后辈吐艳扬葩，青史居功堪慰老；

白头改籍，过数月归真反璞，北庭想像倍怆神。

——谢家荣率矿产测勘处同人拜挽

化雨记东来，讲习廿年，薪火舟传皆子弟；

燕云愁北望，宫墙万仞，钵衣私淑失宗师。

——中大地质系同学会拜挽

学术果天下为公，航海远来，群钦廿六年善诱新材，教泽与神州比寿；

战争信不祥莫大，哲人竟萎，差幸东西陆竞传巨著，真知共大块长存。

——地质研究所、地质学会、地质调查所同挽

治学有方名一世；

诲人不倦是良师。

——苏孟守挽

述作最丰，伟著共欣传后学；

论交至笃，同仁齐恸失宗师。

——李四光挽

著述早名山，海内群贤尊祭酒；

桃李叨末座，公门遗爱系哀思。

——俞建章挽

杏坛主讲，不厌不倦，世间真罕觏，懿范同怀挥涕泪；

东巷羁囚，且勇且忠，天上倘有知，灵旗遥降护宫墙。

——重庆大学地质系全体同人挽

学贯人天，万卷文章悲彩逝；
胸罗今古，满门桃李泣春风。

——四川地质调查所挽

哲人其萎。

——中央研究院植物研究所

坐春风沐化雨，廿年前事；
著地质倡新说，一代宗师。

——侯德封、王炳章同挽

悼葛利普师

学府开先觉，熊光炯炯燃。师来新大陆，先觉觉后贤。地层古生物，精研复广研。岩矿及种种，师也实开先。奠此良好基，前后复经年。衰政久失轨，欠薪顾怡然。北地天寒早，高年步履艰。学海多风波，就教列门前。不辞辛辛苦，指讲答当然。何期我愚鲁，仰钻叹高坚。一别三数载，曾亦瞻师颜。矍铄应犹昔，神思若壮年。焉知三载后，国步际迍邅。举世殇烽火，凶锋肆北燕。幸鲁灵光在，犹堪作少年。国土重光日，长歌巨乐天。不图风■[1]悴，奄忽去人间。匆匆七六载，小半住幽燕。是第二家乡，有弟子薪传。良师终不朽，古道白云端。秉笔情难尽，千秋亦泣然。

乃戢兵戈乱若丝，愁闻天畔失良师。
光阴念载埋研讨，华发婆娑乐教施。
曾痛高材悲入室，勉支晚景苦维持。
太平洋上文星暗，第二家乡焕好枝。

——学生李陶敬挽

中国地质学会花圈一个
地质研究所花圈一个
中央地质调查所花圈一个
翁文灏花圈一个
心理研究所花圈一个
中央研究院花圈一个

---

1 此字在原刊中印刷不清，无法辨认。——编者注

生物研究所花圈一个
朱家骅花圈一个
物理研究所花圈一个
动物研究所花圈一个
地理研究所花圈一个

北碚地质调查所：敬悉追悼葛师，倍增哀思，谨肃短电，藉表痛沉。

——北平师大地学系师生同挽

地质调查所葛利普先生追悼会：哲人殒落，痛悼同深，谨唁。

——浙江大学

中央地质调查所李赓阳兄：兹悉在碚订期开会追悼葛师，谨电请代备花圈敬献

——瑸（田奇瑸）

——原载《地质论评》1946 年第 1—2 期

悼葛利普教授

陶 钝

在《消息》杂志第二期上看到了葛利普教授在北平逝世的记载，使我顿时感觉到深刻的凄怆，为了这个民主中国的朋友，有辉煌成就的老地质学家的死，我心中说不出的悼惜。

葛利普教授是美国人，年幼时当过工人，依靠自己的苦学才成名的，他来中国的时候已经是世界知名的地质学家，他在国立北京大学当教授垂三十年，由于他那种孜孜不倦埋头苦干的精神和慈爱优容的作风，在中国得到了广大青年学生的尊崇和爱戴，在他这种鼓励和培养之下，北京大学地质系出了一批不辞艰苦，愿意献身中国地质研究的学生，他们背起了帆布包，携带着铁锤和铁锹爬上人迹罕至的高山，发掘那荒僻的古地，他最得意的学生赵亚曾到云南去考察地质，深入人烟稀少的村镇，夜间被土匪枪杀了，这消息传来的次日，葛利普含着无限的悲痛在上课时把这个不幸的消息传达了，他两只手按着教桌站起来，因为两只残废的脚不吃力，使身子不住的颤抖，他颤巍巍的用沙咽的声音说："为了纪念赵同学，大家起立静默五分钟。"说着，两行无声的泪水像连珠似的滴在教桌上，同学们都被他这种伟大的国际友爱精神感动了，一个外国人，对中国青年学生给予如此热烈的同情和热爱，我真不知道今天那些以屠杀中国青年为职业的国民党刽子手们，他们的心肠是什么造的。

葛利普教授在中国地质学研究上的成就是辉煌的，中国学术研究的落后可以说是无足称道，只有地质学，主要是"北京人"的发现和研究，博得世界上的重视，这成绩和葛利普教授是分不开的，他的两足是残废的，但阻碍不住他学术研究的热情，他从来不缺课，不请假，每次上课都带着饱满的精神和愉快的心情。对于地质学他有丰富的著作，他的地质学巨著已经完成了六册，还有一通俗性的著作《我们居住的世界》（*The World We Live*）[1]没有出版。

"七七事变"后，他困居在北平，太平洋战争爆发，日寇没有因为他是一个终日玩着岩石的残废老人而放松了他，他被关在孤老残废的集中营里，胜

---

1 应为 *The World We Live In.*——编者注。

利后才被从这个地狱解放出来，可是经过了日寇残酷折磨的老人，已经失掉了他的健康，终于被胃出血的恶疾夺去他的生命，年纪七十六岁。

在悼念葛利普教授之余，使我不禁想到中国那些国民党的"学家"们，拿着地质学作了他们升官发财的捷径，无耻的囤积居奇，大发国难财，这些人看到葛利普教授的终身事迹和忠实于科学研究造福人类的精神能不愧死。

葛利普教授也可以称为来中国的美国朋友的好榜样，他来中国教书，帮助中国的科学研究，不附带任何的野心和企图，他尊重中国的独立和民主，他热爱中国的青年，他也热爱中国。他是最忠实的学者，中国最伟大的朋友。这给了那些口头上帮助中国，实际上帮助反动派大运军火，挑动中国内战，想把中国作为殖民地的美国野心家们，是一个分明的对比！

——原载《山东文化》1946 年第 6 期

为中国地质学打下根基的外国人——葛利普

陈 光

亲爱的读者：当你翻开人类活动的历史，你随时随地可以发现许多千古不朽的人，他们以无比的才智品德，照耀着人类发展的前途，他们努力的结晶是人类福利的泉源，他们活动的累积构成了人类灿烂的文化，如果人类历史的浩渺有如幽邃的秋夜，那么他们不啻是嵌缀秋夜的繁星，而我这里所要介绍的葛利普教授，便是其中之一。

我们之所以要在这里纪念葛利普先生，并把他介绍于读者之前，不仅因为他是世界著名的地质学家，而且因为他是中国真实的友人，是他以他的后半生，帮助中国奠定了中国地质学发展的基础。他诞生于美国威斯康星州，林菁茂密的洋杉堡[1]，但他逝世却在中国金瓦红墙的故都——北平；他生于一八七〇年一月九日，殁于一九四六年三月廿日，享年七十有六。在民国九年，即一九二〇年，那时他四十九岁，即已应聘来华，任北京大学地质系教授，兼地质调查所古生物部主任，以一个外籍人士，知名于世界，而犹能以中国为本位，以发展中国科学为素志，廿七年间未尝离中国一步，终至入中国籍，为中国人，死于中国的土地上，有史以来，葛氏实在可算第一人了。

葛氏有犹太血荫，他祖父原来是德藉教士，为了争宗教自由，一八三九年才搬到美国，在水牛城创设宗教学院，自任院长。同行的千多人当中，有一个冯罗德上尉，他的女儿便是日后葛氏的生母，葛氏的父亲迁居洋杉堡，仍为牧师，家境很寒苦，所以葛氏幼年便在教会附设的小学里念书，十一岁时母亲死了，十二岁才进了洋杉堡的公立中学，十五岁的时候父亲继祖父任路德会宗教学院院长，他才跟随着来到水牛城。这一带终年温湿，冬季一月的均温在冰点以上，夏季七月的均温不高于华氏60度，全年雨水大约在50—200cm之间，是寒温带的林地地区，所以植物很茂盛，引起他对植物学发生很大的兴趣，他白天学习订书业，夜晚就在夜校里读书，每逢假日便到野外采集植物标本。后来，因为常常出没于山头谷地，常与风化之土粒，崩解的岩

---

1 即锡达堡，下同。——编者注

石碎片相接触，日子久了，欲窥自然奥秘之心愈切，兴趣才转向地质学与古生物学，因而选习麻省理工学院函授科之矿物学，成绩非常优异，一八九〇年得克劳斯贝教授之介绍，为波士顿自然科学会之助理员，是在那里他逢到了美国当代古生物学大家赫第氏，并在其指导下整理波士顿滨海一带海产无脊椎动物之材料。同时，入麻省理工学院为特别生，并在波士顿公立中学补习拉丁文及希腊文。第二年，一八九一年，葛氏廿岁了，始改为麻省理工正式生，兼任学生助理，半工半读，到一八九六年才毕业，毕业论文即为《水牛城以南十八哩溪地区泥盆纪之动物化石种群》。按纽约州西部，不仅植物繁茂，而且地层发育绝佳，化石种群完备，葛氏日后在古生物地层方面之成就实奠基于此，而以后工作领域之推展亦以此为起点。毕业后任助教一年，一八九七年才入哈佛大学研究院，从戴维斯氏习地文学，并在其指导下著《论波城左近冰期之古湖》一文。一八九八年被选入美国地质学会为会员，一八九九年得到哈佛大学理学博士学位，时年廿九岁，毕业论文为在姜克森氏指导下完成之《腹足类长辛螺及其同属之发生史》，一九〇〇年任仑赛累尔工学院[1]矿物教授，兼密契根[2]州地质调查所所员，时年卅岁，继因兴趣不合，复改任哥伦比亚大学古生物学讲师，兼纽约州地质调查所所员，时年卅一岁。卅二岁为副教授，卅五岁为正教授，此后继续执教以至一九一九年来华以前，其间曾一度离美赴瑞典出席世界地质学会，遍历英德奥诸国，得以比较欧美地质之发育，统计在哥伦比亚大学十九年间共发表著作四十六种，其中以《地层学大纲》及与夏谟教授合著之《北美无脊椎动物化石》最为世所称道，其他如腹足类及冰川地形之赓续研究，下古生代地层之详细划分，沉积表之分类等，均极有价值。在研究的过程中，遇到名词混淆，含义不清的时候，特别注意"正名"，必要时往往还铸造一些新字，以利系统研究的进行，如"假整合""不整合"等都是由他提供出来的，一直到现在都为举世所采用。

　　细细地阅读葛氏的生平，无异在阅读一部人生奋斗的过程，你可以看到：他从呱呱坠地，以至毕业于麻省理工学院，这廿六年间，他是怎样地突破环境的困阨，一步步地走向成功之路。仿佛一个作曲家，如果不饮尽人生的苦汁，便谱不出人生的真谛；如果他没有前此的勤苦准备阶段，他日后也许便不可能有如许的成就，而我们之所以为我们——同时也为他庆幸，并不是单纯地因为他个人能有成就，而是因为他的成就，变成了整个人类成就的一部

1　伦斯勒理工学院（Rensselaer Polytechnic Institute）。——编者注
2　现一般译为密歇根。——编者注

分。他在寒苦中诞生，在寒苦中成长，然而他具有一种先天禀赋的坚毅素质，他血液里奔流着一种不屈不挠的反抗精神，像一株被寒天冻地封闭了的幼芽，在雪溶于阳春的时候要突破地面开花结实，终于他胜利了。从他学成以至来华以前，廿三年间，他无时不在散发一种成熟的芬芳，他长期努力的成就，累积起来有如满树压枝的果子。其间，虽然他失去了一条腿，虽然不尽如人意的婚后生活，在他心上抹了一层不可磨灭的暗影。但由于他强烈的生的意志，战斗的意志，与顽强的向上意志，他并没有心灰气馁，他反而在他四十九岁的中年，远渡重洋，投身异国，把他成熟的芬芳传播到太平洋的这边，带到中国这片贫瘠的土地上来。在这片过分硗薄的土地上，科学的幼苗是很少不中途夭折的，甚至我们往往还可以看到许多一心献身科学的人，在多年的涵蕴以后，到了成熟结实的时节，不幸不及开花结实而便与世长辞。葛氏来中国以后，不是不曾遇到过频频的政局变动，社会的纷乱，与生活的不安，然而像一棵长成了的树木，不再惧怕外界的摧残，他仍能屹立在不利的环境中，坚守他的岗位，继续他未完成的工作。葛氏在中国廿七年，所完成之著作有一百五十七种之多，其中以脉动学说、中国地层学、古生物志等为主，其他则散见于《中国地质学会志》及单行本。他往往深夜不眠，烟酒不绝，辄夜直书，天明脱稿。于古生代珊瑚之研究，则特倡导作连续之切片，以明察人体发育与种族系统之关系。

但葛氏对中国地质界之贡献，初不只限于著述方面。在他来华以前，国内古生物学一项素无专人执教，而化石之鉴定则悉赖寄运外国，请国外专家鉴定，及他来华以后，此类问题乃均迎刃而解。在教学方面，他不仅是一位热心负责的教师，而且还是一个善于奖掖后进启发后学的导师。古生物学之教学，每因纯型态之记叙，而易流于枯燥支离，但他于记叙中诱悟生态演化之理论，济以绝世的口才，每能引人入胜，使学者的兴趣与日俱增，而且此种深切的兴趣，不因为毕业离校而日减，所以从他学习的人很少有中途易辙的。到现在，国内古生物学专家不算缺少，可以说都是葛氏的功绩。于校内教学以外，葛氏抵平的头几年，又以地球与生物演化十二讲，为通俗讲演，使社会人士对陌生的地质科学有所了解。讲演概在星期日下午，每两周一次，每次二小时，然而往往超过三小时，由此可见葛氏对于通俗普及教育之热心，此后国内学习地质的才越来越多。葛氏又与丁文江、翁文灏诸氏倡议组织中国地质学会，联系国内地质学者，藉以砥砺切磋；又倡议刊行《中国地质学会志》及《古生物志》，用外文刊发著作，公诸国外同好，藉以观摩交换。自此中国地质学的发展才蒸蒸日上，而国际地质界才知道在一切都落伍的中国土地上，地质科学的幼芽正欣欣向荣与日俱长。

葛氏全名为 Amadeus W. Grabau（Amadeus 与昔西班牙无敌舰队之 Armada 发音有相似在），所以生前葛氏友好，因为他富于勇往直前的精神，每每欢喜以此相戏。现在葛氏不幸因为患胃溃疡仙逝北平，可是他那勇往直前的身影，将永远活在我们的记忆里，他对地质学的贡献将使他永远不朽，他后此的功绩将使他在中国地质学的发展上永远闪烁照耀如同东方的启明星。

——原载《科学时代》1946 年第 8—9 期

中央地质调查所古生物研究室主任葛利普褒扬案

国民政府 三十五年九月十六日褒扬葛利普令

中央地质调查所古生物研究室主任美籍葛利普，以地质学专家来华服务，垂二十余年，研究精深，著述宏富，傍搜博采，有本有原。历任北京大学地质系教授，诱掖人材，造就綦众，对于我国地质工作，实为极大贡献。七七事变以后，疾病羁缠，未克南来，虽处境恶劣，艰苦备尝，然犹闭户潜修，写作不辍，勤劬笃学，尤堪敬佩。兹闻溘逝，悼惜良深，应予明令褒扬，以彰贤哲。此令。

抄原呈

据本教育部所属国立北京大学及本经济部所属地质调查所陈称，该校地质系教授及该所研究员曾任古生物研究室主任葛利普，在北平逝世以及其身后萧条情况，呈请优恤及拨发丧葬费等由到部。据此查葛利普（Amadeus[1] W. Grabau）为美籍知名地质学家，在未来华之前，曾任哥伦比亚大学教授将近二十年。著作宏富，如《地层原理》、《北美标准化石》、《非金属矿床》及《普通地质与地史学教科书》，皆系世界闻名巨作，学问渊博，声誉卓著。民国九年，应中国之聘，来华任北京大学地质系教授兼中央地质调查所古生物研究室主任。二十余年来，作育生徒，为数甚多，吾国专门研究地史及古生物学人士，出其门下，实居多数。研究著作，与年俱增，如《中国地史学》二卷、《蒙古之二叠纪》一卷、《中国古生物志》十一卷[2]以及《地球脉动说》[3]五卷等，皆其来华后重要作品，对于中国地质工作，实有极大功绩。七七事变，该员以疾未能南来，太平洋大战以后，尤备尝艰苦，然犹闭户研究，写作不辍。惟终以营养缺乏，日渐羸弱，竟于今年三月二十日下午五时半，因胃出血逝

1　原文为 Amadem，误。——编者注

2　实际应为 7 卷。——编者注

3　即《脉动理论下的古生界地层》（*Palaeozoic Formations in the Light of the Pulsation Theory*），一共出版了 5 卷。——编者注

世。该员在华薪俸为数无几，然其历年所收藏之地质书籍约二千余册，则已于民国二十一年预立遗嘱，于其死后全部捐赠中国地质学会，是其在华终身工作，已早具决心。综其前后二十六年，精勤工作，未尝稍闲。此种服务精神，尤堪嘉尚，实非其他人员所易比拟。恳钧院转呈国民政府命令褒扬，以慰贤劳。似此学者，如在美国服务如此之久，则依照外国定例，自可优厚抚恤。今在华服务死后，尚有妻室留居美国，另有秘书吴兰芝女士（Miss Volange[1]）随侍在平，艰苦举　葛君遗体亦待安葬，并恳钧院特予抚恤葛君遗属美金五千元，并拨治丧费国币二百万元。当此与国外技术合作之际，更可籍以慰死者而励来兹。是有当理合会同备文赍呈，付乞鉴核施行。谨呈行政院院长宋。

教育部部长朱家骅
经济部部长翁文灏

　　——原载国史馆典藏：《中央地质调查所古生物研究室主任葛利普褒扬案》，1946 年 9 月 16 日，现藏中国第二历史档案馆；另载"国史馆"编：《中华民国褒扬令集初编》（九），台北：台湾商务印书馆，1985，5269—5270 页；《褒扬令》刊于《教育部公报》第 18 卷第 9 期（1946 年 9 月 30 日）。

1　原文为 Volqwge，误。——编者注

Obituary Amadeus William Grabau: An Appreciation

By Hervey W. Shimer

The death of Amadeus William Grabau in Peking, China, March 20, 1946, removes one of the great geologists of the world. Although he had lived in China during the last 26 years of his life, he still had many friends in this country who will mourn him personally and many more who will be saddened by this loss to science.

Dr. Grabau's research work lay principally in the fields of stratigraphy, paleontology, and evolution. He published a 2-volume *Textbook of Geology*, a large volume on *Principles of Stratigraphy*, and collaborated in *North American Index Fossils* (2 volumes). He wrote many papers in pure paleontology, especially on mollusks and brachiopods, describing a very large number of new species and many new genera. His descriptions of these are full, carefully done and well illustrated.

He was closely associated with Alpheus Hyatt for eight years at the Boston Society of Natural History. Hyatt during these years was brilliantly applying the ontogenetic theory to the evolution of invertebrates. And Grabau's contributions to this line of research are outstanding. In gastropods, corals and crinoids, in brachiopods and cephalopods, he shows that the ontogeny of an individual repeats the phylogeny of the group to which it belongs, that the geological sequence of these ancestral forms is in the order indicated by the development of the individual.

During his later years Dr. Grabau was especially interested in the development of his Pulsation and Polar Control theories. According to the former the evolution of the crustal features of the earth is brought about by the rhythmic rise and fall of sea-level, expressed in its transgressions (pulsations) and regressions (interpulsations). All continents are thus affected at the same time. During regressions there occur erosion, continental sedimentation, mountain folding and vulcanism. With this Pulsation theory is linked the Polar Control theory. According to this there is a "periodic shifting of the earth's crust (sial-sphere) through the impetus given it by the rotation of the earth on an axis of essential constancy of position." In other words, the poles are constant in position but the sial crust is not. Thus the continents shift to the poles, accounting for glacial climates, and away from them, accounting for mild climates. As the sial is partly submerged in the underlying sima, its movement produces mountains at its forward edge and fissures, with consequent vulcanism at its rear edge.

Dr. Grabau published four large volumes on the Paleozoic pulsations, but did not have time to complete his volumes on the Mesozoic and Cenozoic. However, in his *The Rhythm of the Ages*, which appeared in 1940, he not only sums up his theory as applied to the Paleozoic, but also outlines his evidence for the succeeding eras.

Although many geologists will not accept these theories, they will remain an incentive to research, and the mass of data which he collected in their support is a storehouse of information upon which future workers will draw.

——原载 *American Journal of Science,* October 1946, 244

浪淘沙
——葛利普教授纪念刊题辞

君去已多时!

梁坏! 山颓!

门墙桃李尽含悲!

留得神州新地史, 星日同辉!

才把凯旋卮, 一笑长辞!

名山事业后人思!

廿载他乡成故国, 魂也依依!

民国三十六年双十节

章鸿钊敬撰

——原载 *Bulletin of the Geological Society of China*, 1947, vol. 27

Professor Amadeus William Grabau: Biographical Note

By Y. C. Sun

Professor Grabau was one of the most eminent American palaeontologists and geologists of this century. The biographical note of him was written by late V. K. Ting in Grabau's 60th Anniversary Volume of the Geological Society. The present note is to give more details regarding his work and life during the recent 15 years. He was born in Wisconsin in 1870. He studied first in the Masachusetts Institute of Technology where he served as instructor in 1892–1897. After obtaining his Doctor degree at Harvard in 1900, he served as adjunct professor at Columbia University 1902–05, then as professor from 1905–09. In America he published in collaboration with A. W. Shimer a widely used book on *North American Index Fossils*. Later came: "Succession of Faunas in the Middle Devonian in U.S.A."; "The Hamilton Fauna of Michigan"; "The Phylogeny of Invertebrates, Chiefly Gastropods"; *Principles of Stratigraphy*; *Text-Book of Geology* (2 volumes, 1921) and *Non-metallic Deposits* (2 volumes, 1922).

From the beginning he accepted the famous American palaeontologist Dr. C. D. Walcott's suggestion and selected "Brachiopoda" and "Devonian Stratigraphy" for his research field, although his contributions cover subject ranging from Sinian to Quaternary stratigraphically and from Graptozoa to Pisces palaeontologically. By influence of J. Walther[1] he drew attention particularly to the problem of bionomy and lithogenesis, and he systematized all these principles in his great book *Principles of Stratigraphy*. His *Text-Book of Geology* is the first American Text-Book of Geology in which many European sections and illustrations are introduced. It is most popularly used in America and abroad, but at the time shortly before the Japanese invasion he felt that he wrote that book little too early because of the lack of Asiatic material and hoped to reprepare a new one in which

---

1 原文为 Walter。——编者注

new data of Chinese Geology will be given. His contributions are fundamental and exhaustive. He touched not only American and Asiatic palaeontology and geology, but also impinged forcibly on the fundamentals of world geology.

In 1920 he came to China and accepted the post of the professorship of palaeontology in the National University of Peking and Chief Palaeontologist to the Geological Survey of China under the directorship of V. K. Ting and W. H. Wong. In the Survey, he was in charge of the Palaeontological Laboratory and began to issue palaeontological monographs—*Palaeontologia Sinica*—over hundred numbers of which have already been published. He took the complete responsibility to train Chinese palaeontologists and geologists. He gave lectures not only on palaeontology, stratigraphy, but also on zoology for the college students and research workers. He also prepared Syllabuses in palaeontology and historic geology. His task had been successful and a dozen of Chinese palaeontologists had worked in their field under his direction and now occupy leading positions in geological institutions (Y. T. Chao, C. C. Tien, for brachiopods; Y. C. Sun, T. K. Huang, C. C. Yu, S. S. Yao, Y. S. Chi, S. Chu, for corals; Y. C. Sun and K. Hsu, for graptolites; Y. C. Sun and C. C. Tien, for cystoids and crinoids; S. Chen for Fusulinids; T. H. Yin and. K. Hsu and Y. T. Chao, for gastropods and pelecypods; Y. C. Sun and H. F. Sheng, Y. Wang and Y. H. Lu, for trilobites; C. Ping for insects; Y. C. Sun, C. K. Chao and C. C. Yu, for cephalopods). By his influence, J. S. Lee published an important monograph on *Fusulinidae of North China*. Professor Grabau's publications are very numerous and his achievements on geology and palaeontology of China had been known all over the world. He first published 2 volumes of *Stratigraphy of China*, later came 2 big monographs on *Devonian Brachiopoda of China*, also 2 monographs on *Palaeozoic Corals*. One monograph on *Ordovician Fossil from North China* and another monograph on *Silurian Fauna of Yunnan*. Besides these, he wrote "Problems in Chinese Stratigraphy" and published them in National Peking University Press. After he returned from Washington Congress, he had devoted himself to the Pulsation Theory and finished seven big volumes on *Palaeozoic Formations in the Light of Pulsation Theory*. Volume 6 is partly printed and vol. 7 will be printed. Since 1920 he published 146 papers (while his total publications numbering 291) which were issued mainly in *Palaeontologia Sinica* and the *Memoirs of Academia Sinica*[1], *Bulletin of Geological Society*, *Quarterly Journal of Science of University* as well as in other Chinese and foreign publications.

His great book on *Palaeozoic Formations in the Light of Pulsation Theory* is comprehensive and valuable for stratigraphers and palaeontologists. Many new ideas in his papers "The Palaeozoic Centers of Faunal Evolution and Dispersal", "The Rhythm of the Ages", "Foundamental Concepts in Geology and Their Bearing on Chinese Stratigraphy", "Significance of the Interpulsation Periods in Chinese Stratigraphy", "Classification of Palaeozoic System in the Light of Pulsation Theory", "Huangho River Plain of North China" are of great significance and have been almost accepted by many well-known palaeontologists and geologists. After the Pearl Harbour incident, he was sent by Japanese to the internment Camp in Peiping, but the lack of food and attention and his utter hatred of the Japanese agression toiled heavily on him, therefore he was prevented from completing the remaining volumes of his pulsation theory and the revision of his *Text-Book of Geology* (2 vols). Besides the above stated, two valuable monographs on *Fengninian Brachiopoda of China* and on *Lower Permian Brachiopoda of China* were prepared by him in his late years which will be published in the *Palaeontologia Sinica*.

He was always happy in life and welcome by all his friends and students. He took active service in organizing the Geological Society of China, the Palaeontological Society of China and the Peking Society of Natural History. He was the first recipient of Grabau Medal of Geological Society and also the recipient of Thompson Medal of the National Academy of Sciences of United States. During the war he worked hard all the time at No. 5, Tou Ya Tsai Hutung, West City, Peiping and wrote to the writer at Kunming in June 27, 1941 with the following words:

"Meanwhile I hope you will continue to thrive and carry on the old work as we did in the past but you will probably find very little change at Tou Ya Tsai. I have not seen the University since it was closed.

With best wishes to you and hoping that we will all meet in happier days that will come."

1　应为 *Memoirs of the National Research Institute of Geology, Academia Sinica*。——编者注

He had great hatred for Japanese and asked all his friends and pupils to leave Peiping and go to free China to continue their work. By his will, he contributed his private library to the Geological Society of China and the will was written in the year of his 60th Anniversary with late V. K. Ting and American Ambassador as witnesses. This shows that he had been always faithful to Chinese government and Chinese geological circle as well. He was very ill physically and mentally when he was liberated in September, 1945 shortly after victory and in spite of the care of the authorities of the Geological Survey and the National University of Peking, he died on March 20, 1946. He was buried in the compound of the Geological Department of National University of Peking according to his last will. The Geological Department of National University of Peking has decided to name the departmental library as Grabau's Library in honour of him. At the same time he has been awarded a testimonial by the Chinese National Government through the request petitioned by ministers of Economic Affairs and of Education. In addition, the council of the Society has also decided to issue a Memorial volume this year in recognition of his great achievements on geological science of China. His wife Mary Antin, daughter Josephine and two granddaughters Margaret and Elizabeth Ross survive him.

Scientific Career and Honours

1890.   Appointed assistant in the Mineral Supply Establishment at the Boston Society of Natural History.

1891.   Matriculation at the Massachusetts Institute of Technology.
           Appointed student assistant in the Geological Department.

1896.   Took the degree of Bachelor of Science.
           Appointed assistant in the Geological Department.

1897.   Entered Harvard University.

1898.   Took the degree of Master of Science at Harvard.
           Elected fellow of the Geological Society of America.

1899.   Appointed lecturer on geology in Tufts College.

1900.   Took the degree of Doctor of Science at Harvard.

Appointed Professor of Geology at the Rensselaer Polytechnic Institute.

Appointed member of the Geological Survey of Michigan.

1901. Appointed lecturer in palaeontology at Columbia University.

Elected Fellow of the New York Academy of Science.

Appointed member of the Geological Survey of New York.

1902. Promoted adjunct Professor at Columbia.

1905. Promoted Professor of Palaeontology at Columbia.

1910. Travelling in Europe.

1920. Appointed Chief Palaeontologist of the National Geological Survey of China and Professor of Palaeontology of the National University of Peking.

1922. Elected Fellow and Councillor of the Geological Society of China.

1923. Appointed Research Associate of the Central Asiatic Expedition.

1924. Elected Honourary member of the Science Society of China.

1925. Elected Foreign member of the Kaiserlich Deutsche Akademie der Naturforscher zu Halle.

Awarded the Grabau-Medal by the Geological Society of China.

1927. Elected Honorary member and Life Councillor of Peking Society of Natural History.

1928. Elected Correspondent of the Philadelphia Academy of Natural Science.

1929. Elected Research Associate of Academia Sinica.

1933. Participated the 16th International Geological Congress, Washington, U. S. A.

1934. Appointed Director of the Geological Department of National University of Peking.

Awarded King-Medal by the Peking Society of Natural History.

1937. Awarded the Thompson-Medal by the National Academy of Science, U. S. A.

1940. Received the Honorary Degree of Doctor of Science from Coneseus College, U. S. A.

1946. Deceased in the Compound of the Geological Survey of China, Peiping Branch, Peiping.

1947.   Buried in the Compound of The Geological Department of National University of Peking, Peiping.

葛利普著作目录（从略）。

——原载 *Bulletin of the Geological Society of China*, 1947, vol. 27

Amadeus Grabau: In Memoriam

By Sven Hedin (Stockholm)

It was with pain and grieved emotion that I received the news of Professor Amadeus Grabau's decease some time ago; and I felt, deeply and warmly, that I had lost a real and faithful friend, one who had been close to me in bright and gloomy periods for very many years. Around his bier were assembled in sorrow and regret the foremost palaeontologists of our time in China, America and Europe, and I am infinitely grateful that the gentlemen—Grabau's and my friends in China—who have wished to honour him with a special obituary memoir should have had the kindness to turn to me with a request for a pronouncement about this great man. Many scholars and great palaeontologists who have collaborated with Grabau in the National Geological Survey of China, or who have sought and found support in his wide and profound scholarship will in this memoir bear witness to his pioneering importance, and to the fruitful results of his indefatigable and solid scientific labours. For my own part, I can only make a pilgrimage in thought to his grave, with feelings of reverence and admiration, and devote a few simple words of gratitude and homage to his unforgettable memory.

During the years 1927, 1930, 1933 and 1935, when for longer or shorter periods I was staying in Peking, I very frequently had a real home in Grabau's house in the western part of the city. Never has a scholar shown a greater and more genuine and cordial hospitality to all his friends than did Grabau. One felt and knew that one was always welcome in his home. Leaning on his crutches and with his jovial smile he met his guests on the threshold, and from the very first moment one had the definite feeling that he was glad for the visit. If one was seeking good advice in difficult situations or for the drawing up of plans for a new journey of exploration, one might rest assured that with his clear and keen understanding he would find the right solution. It was a rare pleasure to hear him speak of new discoveries he had made in collections that had been handed over to him or to listen to his comments upon the advances made by palaeontologists in other countries, whose import he often interpreted in his own original and shrewdly accurate way. How often did I not sit in his study listening to his explanation of why he believed that the skeletons and crania of the oldest *primaeval humans* would one day be found in Central Asia. With an imagination so vivid that one might think he saw them himself he would describe how the Hominidae with the help of the branches of trees first learned to be "erecti" and walk on two legs; and how world-encircling was not the flight of his thought when he described how the migration of the continents was due to movements in the earth's crust, a thought which he referred to as the oscillation theory and which has in some, quarters, also here in Sweden, gained adherents.

Amadeus Grabau was a happy man. He was happy in his science, which opened for him limitless and grandiose perspectives over the history of our earth; and he was happy in the circle of his friends, where in his inimitable and inspiring way he gave expression to his bold thoughts and to a never failing wit. When one was a guest at Grabau's table it was less a matter of enjoying all the finest dishes the Chinese kitchen could produce, the delicious fruits or the sparkling wine, than of listening to the host's witty epigrams or his brilliant eloquence. But one had to be constantly *en garde*, for when one least suspected it the host might tinkle on his glass and declare: "Now Mr. X is going to make a wonderful speech to us". And then one was obliged to deliver a speech. Those who knew Grabau's habits at the dinner-table were therefore always ready with some anecdotes or stories that might with advantage be used in a speech. Thus what he most esteemed at table was wit—without, however, forgetting the pleasures afforded by a good kitchen. One always went to Grabau's dinners with tense expectation, and always with the conviction that one would have a good time. But what above all drew us to Grabau's table was the knowledge that we should there meet extremely interesting and celebrated men. Among the more regular guests were Dr. Wong Wen-hao, Dr. V. K. Ting, Dr. Hu Shih and Père Teilhard de Chardin. During the meal gaiety and high spirits prevailed; but afterwards the guests frequently formed small academic groups to discuss scientific problems. Grabau's home became a focus, a salon, for the academic circles in Peking, and his hospitality knew no bounds.

About midsummer in 1933 Grabau, accompanied by his housekeeper, the amiable Mrs. Woodland, V. K. Ting and others, journeyed over the sea and the continent to Washington to take part in the Geological Congress of that year. He stayed for a long time in his native country; and when he came back to Peking he brought with him a good deal of homesickness—for America, and it took some time before he could settle down to his old accustomed ways. But when on my way home to Sweden I spent a couple of weeks in Peking to arrange the affairs of the expedition with my friends Hu Shih and P. L. Yuan. I found Grabau glad, winning and full of ideas as of yore, and the doors of his home stood open for new guests.

Altogether unchanged, however, he was not. He had aged outwardly and was suffering from troublesome ailments. But in his soul the sacred fire burned with as clear a flame as ever, and he was looking calmly forward towards eternity and the solution of the greatest of all riddles.

I think when I pressed his hand at parting that I had a premonition that it was the last time; and I believe that he, too, had a feeling that we should never meet again, for his eyes were moist and lit with a melancholy sheen. Yet a decade was to pass before the news of his death came. Now he is gone, and it is silent in the once so gay and hospitable home in the western city of Peking.

But the memory of this great, good and noble man lives on, and it will go down blessed, honoured and loved, to a remote posterity.

——原载 *Bulletin of the Geological Society of China*, 1947, vol. 27

Minute on the Life and Scientific Labors of Amadeus William Grabau (1870–1946)

By William King Gregory (American Museum of Natural History)

Amadeus William Grabau was born of German stock at Cedarburg, Wisconsin, January 9, 1870. His grandfather was a protestant Lutheran clergyman and so was his father. At the age of fifteen he was apprenticed to a book-binder in Boston. He received his professional grounding as a geologist and palaeontologist in the Boston Society of Natural History (1890), the Massachusetts Institute of Technology (1892–1896) and Harvard University (1897–1900). In his formative period he also acquired a keen interest in the bionomics and evolution of marine invertebrates, under Alphaeus Hyatt and R. T. Jackson. He taught geology and palaeontology at Rensselaer Polytechnic Institute, Troy, New York (1899–1900), Tufts College, Massachusetts (1900–1901), Columbia University, New York (1901–1919) and National University of Peiping, China (1920–1946). In the New York Academy of Sciences he was a fellow, vice-president (geology) and life member.

A brief suggestion both of the wide variety of his researches and of the gradual development of his interpretations of facts may be attained merely by noting a few of the titles of his published writings, with their dates; while citation of the number of pages and illustrations in his major works may give a rough measure of his enormous productiveness.

Many of his investigations on the geology and palaeontology of the Eastern United States quite naturally dealt with specific or regional topics: such as the sand plains of Cape Cod (1897); the palaeontology of Eighteen-Mile Creek and the Lake Shore section of Eric County, New York (1899); the Siluro-Devonic contact in Erie County, New York (1900); Lake Bouvé, an extinct glacial bed in the Boston Basin (1900); preglacial drainage in Central Western New York (1900);

stratigraphy of Becraft Mountain, Columbia County, New York (1903) and the like. But these localized studies abounded in far-reaching implications and were interspersed with more general titles, such as: *Phylogeny of Fusus and its allies* (1904)[1]; Value of the protoconch and early conch stages in the classification of Gastropoda (1907); Mutations of *Spirifer mucronatus* (1907); On the classification of sedimentary rocks (1904); Types of sedimentary overlap (1906); Types of crossbedding and their stratigraphic significance (1907); A review and classification of the North American Lower Palaeozoic (1909); Continental formations of the North American Palaeozoic (1910).

North American Index Fossils, the first of his *magna opera*, was prepared with the collaboration of H. W. Shimer. Volume I (1909) comprised 853 large octavo pages, with no fewer than 1201 text figures; Volume II (1910) included 909 pages, with 726 figures. His second great work, *Principles of Stratigraphy* (1913), contains 1185 large octavo pages and 264 text figures. The third was his *Text-book of Geology* (1920): Part I, General Geology, with 864 pages, 735 text figures; Part II (1921), 976 pages, 1244 figures. A smaller work of the year 1920 was his *Geology of the Non-Metallic Deposits other than Silicates*: Volume I, *Principles of Salt Deposition.*

In 1920 he was appointed chief geologist to the Geologic Survey of China and professor of palaeontology in the National University of Peking. Thus opened the second period of his scientific productiveness, in which his center of observation was shifted half-way round the world, from North America to Eastern Asia. But his major objective remained the same and it was always to integrate his vast stores of knowledge in a comprehensive and generalized but accurate view of the geologic succession, palaeogeography and succession of life of the entire world.

In China, his predecessors, especially von Richthoven, Willis and Blackwelder, and many others, had left him a rich heritage to begin with, while his position as palaeontologist of the Geological Survey and professor at the National University of Peking not forgetting his own genial and vigorous personality brought him able students and colleagues. Unfortunately he was hampered by a long and losing fight against arthritis; this eventually laid him low but not before he had produced

---

1　此书为葛利普 1900 年完成的博士学位论文，1904 指的是出版年。——编者注

numerous special papers and a second series of major works, dealing at first with
the geologic history of China as its focus. "The Sinian System" (1922) was a short
but effective summary and critique of the observations and conclusions chiefly of
von Richthofen, Willis and Blackwelder on a great rock system, mostly of
metamorphic rocks, of northern and eastern China. This system lay beneath the
Cambrian and rested disconformably upon much older crystalline rocks,
presumably of Algonkian age.

Next followed his *Palaeozoic Corals of China* (Parts I and II, 1922–1928), one
of his numerous taxonomic works. Then the *Stratigraphy of China*: Part I
(1923–1924), "Palaeozoic and Older," with 528 pages, 306 text figures, six plates;
Part II (1928), "Mesozoic," with 774 pages, 449 text figures and four plates.

Meanwhile his "Palaeogeographic Maps of Asia" (1925) set forth in thirty-six
maps the inferred successive stages in the relations of the lands and invading seas
from the Lower Cambrian to the Upper Pliocene, inclusive. For the attentive
student only a moderate effort is necessary to correlate them with the descriptive
passages in the text and to imagine the uplifts and thrusts from below and the
gradual invasions and retreats of the sea, with all their terrific revolutionary
impacts upon the slowly yielding faunas and floras.

Part III of the *Stratigraphy of China: A Summary of the Cenozoic and
Psychozoic Deposits with Special Reference to Asia* was issued in 1927. His next
major work (1930) was 'The Permian Fauna of the Jisu Honguer Limestone of
Mongolia and its Relations to the Permian of other Parts of the World.'[1] Its 665
pages in quarto, with 34 beautiful plates, resulted in part from Professor Grabau's
steady cooperation with the Central Asiatic Expeditions of the American
Museum of Natural History under Roy Chapman Andrews.

The same year 1930 brought with it a brief but significant article, "Asia and
the Evolution of Man," summarizing certain great geologic events in Asia, including
the upheaval of the Himalayan Mountains. On one hand, this upheaval had, he
inferred, driven the forest-living anthropoid apes to the well watered forests of
the South, and on the other hand, it had forced the northern progenitors of man

---

1 应为出版于 1931 年的 *The Permian of Mongolia: A Report on the Permian Fauna of the Jisu Honguer Limestone
of Mengolia and its Relations to the Permian of other Parts of the World.* ——编者注

to give up the arboreal life and come out into the open plains of Central Asia.

Although painstaking in his records of fact, Grabau was always in search of generalizations of wider and wider scope and he vigorously defended his theses against all opponents. Beginning with his lecture on the "Rhythm of the Ages" (1933), which was later expanded into a work of 561 pages (1940) and 25 maps, he developed the theory of Oscillation or Pulsation, according to which the catastrophic rising and sinking of subcontinental areas was due in part to the rhythmic accumulation and release of pent-up forces of gravitation, which gradually brought about the shifting of the earth's poles. In order to pull all the lands into a great antarctic Pangaea, with subsequent disruption and drift apart, in accord with Wegener's theory, Grabau invoked the aid of a distant passing star. These stupendous themes were developed in his four-volume work *Palaeozoic Formations in the Light of the Pulsation Theory*, totalling 3,223 pages and published in 1936–1938.

In his later years this great adopted American-Chinese sage was more and more painfully restricted in his bodily movements; but as long as the life flame burned on, he continued to discuss eagerly with his friends the great problems of geology. He suffered privation and grave illness during the Second World War and finally succumbed peacefully in Peking, March 26, 1946.

For all his vast and productive labors, and for his vigorous, genial, and helpful nature, we honor his memory.

——原载 *Bulletin of the Geological Society of China*, 1947, vol. 27

Memorial to Amadeus William Grabau

By Hervey W. Shimer[1]

Ancestry and Early Life

Amadeus was the third child in a family of ten children. His father was the Rev. William H. Grabau (1836–1906), who was born in Erfurth, Saxony, and came to America in 1839 with his parents.

Grandfather A. A. Grabau was a Lutheran pastor in Erfurth, Saxony. About 1830 King Frederick William the Third of Prussia made an effort to unite all Protestants in his kingdom under the Reformed State Church. Many Lutherans objected to this, and under the leadership of Rev. Grabau some thousand of them emigrated to America in 1839. A certain Captain in the King's Guard regiment in Potsdam, Henry von Rohr, had become a friend of the fleeing pastor and his flock. He resigned his commission and joined them, taking over the organization and business affairs of the whole group. The pioneers came over in five sailboats which had been furnished by the Queen of Holland. Landing in New York, they went from Albany to Buffalo by boats on the Erie Canal.

In Buffalo Rev. Grabau founded the Old Trinity Lutheran Church, and also the Buffalo Synod, College and Seminary. It was here that young William Grabau, who had come as a boy of three with his parents from Erfurth, studied theology. As a young pastor he was sent to a small group that had gone west, settling in

---

1   In addition to the helpful co-operation of Doctor Grabau's wife, Mary Antin, the writer wishes to acknowledge his indebtedness for information on family background and youthful days to Doctor Grabau's sisters, Adele (Mrs. Adele Zeimer) and Lucy (Miss Louise Grabau). Mrs. Frank Grabau (the former Mrs. Alice Woodland) contributed an understanding of the conditions under which Doctor Grabau worked in China. In addition to long days of teaching in Peking, she for some years assisted him in his researches and continued to do this by correspondence after her marriage in 1936 here in Massachusettes. Professor Frederick K. Morris, one of Grabau's former students, who visited him at various times in China, furnished additional data and also the photograph used in this memorial. The librarians of the Museum of Comparative Zoology and of the Harvard College Library (Reference Department) have been most helpful in looking up needed material.

Wisconsin around Cedarburg and Milwaukee. With William to this new home went his wife Maria (born in Buffalo in 1846), the daughter of his father's old friend and fellow immigrant, Henry von Rohr. The couple lived many years in Cedarburg. Here were born three sons and three daughters. In 1885 Pastor William was recalled to Buffalo as Dean and Professor of Theology, also of Greek, Latin, Hebrew, and History, at the Buffalo Seminary.

Amadeus William Grabau, the third child of William and Maria, was born in Cedarburg, January 9, 1870, and received there his early education in parochial and public schools. When his parents moved to Buffalo in 1885 he entered the High School and later went to evening school to prepare himself for college.

His mother died when he was six years old. The following year his father married again. It was especially his stepmother who recognized the boy's great intelligence and longing for knowledge, and who encouraged all worthwhile undertakings in her children. Amadeus always gave grateful tribute to the good influence of his stepmother. During his later life in China, he celebrated her birthday each year with a gathering of important people, each of whom signed a greeting to her. In a handwritten letter he always told her about the party and of his great love for her.

It was in the woods and along the creeks of Wisconsin that his love for nature began to develop. While other boys of his age played ball or marbles he loved to roam the wilder parts of the countryside, picking up rocks, bugs, and wild flowers. He had a quick inquiring mind and a passion for reading. He also had a gift for music. So when his parents saw no hope of his becoming a pastor they tried to make an organist of him and had him practice many hours on the church organ. By the time, however, that he was preparing for college, his family could see that his principal interests were in science. His mother encouraged him in this, and his father put no obstacles in his way since he was himself a student and a collector, and his interest in astronomy had led him to put a telescope on the flat roof of the college.

Amadeus' first serious study in science was Botany. He identified, pressed, and mounted a large collection of plants that grew around Buffalo. He became closely associated with the Buffalo Society of Natural Sciences, and with a group of young men he organized the Agassiz Society of that Institution.

Mature Life

His interest in geology grew out of a correspondence course in Mineralogy with Professor W. O. Crosby of the Massachusetts Institute of Technology. Professor Crosby was attracted to him as one of his outstanding students and visited the Grabau home in Buffalo. A close friendship developed between him and Amadeus, and he persuaded the young student to go to Boston to study at M. I. T. and become Assistant in Paleontology. While studying there Amadeus retained this position from 1892 to 1896 when he received his B.S. degree and was promoted to be Instructor in Paleontology and Assistant in Geology, a position he held for one year. In 1897 he went to study at Harvard University, receiving his M.S. in 1898; he was Fellow at Harvard from 1898 to 1899 and received a Sc.D. degree in 1900.

During his entire time in Boston he was guide and lecturer for the Museum of the Boston Society of Natural History (1892–1900), conducting classes to various places of geologic interest and especially to the beaches for studies of the shore life. It was on one of these trips that he became acquainted with Mary Antin, a young Jewish girl who had emigrated to America with her parents from Russia. Her quick and brilliant intelligence attracted the young teacher. At the age of 11 she had written an account of the trip from Russia—first in Yiddish and two years later in her own English translation; this was published in 1899 under the title of *From Polotzk to Boston*, with a foreword by Israel Zangwill. Amadeus and Mary were married in 1901. The brilliant promise of Mary Antin's youth flowered in the widely popular *The Promised Land*, a book of 373 pages published in 1912.

From 1899 to 1900 Doctor Grabau was Instructor in Geology at the Renssalaer Polytechnic Institute (Professor of Geology 1900–1901) and at the same time Lecturer in Geology at Tufts College. He continued his study of the geology and fossil faunas of western New York, this time concentrating on the Niagara Escarpment. This was published in 1901 as Bulletin 45 of the New York State Museum. Its emphasis on the contributions of the geology underlying the scenic features of the Niagara region made it useful as a readable guide for the visitors to the Pan-American Exposition being held in Buffalo that summer. With characteristic thoroughness Doctor Grabau devotes 61 pages to the history of the development of the falls. He describes the stratigraphy of the region and lists and describes the fossils of each bed. His love for languages is shown not only in his quotations but in giving the etymology of the generic names (it is interesting to note that nearly all of these names which had been given by the early workers are from the Greek).

Doctor Grabau went to Columbia University as Lecturer in Paleontology in 1901; he was made Professor at the University in 1905 and continued there until 1919. During the summer of 1901 the author of this Memorial met him for the first time and for the next two years was his Assistant at Columbia. He was also his assistant during this first summer in geologic work in Alpena and other localities in Michigan, in Thedford, Ontario, and in western New York. He thus knew him well, and to know him was to love him.

The writer often recalls with deep affection and gratitude the hospitality with which Doctor Grabau and his wife always welcomed him in their new home in NewYork. It was only a small apartment, but always there was a warm welcome and a pleasant evening assured, including reading aloud by the host and the making of a Welsh rarebit in the kitchen. Doctor Grabau had a marked capacity for inspiring devotion to paleontology in his students. One felt when listening to him or when working in his laboratory that fossils and the ways in which they evolved were the most interesting and important things in the world. Students would postpone their own work to help him finish one of his numerous projects on time. He, in turn, was unstinting of his time to direct and aid them in their work under him. One recalls, also, the generosity with which he shared with his students the great people who came to the laboratory at Columbia. There was opportunity to meet, for example, the physiographer Albrecht Penck, Alpheus Hyatt, the researcher in fossil cephalopods, the biologist Jacques Loeb, and many others.

The nineteen years that Doctor Grabau spent at Columbia were very prolific ones. His published works during those years cover a wide field.

The Grabau family life suffered as a casualty of the First World War, like so many other families with a German-American problem. Mary was lecturing throughout the country in the Allied cause. Amadeus, with his admiration for German science, could see little wrong in the German nation. It was a time of intense feeling. People ceased studying German and even listening to German music. Doctor Grabau expressed his German sympathies rather forcibly with the result of the severance of his relations with Columbia University in 1919. Before this the family life had become impossible, and there resulted an informal separation. When Doctor Grabau left for China in 1920 his wife remained in this country with their daughter Josephine. It is pleasant to record that they later became reconciled, but because of the daughter's years of illness and later her own Mrs. Grabau never went to China. With the healing of the old wounds of the war, correspondence was resumed and much companionship expressed itself in letters across the Pacific. Doctor Grabau dedicated what he considered his most important book, *The Rhythm of the Ages*, "To my wife Mary Antin, my daughter Josephine, and my granddaughters Margaret and Elizabeth Ross". He felt strongly hostile to the Nazis and gave aid to German refugees escaping by way of China.

Doctor Grabau went to China in 1920 as China Foundation Research Professor at the National University at Peiping and Chief Paleontologist of the Chinese Geological Survey. The years in China were years of vast accomplishment, notwithstanding the fact that he was always handicapped by lack of facilities and funds to carry on the researches as he wished. During all this time he was also an active member of various scientific societies in China. During his later years attempts were made by friends in this country, especially by his relatives, to persuade him to come back to America, but he refused, at first because of the work he was carrying on in China, and later because of his increasing ill health. His final years are summed up in the following communication from the National Geological Survey of China.

"We announce with regret the death of Prof. Amadeus W. Grabau, Chief Paleontologist of the Geological Survey of China, on March 20, 1946, in Peiping, in the premises of the Peiping Branch of the Survey. In 1937, when the Japanese took Peiping and North China by force, Prof. Grabau had been unable to leave that city owing to illness and lack of communication facilities. Shortly afterwards, the Geological Survey established its new headquarters in Chungking while the National University of Peking was also forced to move to Kunming (Yunnan). Having been prevented to join either of the two institutions, he remained in the old Chinese capital, keeping himself busy in reading and writing. Shortly before the recent Sino-Japanese conflict he began his researches on his 'Pulsation Theory', and during a period of five or six years he completed four big volumes on *Paleozoic Formations in the Light of the Pulsation Theory*, which were published by Henri Vetch in Peiping. The last volume appeared in 1938, when fierce battles were being fought both in North and Central China. Following the Pearl Harbor Incident in 1941, he was taken by the Japanese who housed him in the old British Embassy in Peiping. Lack of nutrition, comfort, comradeship, and above all the stabilization of Japanese dominance which he hated from the beginning put him in utter despair. His health began to decline and his mind also became enfeebled. And in September 1945, when the Japanese surrendered, we found him a very old and changed man, with a face so thin that we could hardly recognize him. Measures were then immediately taken by authorities of the Geological Survey to give him better accommodation. Consequently he removed from the British Embassy to the premises of the Survey at No. 3, Fengsheng Hutung. This was, however, too late, and he died, at the age of 76 on March 20th, as a result of stomach bleeding."

A dramatic life, —from a pioneer boyhood, a rise to a happy full activity up to the First World War, the severance of relationships in this country, then many years of handicapped activities in China, ending with the Second World War and its intolerable conditions leading to his death.

Contribution to Geological Thought

Doctor Grabau's active research leading to publication began with his coming to Boston in 1892. His first five publications and his Niagara Guide represent the consummation of subjects in which he had become interested while living in Buffalo. His papers on the Cambrian and Lake Bouvé resulted from research while a student in Boston. That he could do all this research during eight years, while teaching, leading numerous field trips, and studying for his three degrees, shows the remarkable character of his mind and his tremendous energy. From his boyhood he was interested in all lines of science, but his early association with the very fossiliferous exposure of Hamilton shales just south of Buffalo at Eighteen-mile Creek inclined him definitely toward paleontology and stratigraphy. At the same time, glacial erosion and deposits around Buffalo whetted his interest in glaciology. These two lines of research remained his dominant interests throughout his life. They led finally to his pulsation theory and his polar-control theory, and the many publications on them. His last published paper, like his first, was on glaciology. Between these were various publications along this line, including the discussions in his two-volume *Text-book of Geology*.

In all of his stratigraphic papers the careful detail of his sections is a joy, and his paleogeographic maps are a spur to thought. These papers include both local sections and broader regions as well as those correlating the formations of continents. The last is especially true of his monographs on his Pulsation Theory. His publications on *Relation of marine bionomy to stratigraphy, Physical and faunal evolution of North America* (during Middle Paleozoic), *Continental formations in North American Paleozoic, North American index fossils, Text-book of Geology*, and many others indicate a mind developing toward the worldwide view represented by his Pulsation Theory. His 1200-page volume, *Principles of stratigraphy*, added greatly to the breadth of view. This book covers the entire field of geology—the composition and movements of the atmosphere and hydrosphere, the formation of present-day sediment, a discussion of all rocks (sedimentary, igneous, metamorphic), the classification of plants and animals and their relation to their environment, leading to a consideration of fossil organisms and the principles of correlation. On all these topics Grabau's references are full. The book is a storehouse of information, with many new ideas and new ways of looking at old ones. His inclination toward exact and detailed classification is well shown in this volume.

The contributions to Economic Geology include discussions of limestones, coal, oil, salt, gypsum, and potash. One of these is the 435-page *Principles of salt deposition*.

It was natural that the early interest in shells, fossil and recent, should lead to thoughts about their origin. His boyhood interest was given depth and impetus by his close association with Alpheus Hyatt at the Boston Society of Natural History where Hyatt was brilliantly applying the Ontogenetic Theory to the evolution of the cephalopods. In his study of many of the invertebrates Grabau applied many of these same principles—that the ontogeny of an individual repeats the phylogeny of the group to which it belongs; that the geological sequence of these ancestral forms is in the order indicated by the development of the individual.

Doctor Grabau's publications on evolution include six large papers on gastropods; the titles, *The phytogeny of Fusus and its Allies*, and "Value of the protoconch and early conch stages in the classification of gastropods", indicate his line of approach to the problem. He wrote six papers on the evolution of corals, three on crinoids, five on brachiopods, and two on cephalopods. Besides these there are descriptions of the evolution of many species of invertebrates included in his papers on stratigraphy and paleontology. There are also three papers on the evolution of man in Asia. In his numerous paleontological publications he described a very large number of new species, many new genera, and a number of new families. His descriptions are full, carefully done, and well illustrated.

The Pulsation and Polar Control Theories

According to Grabau's Pulsation Theory the evolution of the crustal features of our earth is brought about by the rhythmic rise and fall of sealevel as shown by its transgressions (pulsations) and regressions (interpulsations). Thus all lands are affected at the same time since these alternating changes are presumed to be due to a decrease and an increase in the depth of the ocean basin. Grabau treats the breaks in marine sedimentation, the interpulsations, with their continental sediment or absence of any sediment, the evidence of erosion and the folding of mountains, as of equal significance with the transgressive marine sediments (pulsations). Geologists have long been aware of the alternation of marine floodings of the lower lying areas of continents with the elevation of the land and the formation of mountains and a consequent withdrawal of ocean waters. Some of these transgressions have been long known to have been worldwide. Grabau suggests this to be true of all. He publishes lists of fossils in evidence of this correlation. This listing and the discussion of the evidence for his correlations form a most valuable portion of his contributions.

Doctor Grabau links this Pulsation Theory with his Polar Control Theory. He makes of the two theories a unity to account for variations in climate, and hence in faunas and floras throughout geologic time and on all parts of the earth's surface. According to the Polar Control Theory there is "a periodic shifting of the earth's crust (sial-sphere) through the impetus given it by the rotation of the earth on an axis of essential constancy of position", *i.e.*, the poles are constant in position, but the sial crust is not. He shifts the continents to the poles to account for the glacial climates and away from them for mild climates. He also shifts continents to bring plants and animals of each land into contact with related organisms of lands now far apart. According to this theory there was never any great ice age. The lands shifted their positions, and polar glaciation took place in normal fashion whenever the lands were at the ice centers associated with the poles. Since the sial crust rests upon the sima and is partly submerged in it, these movements produce mountains at its forward edges and fissures with consequent vulcanism at the rear edges.

Four volumes have been published on the Paleozoic pulsations. Volume five, on the Mesozoic, was partially completed and remains unpublished. In his volume, *Rhythm of the ages*, Doctor Grabau summarizes the Pulsation Theory as applied to the Paleozoic and outlines the evidence for the Mesozoic and Cenozoic.

Doctor Grabau possessed a detailed knowledge of the formations of the earth and of their faunas. Many geologists will not accept the threads with which he has bound these scattered facts together; but all acknowledge their very great indebtedness to him for his persistent labor in gathering the data and publishing them. The threads will remain an incentive to research, and the mass of data constitute a storehouse upon which workers will draw for generations to come.

Societies and Honors

Doctor Grabau became a Fellow of The Geological Society of America in 1898 and was a constant and enthusiastic attendant at its meetings until he went to China in 1920. His only visit to the United States was at the special invitation of the Society to attend the 16th International Geological Congress at Washington in 1933. He also belonged to the Paleontological Society (Fellow), the American Association for the Advancement of Science (Fellow), the New York Academy of Science (Vice-President of Geology and Mineralogy section 1906–1907), the Philadelphia Academy of Science (Corresponding member), the Deutsche Akademie der Naturforscher, the Geological Society of China (Councilor 1922–1925; Vice-President 1925), the Scientific Society of China (Honorary member), the Peiping Society of Natural History (Honorary member; Councilor), the China Institute of Mineralogy and Metallurgy (Honorary member), the Academia Sinica, the Academia Pepinensis, the Paleontological Society of China (Fellow), the Scientific Society of China (Honorary member), the Kaiserlich-Deutsche Akademie der Wissenschaften[1] (Foreign member). He was Research Associate for the Central Asiatic Expedition of the American Museum of Natural History and Research Fellow for the National Research Institute of the Chinese Government. He served as Honorary Advisor of the Ministry of Mines of the Chinese Government and Honorary Advisor of the Feng Memorial Institute of Biology. He was Chief Paleontologist and Director of the Paleontological Laboratory of the National Geological Survey of China, and Dean of the Peking Laboratory of Natural History.

He received the first Grabau Gold Medal in 1925. This award was founded by the Geological Society of China in 1925 to be awarded biennially for outstanding contributions to Chinese science. He was awarded the Mary Clark Thompson Medal by the National Academy of Science (Washington) in 1937.

——原载 *Proceedings Volume of the Geological Society of America Annual Report for 1946*, 1947

---

1 应为：Kaiserlish Deutsche Akademie der Naturforscher zu Halle。——编者注

追念地质大师葛利普先生

——安葬纪念日献辞

孙云铸

今天为已故北大地质系教授葛利普先生逝世一周年纪念，同时也是先生骨灰安葬日期，先生生前友人，同事及学生齐集于北大地质馆前，亲视先生遗体入土，长眠地下，在此黯然永别之一刹间，实令人有无限之回忆与感触。

先生为驰名世界之地质学家，在北大任教授 26 年。先生研究地层学及古生物学，先生具世界知识，了解世界各处之地质情形，为地质学界一代之典型人物，从先生之著作和言论，可以知道先生伟大。

先生生于美国维斯康新省，祖父原为德籍，对于英德文字，均极精通，对于自然科学发生浓厚兴趣。在麻省工业学校毕业后，即从事于地质工作，凡 50 年，任 46 年教授。

先生对于纽约省古生代地层及古物，有特殊贡献，美国地层因此亦以纽约省研究最为详细，并著有《北美标准化石》两巨册（与夏谟合著）。民国九年来华，对于中国古生代地层及化石，精详研究，贡献尤多，在《中国古生物志》及《中国地质学会志》登载，尤以珊瑚类、腕足类、头足类三类贡献最大。出版各书均为世界名著，获得国际荣誉。

先生与耶鲁大学已故地质系主任舒可特，同为一代地层专家，不独对于美洲地质及中国地质研究详尽，抑且对于世界地质，造诣甚深。于逊清末年，先生赴欧参加世界地质学会，遍游欧洲，返美后著《地质学教科书》两册及《地层学原理》一巨册。其《地质学》一书有不少欧洲材料加入，为当时美国教科书中所无，以是与舒氏所著之《地质学》同为美国教本。《地层学原理》一书出版已历 34 年，至今仍为重要范本，尚少有其他著作可以媲美。来中国后，著有《中国地质史》，对于整个世界地质更为了解。先生研究范围之广，著作之多，美人无出其右，可与英国之司密士（W. Smith），德国之凯塞（Kayser）教授先后媲美。

先生思想，俱有革命精神，从不守旧。50 年地质工作如一日，治学之毅力如此。先生深信各地质时代海之进退，为世界普遍性。因创脉动学说——为地质学主要学理之一——解释中国及世界地层问题，并用此学说来划分中国地层，在最近 15 年中，先生根据其学说著成《古代地层》8 部，不但对于中国地层，详加记载，欧美各地地质，靡[1]不搜罗，实为地层学之名著。

先生幼为订书工人，晚入夜班，习自然科学，抗战期间困处北平，辛苦备尝，以残废之躯，单独任此烦难工作，共成著作 262 篇，逾两万页。其著述之精神，至堪欣佩。4 年前本人接其来书谓"北平沦陷后即未再见北大校门"。临终遗言愿葬北大地质馆，其忠于北大有如此者。本日北京大学遵其遗嘱，安葬北大地质馆前，立碑纪念。

中国地质基础，虽经章鸿钊、丁文江、翁文灏三先生奠定，而先生协助之功尤大。现时主持中国各地质机关者，如清华地学系、北大地质系、北洋地质系、中山大学地质系、中央大学地质系、经济部中央地质调查所、北平分所、西北分所、湖南地质调查所等均为先生学生，将来定能继续先生之未竟工作，先生虽逝，先生之精神永不朽矣。

——原载《新生报》1947 年 3 月 20 日
"葛利普教授逝世周年纪念特刊"

---

1 原文为弥。——编者注

我对葛利普先生之感想

王竹泉

民国肇兴，政府始有调查地质之议，并由工商部创立地质研究所，以造就专门人才，民国五年改称地质调查所，以原研究所毕业生充任调查员。中国人调查中国地质，实自此开始，但外国人调查中国地质，则远始自距今 75 年以前。德人李希霍芬氏曾简略将中国地质调查数处，并著有《中国》一书，所谓山西一省所储之煤可供世界数千年之用，即李氏调查结果之一部。据泉于民国六、七年间在山西调查所得之结论，知李氏之地质工作，尚多错误，尤其关于山西煤矿储量之估计，实嫌过于夸大。泉之重新估计山西全省储煤总量约不过 1271 亿吨，其中无烟煤占 353 亿吨，烟煤占 518 亿吨，若以山西储煤量与美国全国储煤量相较，仅占 4%，若与英国储煤量相较约等于其 94%，若与德国储煤量相较约等于其两倍，若与法国储煤量相较约等于其 28 倍。若以山西全省人口 1100 万计，每人储有煤量达 11 555 吨，即以之供给全国 4 亿之人口，每人亦可得 317 吨。故山西之煤藏，虽不若李希霍芬氏估计之多，在中国各省中实亦首屈一指，即在欧亚大陆煤矿区中，亦占重要之位置。关于地质方面李氏曾谓山西大同煤田之地质时代，应纯属于侏罗纪，盖李氏当时意见以谓中国煤田非石炭纪即侏罗纪。嗣泉于民国七年重测大同煤田之结果，发现两种煤系，一属侏罗纪，一属于石炭——二叠纪，自此次证明两种地质时代不同之煤系可存在于同一煤田后，继于绥远大青山、陕北沿黄河一带亦发现同样之煤田。以上所举数例乃表示外国人初次调查中国地质之困难，并说明中国地质事业决非一人一手一足之力所能成功。继李氏之后，外国人来中国调查地质者，为 40 年前美人维理士氏。维氏遍历中国南北各省，尤其在山西五台及山东泰安，曾作较详密之考察，将李氏在中国北部所分之地层，多所纠正，如将李氏之太古代五台系及下震旦系，改属于元古代，并将李氏之所谓石炭纪石灰岩，用化石确定其为奥陶纪，乃其重要之供献。民国四年中国政府拟与美国美孚石油公司合办陕北石油矿，延聘美国石油技师赫乐德、马栋臣、王国栋等考察陕北地质，在肤施、延长、延川一带曾作精细之勘测，结果完成陕北百万分之一地质图，并将所见之地层，分类如左：

一、山西系　灰色砂岩、页岩及煤层

二、汾河系　红色砂岩

三、陕西系　灰色砂岩及页岩，含薄煤层及石油，以上三者之地质时代均属于石炭纪

四、红色页岩、砂岩及薄层灰岩二叠纪

五、红色砂岩及二叠—三叠纪

民国十二年，泉重调查陕北地质，曾采得大量植物化石及鱼类化石，因证明汾河系应改属于二叠—三叠纪，陕西系应改属于下侏罗纪。又美国煤油技师之所谓二叠纪红色页岩、砂岩及薄层灰岩实应属于中侏罗纪及上侏罗纪，彼等之所谓二叠—三叠纪红色砂岩实应属于上侏罗纪之上部以至白垩纪。民国三年，中国政府聘瑞典人安特生氏为矿务顾问，安氏居中国约 10 年，平时除勘测矿床外，颇致力于新生代地质之考察，于脊椎动物之化石、石器时代之陶器，尤尽力采集。斯时泉方携斧裹粮，终年往来于秦、冀、绥各省山中，思对于华北地质有所补助。每次旅行归来安氏辄邀至其家，详询各地见闻，孜孜不倦，有时告以某地产龙骨（脊椎动物化石，华北乡村呼为龙骨，为国药之一种），或赠以石器时代所产石斧一件，安氏即欣喜若狂，必详记其产地，继派人前往大量采掘，如其所著《中国北部之新生代》一书中，关于山西保德、河曲龙骨化石，即用此法发现。其他若所著《中华远古之文化》《甘肃考古记》等书，尤为中国考古之先导，即周口店猿人洞穴遗迹，亦系安氏首先发现。总之安特生氏对于中国北部之新生代地质，实有不可淹没之功绩。观其于林斯顿所著《中国山西新发见之犀类化石》一文中，安氏曾附言谓"当 1917 年吾人与中国地质调查所立有契约，共同采集及研究中国哺乳类动物化石，数年以来，成绩卓著……"。又安氏于其所著《甘肃考古记》导言中曾谓"中国地质调查所对于中国远古历史之探考，于民国八年即着手进行，彼时已于华北各地屡次发见石制器物……"。由此两次宣言中，益可窥见安氏对于新生代地质之如何努力矣。

以上所述各外国人之研究中国地质，率由其本国派遣或由中国政府聘请，工作时间少者 1、2 年，多者达 10 年，悉按其派遣目的所在或个人兴趣所寄，一俟目的兴趣完成，即束装返国。惟葛利普先生原系美国之名地质学家，在美国任各大学教授有年，其所著《地层原理》《地质教科书》及《非金属矿床》等书早已风行全球。自民国九年由美洲来中国，即视中国为其第二故乡，一方面在地质调查所从事地质研究，同时在北京大学地质系充任教授，为中国训导后

进，历 20 年如一日，每与谈及地质问题，恒引证发论，滔滔不绝。在葛氏未到中国之前，古生物人材颇感缺乏，犹忆民国七年泉调查江西吉安、安福、永新一带煤田地质返平，曾采有大量化石，可称满载而归，但因无专门鉴定化石人员，徒呼负负。乃进而商之地质调查所丁文江所长，据云不久所中总要想法将化石鉴定，故民国九年葛氏之来中国与七年江西化石之采集，不无因果之关系。葛氏到中国第一批鉴定材料，即系民国七年泉自江西所采者。厥后中国调查地质人员对于化石采集，益感浓厚。民国十年泉曾在山西保德附近于石炭——二叠纪煤系下部，采得化石 20 余箱，雇骡 10 余头运往大同，由平绥铁路转北平，亦悉由葛氏一一鉴定。继则中国有生物学者，因受葛氏之熏陶，竞起研究，故《中国古生物志》在最近 20 年中著述特富，悉葛氏倡导之力也。每与葛氏遇于研究室，恒以"山西王"呼余，盖因泉曾完成山西全省地质图，对于山西地质较熟，因而特一词绰号呼之。民国二十六年，卢沟桥事变起，葛氏因腿疾留于北平。二十九年秋，泉因事自昆明返平，是时伪师范学院及伪北京大学屡遣人来劝泉前往执教，甚至以如执拗不允将处危境之言辞相恫吓，终嗣泉告以旧病复发，闭门谢客，始获见谅。一日晤葛氏于起所住豆芽菜胡同寓所，倾谈之下，为以谢绝教书之经过相告，氏两目忽注视余良久，骤前握余手，连呼"好"、"好"。此时泉欲南下而交通已阻，居家则经济窘迫，质典渡日，其困难有非言语之所能形容者。未几太平洋战争爆发，葛氏被拘于东交民巷收容所，泉数次拟往探视而未果。日本投降曾晤葛氏于德国医院，见其面容焦悴，知其身体已不易支持矣。统查葛氏之所以于地质，尤其于古生物，能有如此伟大之成就，固由于其天资之独厚，而终身专其业，不因权利而易初志，乃为人最可钦佩之处。今葛氏已矣，吾地质同仁应如何继其遗志，对于地质、对于古生物从而扩大发展之，是又地质同仁之共同责任也。

——原载《新生报》1947 年 3 月 20 日
"葛利普教授逝世周年纪念特刊"

葛师利普千古

袁复礼

猗与夫子，新陆钟灵。再世积德，学出名门。是天之纵，质敏性贞。是人之英，烛照幽明。远征风流，谈无与伦。太学历聘，安驷盈庭。诲精研笃，久而弥新。横涉三洲，誉载寰瀛。浩然永息，归宅还真。老成邈矣，犹有典型。呜呼哀哉！遗泽长存。

—原载《新生报》1947 年 3 月 20 日
"葛利普教授逝世周年纪念特刊"

葛利普教授逝世周年纪念

王 烈

　　先生逝世时，享年 76 岁，生平事绩可分 3 期，每期各 25 年。青年期为求学时期，中年期及老年期均为教授时期。中年期执教于美国哥伦比亚大学，老年期执教于中国北京大学。在北大执教期内，余与之同任地质学系教授，朝夕与共，知之最稔。其丰富之著作，精深之研究，服务之精神，高纯之人格，足资吾辈矜式者，系中同人多已，如孙云铸先生等论文述及。然余仍有不能已于言者，盖先生自民九入北大，以迄七七事变，共 17 年，平日教诲诸生，循循善诱，孜孜不倦，始终如一，从未少懈。吾辈时相过从，闲谈讨论，不离地质，常以发展中国地质学相劝勉，到校未及一月，深虑吾国地质教育不能普及，于是：（一）发起星期日公开讲演，使地质教育不仅限于地质系内学生，而兼推及于本系以外；（二）提议创设中国学会，使校内之地质学者得与校外学者有接触讨论之机会，地质知识渐推及于全国；（三）提议由中国地质学会编印《中国地质学会志》《地质专报》以及各种《古生物志》等刊物，使与欧美各地质学会之印刷物互相交换，由是欧美各国始知中国地质学之成绩。中国各种科学每多落后，独地质学在世界各国中具有相当之地位，此无他，先生之力也。最可钦佩者，先生在北大执教 17 年中，系中同人或为境遇所迫，或因见猎心喜，每有见异思迁者，而先生则守住岗位，始终如一，数十年如一日。其服务之精神，实有难能可贵，可为吾辈模范者。先生来华后，吾国政局动摇无定，每逢转变，校薪即无着落（先生虽系外籍，然薪给折扣与本国教授同从，未稍示优遇），而先生则安之若素，毫无怨尤，以外籍教授而能与国人共甘苦，其义侠胸怀有足多者。卢沟桥事变后，先生以足疾难行，未能随校南迁，遂与吾等分离留平，闭户读书，潜心研究，著述脉动论一书，凡 8 厚册。虽经日人威胁利诱，从未再入北大校门。日美宣战，居集中营，艰苦备尝，亦未尝少变其志，殆古人所谓：富贵不淫，贫贱不移，威武不屈者，是耶！先生之人格，吾辈应奉为百世之典型，当与其服务之精神，研究

之学说，丰富之著作，历久常新，并世而不朽，先生虽死不死也。兹值先生
逝世一周年之期，爰掇数语，以为纪念。

——原载《新生报》1947 年 3 月 20 日
"葛利普教授逝世周年纪念特刊"

葛利普先生略历

1885 年　习订书业于水牛城，日充订书工人，夜晚就读夜校。

1890 年　波士顿天然博物院助手，整理矿物标本。

1891 年　入麻省工业学校，肄业，并任该校地质系学生助理。

1896 年　得学士学位，任地质系助教。

1897 年　入哈佛大学，肄业。

1898 年　得哈佛大学硕士学位，被选为美国地质学会会员。

1899 年　美国特辅词学院（Tuft's College）地质学讲师。

1900 年　得哈佛大学科学博士，任雷赛尔工艺学院地质学教授。

1901 年　哥伦比亚大学古生物学讲师，纽约科学院会员、纽约地质调查所所员，与玛丁[1]女士结婚。

1902 年　哥伦比亚大学副教授。

1905 年　升任哥伦比亚大学古生物学教授。

1910 年　游历欧洲参加瑞京国际地质会议。

1920 年　任农商部地质调查所古生物研究室主任兼北京大学古生物学教授。

1922 年　选任中国地质学会会员及评议员。

1923 年　美国中亚考察团团员。

1924 年　中国科学社名誉会员。

1925 年　德国哈勒皇家自然科学院会员，受中国地质学会第一次葛氏奖章。

1927 年　北京博物学会名誉会员及永久评议员。

1928 年　费城自然科学院通信会员。

1929 年　中央研究院研究员。

1933 年　赴美国华盛顿出席第 16 次国际地质会议。

1934 年　北京大学地质系代理主任。

1946 年　在所逝世。

——原载《新生报》1947 年 3 月 20 日
"葛利普教授逝世周年纪念特刊"

1　Mary Antin，汉译应为玛丽·安亭。——编者注

纪念葛利普师

张席禔

　　葛利普教授，于民国九年来华，任国立北京大学地质系教授，兼农商部地质调查所古生物技师。在中国供职计 26 年，去年 3 月 20 日，病故北平，享年 76 岁。兹卜于 3 月 20 日，安葬于北京大学地质馆。席禔昔年在北大攻读时，亲受葛师之陶冶，教泽广被，耿耿不忘于衷，谨缀数语，以表爱戴师长之诚，而志纪念。

　　葛师学问渊博，著述宏富，治学以谨，待人以诚，其研究科学之精神，孜孜不倦，终生如一日，实令人钦佩。葛教授到中国后，一面授课，一面研究，调查所历年野外调查时，采集之标本资料，多经先生整理研究，先后继续出版多种，确定中国各处之地层时代，树立中国地质基础，俾后学者有所遵循。

　　中国自古以来，闭关自守，西洋科学，输入甚晚。即专就地质一科而言，民国初年，国立北京大学，始成立地质系，农商部始创办地质调查所。惟以当时国内地质专门人才，尚甚缺少，即有之，亦属凤毛麟角。因此指导乏人，故进步殊鲜。自葛教授到华之后，对于北京大学地质系，及调查所方面，多所筹划建议，数年之后，始稍具规模。

　　葛师在校，用英语讲授，学生直接听讲，最初尚难免困难，不能透彻暸解之点，彼则循循善诱，教导有方，督促鼓励，面面具到，故一班学子，皆敬而爱之。对于毕业青年，在校方或调查所工作者，常受葛师之督导，使能作独立研究工作，造就中国专门人才。近年来中国地质界后继有人，葛教授之功劳，当不可没灭。

　　今葛师虽死，彼之著作，永垂不朽，彼之门徒满天下，当能继续葛氏之志，为中国地质界效力。葛师有灵，亦当含笑于九泉。

民国三十六年三月十四日谨书于清华园

——原载《新生报》1947 年 3 月 20 日
"葛利普教授逝世周年纪念特刊"

葛利普周年祭

寿振黄

　　古生物学大师葛利普（Amadeus W. Grabau）教授，离开人世，迄今又是十二个月了！这十二个月中，国内自然科学界，均已先后复原，财力有限，人力不足，不景气的情形，较之抗战期内，或更凄惨！回想当年，瞻望将来，适逢先生逝世周年祭日，中心不觉有无限感慨！

　　葛利普教授的生平，详见丁文江所作先生小传（民国二十年《中国地质学会志》第十卷），和拙作《葛利普教授年谱》（去年五月二十日《大公报》副刊）。民国九年，葛利普先生正五十岁，应丁文江之聘，从美国东部的纽约，渡过太平洋，来到北平，担任国立北京大学地质学系古生物学教授，同时兼任农商部地质调查所古生物学研究室主任。先生与丁文江、翁文灏、章鸿钊、李四光等，协力奠定中国地质学的基础！二十余年来，善用他广泛的学识和丰富的经验，献身于中国地质学和古生物学的进展。在没有研究空气的故都，创造了可以研究的环境；对于中国自然科学的各方面，均有巨大的影响！现在中国地质学和古生物学，在国际学术界中，已有相当的地位，葛利普教授就是造成这样地位的一位功臣！

　　民国纪元以来，欧美学者来讲学的，每年均有若干人；有的是来宣传某种学说，有的是来游览名胜古迹，有的是来搜集研究资料。求其努力研究，热心教导，值得大家仰慕，能如葛利普教授的，想不到还有些什么人！每年来华讲学的，不乏当代名师大儒，然而检讨他们对于国内学术界的影响，似乎都很微薄。如今想起葛利普老教授来，实在令人伤悲！

　　第一次世界大战结束之后，先生辞去美国哥伦比亚大学教授，来到遥远的中华民国，他已经选定北平做他的第二故乡。先生爱吃中国菜，爱穿中国马褂。晚年曾穿黑色马褂，照一张半身相片！这一张穿马褂的相片，白发皑皑，神采奕奕！老先生自己十分珍爱，生前常留在他的案头！

　　从日常的行为中，我们可以看出先生是如何热烈地爱好中国。寄信到香港去，他一定要写"中国香港"；乘船过日本时，他曾大骂日本警察，因此不上岸游览，晚年愿入中国国籍，可惜没有办理入籍手续，便因胃出血逝世了！

先生离美时，将全部私人藏书，随身运来；这些书籍，给予故都地质学界人士不少的便利。先生生前曾拣出若干图书，赠予中央地质调查所。临终时先生遗嘱：将全部藏书，捐赠中国地质学会。该项藏书包含书籍六百数十册，期刊一百数十种，和单行本数千册，已于去年年内，交由中央地质调查所北平分所图书馆，暂时代为保管。

葛利普教授运用文字的能力，实在令人惊异！著作之丰，多至二百九十余种；有少仅数页的小型论文、有多到千余页的长篇巨著，其中最重要的，有《地质学教本》（*Text Book of Geology*，2 vols），《地层学原理》（*Principles of Stratigraphy*）和《北美标准化石》（*North American Index Fossils*）均是五十岁以前在美国所著的。五十岁以后在中国所作的，有《中国地质史》（*Stratigraphy of China*）上中两册，《蒙古之二叠纪》（*The Permian of Mongolia*）《中国古生物志》十一册，《脉动学说》（*Palaeozoic Formations in the Light of the Pulsation Theory*）五册（尚有二册，写成而未出版）等等。

七七事变，故都沦陷之后，曾写一本通俗书籍，名为《时代的律动》（*The Rhythm of the Ages*）共四十章，民国二十九年，由北平法文图书馆出版（见二十九年六月十六日英文《时事日报》，笔者所作书评）。民国三十三年，又写一本《我们所居住的世界》（*The World We Live In*），共二十三章，现在尚未出版。

葛利普教授对于地层的研究，十分广博；他对于古生代、中生代和新生代，均发生兴趣；其中以震旦纪、泥盆纪、二叠纪，贡献最大，先生对于古物学的研究，尺度亦相当宽大，从无脊椎动物，到低级的脊椎动物；其中以软体动物最为精深，腔肠动物的珊瑚类次之。

先生行动不便，每逢星期日下午，聚集男女友好，在西城豆芽菜胡同寓所，举行非正式的座谈会。先生健谈，且很幽默！往往从微小的化石，谈到巨大的地球；从欧美文学，谈到自然科学。有时天色已黑，余兴未尽，乃出旨酒美肴，留客晚膳！

先生是国际上有地位的名教授，学问渊博，诲人不倦。他在北京大学学堂中，时常对学生说……你们的老师是葛利普，葛利普的老师是哈佛大学的名教授海阿德（A. Hyatt），海阿德的老师是美国动物学的开山祖师阿伽西（L. Agassiz），阿伽西的老师是法国比较解剖学泰斗居维叶（Georges Cuvier）。现在你们应当记住，居维叶是你们的太太老师，你们是居维叶的徒弟——学生的学生！话未说完，老教授和全体学生，哄堂大笑！这些回忆，愉快而又和谐；引人入胜，好比黑暗中一盏明亮的灯！

先生在国内外最高学府，掌教四十余年，提拔后进，不遗余力；桃李满门，人才辈出。国内地质学家，如王宠佑、叶良辅、袁复礼等，都是先生在美国的学生，古生物学家如杨钟健、孙云铸、裴文中、尹赞勋、赵亚曾、计荣森等，都是先生在北大的学生。至于李四光的研究螆科（*Fusulinidae*），金叔初的研究贝介，遇有疑难的问题，亦当到豆芽菜胡同寓所，与先生商讨，重视先生的意旨！旅者客居故都，垂二十年；每逢对于自然现象，有所请益的地方，先生莫不和颜悦色，详加指示，与以善意的批评！

七七事变以后，先生深切地痛恨并憎恶侵略中华民族的敌人；他们的失败，先生早已预知！一二八事变以后，先生拘留在集中营中（东交民巷英国大使馆），生活艰苦，日益衰弱。但是第二次世界大战的结束，和他的预言相符，这位生长在自由气氛中的老教授，因全面胜利而恢复身体和精神的自由；当时先生心中的快乐，是很难用文字来形容的！

七十六岁的生命，似乎不算太短，在被迫集中之前，他没有浪费过着生命中的任何一刻。他将整个的生命，贡献出来，去搜求自然界的真理！他自己不断地工作，并且尽力援助别人工作。和那些"自己不去工作，却阻止别人工作"的人比较，伟大和渺小，不可同日而语了！

去年三月二十三日下午，我们将这位科学界的领导者的遗体，送到朝阳门外去火葬时，大家觉得有些茫然！时间过的很快，如今又到先生逝世周年忌日，心中有千言万语，不知道从何处说起！我们用沉重的心情，来追忆先生，来纪念先生！我们应当向先生学习渊博和宽大，继续先生的工作！

先生谈话时，旁证博引，风趣横生，听众为之入神；这样可以知道先生的渊博。先生待人，非常宽大，对于中国学生，丝毫不存种族的界限，没有歧视的心理。青年人的文稿，先生总是仔细阅读，把它当作是自己的工作；不挟偏见，不怀私心！先生一生拿古生物学和地质学，当作严肃的终身事业，始终不渝，没有离开过这个岗位。先生的学问好比一座埃及的金字塔，有它宽广的基础。先生不希望后辈做狭小的专门家，越钻越小，钻进牛角尖去！

先生的音容，或许逐渐的被人们忘记；但是先生的精神和业绩，留在自然科学界，永远不会衰减！

——原载《华北日报》1947 年 3 月 17 日

地质大师葛利普入土了

徐 盈[1]

　　中国的真正友人——地质学界权威大师葛利普教授（Amadeus William Grabau）在去年三月二十日经过了被日伪六年的囚居生活，终于在中国胜利后一年，死在地质调查所为他终老所特设的小屋内。过了整整的一年，同一天，在狂风怒吼，沙石乱飞的情景下，遵照了他的遗嘱，由他的女书记捧了他的骨灰，安葬在北京大学地质馆之前的广场，中外友人以无言的沉痛，举行了一个小小的纪念会。

　　这个七十六岁的老头儿，到死也没有留胡子，他留下的二百六十二篇巨著，二万页的黑字，说明了这个人至死仍是青年。追悼他的人，谁也不会想到，他是一个德国种的美国人，谁也不会轻视这一位订书匠的成就，他以德国性格的勇敢，美国性格的进取，用订书的技术，综合了诸子百家，创造了地层学上的脉动总共十四卷，已出版的有五巨册。美使馆文化联络官傅瑞门[2]对我说："他是一位大人物，一位真正的大人物，我们在他的身边小得看不见了。"辅仁大学马德武[3]神父是古植物学家，他说："葛氏的光辉照耀北大，照耀北京，照耀中国及世界。"

　　北京大学的代表人，印度哲学权威汤用彤以地主资格欢迎葛氏能在北大的土地上长眠，申述葛氏成功的历史，足以引发北大师生的学术事业。他特别指出"他比中国人更爱中国，在危境中十年，他不与中国的敌人合作，真

---

1 徐盈（1912—1996），原名徐绪桓，20世纪上半叶著名记者。山东德县人，曾在保定河北农学院、南京金陵大学农业专科学习。因对文学和实地考察极感兴趣，为当时的报刊撰写了大量通讯报道。先在上海《大公报》任练习生，1938年底任重庆《大公报》采访部主任，抗战胜利后担任天津《大公报》驻北平办事处主任。解放后历任政务院文教委员会参事、国务院宗教局副局长、全国政协文史资料委员会工商经济组组长等。编撰有《抗战中的西北》《当代中国实业人物志》等。参见徐盈：《北平围城两月记》，北京：北京出版社，1993年，3—4页。——编者注
2 傅瑞门（Fulton Freeman，1915—1974），美国外交官，曾任美国驻华领事。——编者注
3 马德武（Gregory Matthews，1903—1949），1934—1949年在辅仁大学任教，1948年任农学院院长，在预防伤寒症方面有较大贡献。1949年在北京去世。参见（奥）雷立柏编注：《别了，北平》（汉德对照），北京：新星出版社，2017年，第30页。——编者注

是中国学术界的一大骄傲。"他中道残废,身体虽不健全,但精神上的生活特丰,留下这么多的著述,他不仅是一位教授,而实是一代导师。

民国九年北大地质系同时到了三位教授,一位是丁文江,一位是王烈,一位就是葛利普。当时的助教,今天的地质系主任孙云铸说:"葛利普先生由丁先生请到了北平,他遇到的第二个人就是我。有一天我正在丰盛胡同地质调查所楼上办公,丁在君先生介绍他给我,要我带他到北大去看蔡元培先生和蒋梦麟先生。他那时发黑,面上略有皱纹,路过煤山,疑为矿地,大为惊异。今天我又是沿着这条旧路,过煤山,到丰盛胡同去接他的骨灰,音容宛在,景物依昔,但他已不复存在了。他有五十年的地质生活,作了二十六年的教授,有十七年是在北大地质系。这是空前的创举。"

葛利普不是温室中成长的学者。他诞生在美国维斯康新省,一八八五年他十五岁,在水牛城白天当订书的工人,到夜晚在夜校读书。一八九〇年在波士顿任天然博物院助手,整理矿物标本。翌年,入麻省工业学校[1]肄业,并任该校地质系学生助理。一八九六年得学士位,专任助教,又入哈佛大学,专攻地质,一八九八年得哈佛大学硕士,被选为美国地质学会会员,又任特辅斯学院[2]地质讲师。一九〇〇年得哈佛博士及任雷赛尔工艺学院地质学教授。一九〇二年任哥伦比亚大学副教授,越三年,任古生物学教授。

孙云铸说:"因为在哥大任教,他对纽约省比较详细,他与夏谟合著的《北美化石》[3]两巨册,是集大成的著作。一九一〇年到欧洲游历,参加瑞京[4]的地质会议,与英、德、法、奥的学者颇有接触,归来之后,不仅有了欧洲且有其他地带的材料,归来著有《地质学教科书》两册,及《地质学原理》一巨册,其材料之丰,为同时各教本所无,与耶鲁大学地质系主任舒可特的《地质学》,同为两大重要课本。堪与英国的司密斯,德国的凯沙两教授媲美。但葛氏死后不久,舒可特也继之去世了。"

葛氏任中国地质调查所古生物研究室主任,那时即由孙云铸任助理。在学校,则教地质,亦由孙氏助理。"他帮助别人,自己却永远是退让的,他不作行政事体,古生物学会只肯担任副会长。有一年李四光先生到国外,他只

---

1 即麻省理工学院。——编者注
2 即 Tufts College,现一般译为塔夫茨学院。——编者注
3 即《北美标准化石》。——编者注
4 指瑞典首都斯德哥尔摩。——编者注

作过一次名誉的代理系主任，实际则由孙及谢家荣维持了一年。他不仅对名不居，对利亦然。那时内战循环，北大发不出薪水，只是一折八扣，他只有吃得坏些，让学生到他家里上课。""那时北平大学有薪水，许多人都去了，他是不肯去的。""那时有一位外国教授为了薪水问题，曾向我们请出他们的公使提出严重抗议。而葛利普先生从来没有说过什么。"

这位五十年生活如一日的人，在中国抗战发生，大节来临的时候，他镇静如当，创有脉动学说，深信各地质时代海之进退，有世界的普遍性，以此说来解释中国及世界的地质问题。计划要写十四大部，付印者已有五部。这时候他的身体已半成残废，但写作不辍，仍然能够集中各种材料，从容整顿，给后学节省了不少精力。

葛利普住在豆芽菜胡同里听到说英文《北平时事日报》[1]已为敌伪所接收，他立刻改以几倍的高价改订《华北明星》报[2]。在极困难的时候，他还辗转写信给后方的友人，指出西南的地质空白太多，希望能够在这种机会之下加以努力。珍珠港事件爆发，他便被敌伪因在英国大使馆内的集中营中，因他的身体不能自由行动，所以也就没有送到外地。他在失自由的期间，偷偷告诉友人道："自从北平沦陷之后，我就没有再看见过北京大学的大门。"

没有这样倔强的精神，也就没有这么多伟大的著述，也就不会有这么一位时代定型的人物。北大老地质学系主任王烈说："我们复员回来了，北大重新开了大门，欢迎你回来，但是你已经不能再生了。我们只能遵照你的遗嘱，把你的骨灰葬在这里。"葛氏是把一切献给地质的人，在讲堂上谈地质，在家中仍然谈地质，他的第二生命就是地质，他的第二祖国就是中国。

在胜利之后的那一年，葛利普曾要求入中国籍，这其中也许有他个人不愉快的婚姻问题，但他爱中国的心是坦白无私的。自从一九〇一年他与纽约地质调查所员玛丁[3]女士结婚，但并不愉快，一九二〇年到了中国之后，便与家庭断绝。这二十六年内，在与他的女书记恋爱之前，也有一个时期的放荡，而这些全不影响到他正在学术上的工作。有多少小事情，仍然留在朋友和学生们的脑海中，譬如北平研究院杨光弼先生说："那时的几个科学集会，在欧美同学会开会，平均总是葛利普到的最早，而且每次必有贡献。"又有人说：

1　即 *Peking Chronicle*（*The Peiping Chronicle*）。——编者注

2　1918 年创办于天津的英文报刊，英文名称为 *North China Star*。——编者注

3　葛利普的妻子为玛丽·安亭，而非玛丁。安亭为一著名作家，非纽约地质调查所人员。——编者注

"他在地质旅行时从不以外国人自居，和中国人生活在一起，在街上，遇到熟人，即便是他的学生，他纵然在车上，也总是先脱帽示敬。"孙云铸氏说："他像是一个中国人，而且是一个'温良恭俭让'的良好中国人。"

葛氏自己行动不便，但他却是野外工作的坚持者。学习地质者不见得全能认识化石。王竹泉说："民七在江西吉安、安福、永新一带得到大批化石无人鉴定，民国九年葛氏到中国第一批鉴定的，便是这些化石。"葛氏鼓励同人采集标本，中国古生物学从他的手里才开始光明昌大，《中国古生物志》的为世界看重，中国古生物界的人才较多，则由于他的训练。

新生代研究权威杨钟健是葛氏的学生，他说："我离开北平九年四个月，今天能自南京赶到参加葬礼。我是葛先生的最早一班的学生，当时我是不配听他的讲演的。我还记得他到了之后，第一个公开讲演就是《地球与其生物之进化》，后来由几个人翻译了在商务出版。七七事变之后，我离开北平比较晚，十一月三日到葛氏住宅辞行，他的腿已不能动，说起长别，大家痛哭流涕，当时就怕再见困难，今天我虽赶来，已然不能看到他了。葛先生是不朽的！他有二六二种著作，不必我们纪念他，他已经不朽了，今天的问题，是我们如何能接受他的事业？我们不能继承他的事业，我们的损失就更大了，这一点是不能不注意的。"

杨氏指出葛先生直接间接培养的人才虽多，但今天看到已然空乏。赵亚曾、计荣森、朱森、许德佑等或死于匪，或死于病，黄汲清改攻其他学科，田奇瑰以专家兼任所长，俞建章也不能全离事务工作，其他如王恭睦、戈定邦、潘钟祥也有不专的现象，"古生物学百年不败的基础"实在还成问题。

中国地质学的启发大师死了，大家都在纪念他。但怎样继承他的工作，目前还没有人想到，至少没有着手去作。在中国的外国地质学者，如德人李希霍芬，如安德生，如新常富，在中国人的心目中，谁也不能如葛利普氏一样的完整，这就是说，葛利普完成了播种的工作，但是种子的萌苗，却赖于自己的环境。播种人，他也许不会瞑目吧。（二十日寄自北平）

——原载上海《大公报》1947 年 3 月 27 日和 28 日

地质学家葛利普

杨钟健

　　葛利普是对我生平受影响最深的一位先生。民国八年我自北大预科毕业，升入地质学系。那时葛先生尚未来华。一年后葛先生到校任课，我听他的古生物学、史地[1]学等课，同时他在北平作长期公开讲演，题目是《地球及其生物之进化》[2]，由赵国宾和我笔记，分段在报上发表。后来即成专书出版。在他的课程和讲演中，我得的益处很多。最重要的，还是他的精神上的鼓励，使我对于古生物学和地史学，发生浓厚的兴趣。后来到三年级，地质系分三组，为地史古生物组、矿物岩石组及经济地质组。我特意选入第一组，可以说完全受了葛先生的熏陶。

　　在这大学最后两年中，重要的功课，可以说全是葛先生讲授。并有时到地质调查所作古生物实习，也是亲自指导，民国十一年毕业，发生就业问题。我因那时国内政局极不安定，在国内找不到合适的工作，想去外国。我的父亲也很同意，于是决定秋间去国。葛先生知道这消息，特别兴奋，为我写了三封介绍信。一封给柏林的彭伯士，一封给哈勒的瓦尔特，一封给明兴的白劳里。我到德国后，先到明兴，见了白劳里。白先生十分热心，我就留在明兴，从劳先生研究。所以也是受了葛先生的影响。

　　由德国回国以后，在地质调查所服务。我虽然专习了脊椎化石的研究，和新生代地质的勘查。然究竟还是古生物一行，因而与葛先生过从仍甚密。自民国十七年至抗战前后十年，由师友而变为同事，更形亲密。二十六年夏，卢沟桥事起，北平地质界令人处境困难，我们由普通的朋友，又变成患难的知交。凡设法对付敌伪，及保存文物之工作，无不相与讨论。十一月初，我要离平南下，当时在平外国友人中，一人赞成我南行。但为环境所迫，不得不辞别他南下。当我到他的寓所，与他话别时，他几不能自持，老泪纵横，真是最伤心的一幕。但当时决未料到这就是我和他的永诀。

　　在抗战期间，他在平情形，当然很苦。我们虽在南方，时时关心他的状

1　应为"地史"。——编者注
2　当时演讲和后出版成书的标题均为《地球与其生物之进化》。——编者注

况与安全。珍珠港事变后，但也失去自由，被送至前英大使馆拘禁起来，其精神上之痛苦，可想而知。我们爱莫能助，只有默祝他身体康宁。

民国三十三年我去美国，游踪至佈佛楼[1]，特别访问他的老弟，年六十五岁，那时他尚身体如常，我即以此安慰他的老弟。在那里看到他寄他老弟的书与文章，他的老弟为订书出身，把那些材料装订得很好，还有他一九三四[2]年最后一次到美时，登在报上的新闻和照片，他的老弟，也集留一册，装订起来。凡是在美国遇到地质界的朋友，莫有不问我打听他的消息。可惜我所能知道的，也不过那一点，无以安慰一般人的期望。

但无论如何他的健在是不成问题的。尤其当去年八月，日本投降以后，大家都以欣慰的心情，迎接未来。我们知道他已恢复了自由，想到不久即可在平会晤，重叙以前的悲欢，谁知道这期望，只是一场梦境呢！

三十五年三月底，我由英美回到战后的上海，在一种欢慰的情绪下，竟得到一个极不幸的消息，乃是他于三月二十一日[3]，病逝北平地质调查所内。当他逝世的时候，我尚在太平洋的舟中，计划着如何回北平，如何与他相见，相见后将是如何的欢乐。这消息一来，一切都完了。中国地质朋友，失去了一个挚友，全世界地质界，失去了一位同道。而我呢，回想二十五年来我们相处的情况，既为师生，又为同事，又曾共患难，当然更为悲伤。

葛先生的生平事迹，与夫在学术上的贡献，将有人作为专传，非本文可能详尽。我现在只能说他生平言行最重要的几点：

第一，是他以全付精力，尽瘁学术的精神。他生于一八七〇年一月九日。自一八九六年从麻省工业大学毕业之后，直到他死，整五十年，可以说全过的研究和写作生活。此等记录，在外国固为易见，在中国则至少至目下止，尚无人可与之比拟。我在纪念他六三诞辰时，曾有一短文中间引用了丁文江先生为他到华十年，也是纪念他生辰所说的几句话，最可表现葛先生尽力学术的精神。那是："他十年的工夫，只在北戴河过了个短的假期，但还和金叔初先生，共作了一本北戴河的介壳类，作为副产品。"

第二，是葛先生到中国后，对于中国古生物及地质研究推进的作用。此点凡熟悉中国地质发达经过的人，没有不深切认识。他到平后，一方面在北

---

1 即布法罗。——编者注
2 应为"一九三三年"，葛利普是 1933 年 6 月回到美国，参加 7 月在华盛顿举行的第十六届国际地质大会，会后顺访了一些学术机构，并返回布法罗老家探视亲友，最后于 9 月返回中国。——编者注
3 葛利普实逝世于 1946 年 3 月 20 日。——编者注

大教书，一方面在地质调查所担任研究，后来北大毕业者数位，也入地质调查所，而在北大葛先生的助教孙云铸先生，更亲受他研究指导。计先后亲自与他共同作研究工作者，除孙先生外，如赵亚曾、俞建章、田奇瑪、计荣森诸先生，均成为古生物之专才，尤以赵亚曾先生，亲受熏陶最多最久。此等作风，在今日看来，或不稀奇。但在当时无疑的为破天荒之举。其于地质界，亦自发生了推进作用。因葛先生学识广博，除古生物外，如地层、岩石、矿物乃至关于"北京人"之研究，葛先生均参加讨论。中同人一名，即葛先生所建议，而步达生采用者。他当时在北平，至少有五年之久，无疑的为北平研究的中心，一般作研究工作者，不但地质界，其他方面亦然，均有若卫星，环绕着他。

第三，为他对中国地质工作，不但努力的作，能且向外尽量的宣扬，他不时在各方写通俗文字，或作口头与书信式之报导，把中国地质工作情形，向外宣传，使得世人知中国地质之进展。我们知道，在科学工作初期，外人对中国学者工作，尚多不肯深信。同一说法，中国人言之，觉其可疑。即认为是。此固由西人轻视我民族，然亦信用未立之故，就连中国人自己，有许多人，对一事之评判，一人之好恶，一学理之当否，全以洋人之言为言，幸而中国地质界初年，有一葛利普，故能迅速的树起根基。此点自然有人或稍怀疑，然一悉其发展经过，当知若言之不谬矣。

第四，葛先生学术之贡献，浩如烟海，难以详举。但我今可得而言，以概其余者，为其早年之地层原理学说，三十年前，无人置信，北美地质家，作地层工作，多忽视海侵及地层相之变迁种种迹象。今则知旧路不通，多以葛先生之观点研究。故葛先生之声誉，久而弥高，其晚年努力于脉动学说，人多识其妄，不幸大著尚未能及身完成，此为可憾者。然其要节，固已宣布。关于葛先生之脉动学说之未来地位，可以葛先生已之言评判之。彼曾云，彼之学说，本不望及身得到世人之公认。然三十年后，必有为世人公试公认之一日。（此葛先生曾亲为作者言过。）此等目光抱负，与为时代先驱之豪气，环顾世上学者，能有几人？葛先生可以不死矣。

葛先生已矣，然葛先生之著作，与其治学精神，将永留人间。我希望不久中国地质界，可以有几位像葛先生那样精神、学问与修养的人才出现，则葛先生在中国二十五年[1]之工作精神，更为光芒，更为发扬。

——原载《人物杂志》1947 年第 2 卷第 2 期

---

1 应为二十六年。——编者注

葛利普教授

孙云铸

葛利普先生（Amadeus W. Grabau）原属德籍，其祖父于 1839 年始迁至美国。祖及父均为教堂牧师。1870 年先生生于美国威斯康星州席达堡村（Cedarburg）。年十五，在水牛城充订书工人；晚入夜校补习，对于植物学及地质学尤感兴趣。最初由函授从麻省工业学校克罗斯贝（Crosby）教授习矿物学，至 1890 年始充该系特别生。自此以后，先生得与当代古生物学家赫弟（Alpheus Hyatt）及姜克森（R. T. Jackson）两氏学习古生物学，学业大进。至 1896 年毕业。次年，入哈佛大学；1900 年得科学博士学位；又次年，任哥伦比亚大学讲师，至 1905 年升为教授，担任古生物学及地史学课程。

1910 年，先生游历欧洲，赴英瑞德奥等国实地研究古生代地层，并与当时欧陆诸地质大师相接触［为德国克塞（Kayser）、瓦特（Walther）两氏，英国马尔氏（Marr），法国奥格氏（Haug）］。对于华特氏著作，尤为悦服。返美后，曾注意研究古生物之生活环境，（特别注重珊瑚类、腹足类及笔石类）及各种属之发生史，并著有欧美下奥陶纪地层之比较，及大地槽迁移问题。先生在美研究范围至广，零星著作尤多，最重要者有《地层学原理》一巨册，《北美标准化石》两册（与夏谟博士 Shimer 合著）及《地质学》两册，均为模范教本。所著地质学内容新颖，含有不少欧洲材料，为本书特色。先生在美国地质界颇负盛名，弟子遍全国。当时中国留学生如王宠佑、王臻善、叶良辅、袁复礼诸先生均系先生弟子。

第一次欧战开始，美国对德宣战，学校禁用德文。先生以德国科学重要，兼反对战争，是以有辞去哥大教授之议。适丁文江所长托美国华提氏（White）代聘古生物学专家来华协助工作，当时美国地层学家舒可特（Schuchert）、露德门（Ruedemann）两氏均有来华之意。嗣后各方因先生较为适宜，决定请先生赴华任职。

1920 年 9 月，先生到平，协助丁文江所长奠定中国地质学基础。除在大学地质系任地史学、古生物学、欧美地层比教学三课外，并于每星期日下午公开讲演，讲题为"地球与生物之演化"，由王烈、李四光、龚安庆、谭熙鸿诸教授任翻译。1920 年，农商部地质调查所设古生物研究室，先生任主任，定研究计划，并发刊《中国古生物志》，由丁文江、翁文灏、孙云铸先后任编辑之责。1922 年，先生著《中国北部奥陶纪动物化石》，1924 年《中国北部寒武纪动物化石》（孙云铸著），1926 年《中国石炭纪海百合化石》（田奇㻜著），1927 年《中国北部蟆科》（李四光著），1927 年《中国北部长身贝》卷上（赵亚曾著），相继出版。此项刊物至今已出版百余册，为世界重要出版物之一。当时北方大学、山西大学、北洋大学、北京大学尚少有研究所之设立，中央、北平两研究院亦均未成立，斯时地质调查所之古生物研究室尚属首创，在中国科学史上实占重要之一页。

先生来华二年后，即与中国友人章鸿钊、丁文江、翁文灏、李四光、王宠佑诸氏等创中国地质学会，并发刊《中国地质学会志》，至今已继续出至第二十六卷，为中国科学重要出版物之一。先生不辞劳苦，时任校对修改之责。先生在华廿六年中，贡献尤多；终日研究，毫无倦容，著作丰富，共二百九十二种（每种自十余页至千余页不等），最重要者有《中国地质史》两册，《蒙古之二叠纪》及《中国古生物志》十一册。最近十余年，先生首创脉动学说及地极控制说，并用脉动学说划分古生代系统，已先后成巨著七大册（已有四册出版）[1]。内容丰富，材料新颖，颇有独到之见识，为研究地层学之范本。又于 1939 年引用欧美材料，另著《各纪之脉动》[2]（*The Rhythm of The Ages*），尤为欧美所重视。先生思想纯洁，具有德国血荫，最恶守旧，具改革精神，常用脉动学说解释中美地层问题。即先生深信世界各期海浸与海退大致均系同时，且属普通性，并以之划分古生代为十四个脉动期，颇为地质学界所重视。先生在中国及美国之成就，尤为伟大。

先生之古生物研究范围颇广，尤以无脊椎动物之腕足类、珊瑚类、腹足类及头足类为最精。地层研究包括古生代、中生代及新生代，其中以震旦纪、泥盆纪、二叠纪等纪贡献为最大。地质学科在中国科学地位尚不算落后，实由于中国古生物学及地层学有不少贡献，并已在《中国古生物志》及《中国地质学会志》发表。而此大部分工作，均为先生本人及先生弟子所担任。

先生昔日在美，与俄雷希（Ulrich）、傅斯第（Foerste）两氏彼此常因学术问题时常争执，先生从不介意。1933 年先生赴美参加国际会议又与两老友

---

1 应出版了 5 卷。——编者注
2 现在一般译为《时代之律动》或《年代之节律》。——编者注

相晤，彼此均甚和睦。著者于两年后过美，再遇俄雷希等氏，俄氏深赞扬葛氏在华之成绩，并谓亦即美国地质学会之光荣云。先生感人之深有如此者。

先生对于国际地质会议参加两次，并时常提出重要论文。1923 年比京[1]第十三次国际地质会议，与翁文灏所长提出"中国之石炭纪"，1926 年西京第十四次国际地质会议，与孙云铸提出"中国之寒武奥陶及志留纪"，1933 年与丁文江教授提出"中国之石炭纪"及"中国之二叠纪"。先生对于中国地质学会协助尤大，1925 年创设葛氏奖章以表先生之功绩，而先生又为葛氏奖章第一次受奖人。其他如北京博物学会，中国矿冶学会等亦常往演讲并撰稿；先生推进中国地质学科及其他科学之精神有如此者。

先生在华二十六年，训练不少古生物学人才，且均能独立研究：赵亚曾之于腕足类及石炭纪地层，杨钟健、裴文中之于新生代及脊椎化石，黄汲清之于二叠纪，斯行健之于古植物，计荣森之于珊瑚，许杰之于笔石，尹赞勋、赵金科之于头足类，田奇璃、乐森璕之于泥盆纪，俞建章之于下石炭纪珊瑚，朱森之于石炭纪，陈旭之蜓科，张席褆、丁道衡之于地层，王鸿祯之于云南地层，卢衍豪之于三叶虫，孙云铸之于下古生代，均系直接受先生指导；其他如科学社秉志先生，初有中国白垩纪昆虫之著，亦系受先生鼓励之影响。

先生在大学任教期中著有《中国地质史》两册，《中国地质问题》（登载《北大科学季刊》）及《腕足类种属》，并印有《古生物学》及《地史学》讲义；对于学生发问，回答精细，俟学生完全了解而后已。其诲人不倦之精神有如此者。

先生在地质所[2]协助丁翁两先生推进一切，因翁先生任所长职最久，先生与翁先生相处亦最久。凡关于研究工作，地质学会工作，及一切推进事宜，先生均极热心。翁先生偶有所托，先生立即完成。忆昔日每次中国地质学会评议会开会，当时全体评议员九人，出席者常仅半数或不及半数，先生每次必到；又 1934 年北大地质学系主任李四光先生赴英讲学，北大系务由先生代理，当时因地质馆方完成，系中经费极端困难，地质馆器具费学校仅充拨一千五百元，但各室器具非四千五百元不办，先生一面劝告研究教授捐助本年及下年研究费，一面函前校长蒋梦麟先生请预支下学年设备费一部，并请学校增加二千元。学校允许后，先生协同教授计划图样，不出三月，地质馆各室设备就绪，下学年地质系得按时迁入地质馆工作。先生服务精神及任事之

1　即比利时首都布鲁塞尔。——编者注
2　即地质调查所。——编者注

勤有如此者。

先生于 1933 年由美返平后，觉其所创脉动说深得美国地质学家之欢迎与认识，仍继续努力研究脉动说及世界性地质问题。不幸日寇侵略，北平陷落，北大与地质所南迁，先生因行动不便，仍留平寓，继著《脉动说》[1]；同时催促校所同人悉数离平，并大骂日本军阀及希特勒之疯狂，且料侵略者必败。先生一人留平，继著《脉动说》五六部（共写七部，仅有四部出版）。并著有世界性地质问题，先后寄长沙、重庆中国地质学会，在《会志》[2]发表，最著名如《中国地层区间脉动纪之意义》（1938 年在长沙出版），《地球发育之地极控制说》（1939 年重庆《中国地质会志》），《中国中部几个下古生代剖面之改正》（1940 年《中国地质会志》）、各纪之脉动一册（1940 年）及地史之新解释（1941 年北平《北京博物学会志》）。先生于 1941 年曾函昆明北大地质系，谓："在寓工作情形，一如往昔；北大自南迁后，即未见伪北大之门。"但自珍珠港事件发生后，太平洋战争开始，中华文化基金会[3]亦劝先生离平。先生因风湿症不便，仍留平寓，自此经济来源断绝，生活艰难，不久即被送入前英大使馆集中，即少有著作。身体日坏，憔悴不堪，有时神志不清，状至可悯。当时仅有德籍助手吴兰芝女士随侍在侧。

胜利后，地质所及北大均克来平，先生旋迁入地质分所。中华文化基金会恢复先生薪俸，北大代校长傅斯年先生亦即汇款接济，朱家骅部长两次往谒，甚叹先生神经之失常。地质界同人均冀先生早日恢复健康，卒因胃出血，竟于 1946 年 3 月 20 日在平逝世。遗言"愿葬北大地质馆。"昆明获电后，北大校务会即议决遵先生遗言办理。先生一生系一部奋斗史。在敌伪时期，先生不愿见伪北[4]大之门，遗言愿葬北大地质馆，先生全部地质杂志及书籍遗嘱捐赠中国地质学会。先生之忠于中国政府及中国地质学会有如此者。

先生人格伟大，亦足为人师表。先生以刻苦始，以困苦终。1920—1928 年平津教育经费无著，欠薪达数月之久，同事离平者颇多，先生刻苦如故，毫不为动。又在 1928 年，北大因反对取消北大名称，致薪俸停发数月，先生在寓授课，毫无怨言。先生初来中国，首赴冀东调查，乡儿嬉笑追随，先生不瞋怒；濒晚散归，犹颔首曰"再见""再见"。先生赴校授课，途遇相识同

---

1 即《脉动理论下的古生界地层》（*Palaeozoic Formations in the Light of the Pulsation Theory*）。——编者注
2 即《中国地质学会志》。——编者注
3 中华教育文化教育董事会，下同。——编者注
4 原文缺"北"。——编者注

事，先生首先脱帽为礼。著者追随先生二十六年，未见先生发怒一次。先生为人之伟大有如此者。

先生享年七十六岁，遗有夫人并一女及两外孙女。先生亲见中美胜利，地质所及北大同事及学生返平；又确知国外地质界对其学说之推崇，以及其所协助中国友人丁、翁、章、李诸先生奠定中国地质基础事业，业已成功。先生之素志已遂。先生虽死，先生对于世界之贡献已不朽矣。关于纪念此一代地质大师办法，除已由经济、教育两部呈请国府明令褒扬，北平地质分所立有纪念碑并纪念堂，北大将于先生逝世周年纪念日安葬骨灰并立墓碑，中国地质学会业决定发刊纪念，年内出版；此外教育部及中华文化基金会及北京大学亦拟在北大地质系设葛利普先生纪念讲座以资纪念。

——原载《科学》1948 年 30 卷第 3 期

古生物学大师葛利普教授年表

寿振黄

中央地质调查所北平分所研究员

　　葛利普（Amadeus W. Grabau）教授，为当代地质学及古生物学权威，旅华二十六年，桃李满门。晚近中国地质学及古生物学之业绩，能在国际学术界中有相当地位，非先生热心教导，曷克臻此。先生学问渊博，著作等身（论文 266 篇，书籍 41 种）。平津之治自然科学者，均不时请益于先生。珍珠港事变发生之后，先生被迫集中，衰老之躯，艰苦备尝。胜利以后，方冀其恢复健康，奈天不假年，哲人其萎，抚今思昔，悲从中来，爰集先生生平事迹，兼及近年自然科学界之进展，草先生年表。遗漏之处，知所不免。尚希海内贤达，有以正之！

　　同治九年（1870）　　一月九日，生于美国维斯康辛省之席达堡（Cedarburg）村。

　　同治十年（1871）　　一岁（年龄按西法计算）。

　　同治十一年（1872）　　二岁。

　　　　李希霍芬（Ferdinand von Richthofen）由华考察地质回国。

　　光绪二年（1876）　　六岁，章鸿钊（爱存、演群）生。

　　光绪三年（1877）　　七岁，清室送学生至英法留学。

　　光绪六年（1880）　　十岁。

　　光绪七年（1881）　　十一岁，先生丧母。

　　光绪八年（1882）　　十二岁，达尔文卒，德日进（Pierre[1] Teillard de Chardin）生。

　　光绪十年（1884）　　十四岁，步达生（Davidson Black）生。

　　光绪十一年（1885）　　十五岁，随父移居纽约之水牛城[2]（Buffalo），日间充订书铺学徒，晚间进夜校补习。

1　原文为 Pere。——编者注

2　即布法罗，水牛城为意译。——编者注

光绪十二年（1886）　十六岁，秉志（农山）生。

光绪十三年（1887）　十七岁，丁文江（在君）生。

光绪十五年（1889）　十九岁，李四光（仲揆）生，翁文灏（詠霓）生。

光绪十六年（1890）　二十岁，任波士顿博物学会矿物组助理员，进麻省理工大学[1]充地质学系特别生。初次发表研究论文（二篇）。

　　光绪十七年（1891）　二十一岁，在麻省理工大学改正式生，同时任地质学系助教。

光绪十九年（1893）　二十三岁，发表研究论文一篇。

光绪二十年（1894）　二十四岁，发表研究论文二十一[2]篇。胡先骕（步曾）生。

光绪二十一年（1895）　二十五岁，发表研究论文一篇。

光绪二十二年（1896）　二十六岁，麻省理工大学毕业，得学士（S. B.）学位，任母校地质学系助教。发表研究论文二篇。

光绪二十三年（1897）　二十七岁，得补助金，进哈佛大学研究院。发表研究论文一篇。

光绪二十四年（1898）　二十八岁，得硕士（S. M.）学位，被选为美国地质学会会员，发表研究论文五篇。中国京师大学堂成立。

光绪二十五年（1899）　二十九岁，任塔甫茨学院（Tufts College）地质学讲师，任雷塞拉（Rensselaer）高等工业学校[3]矿物学及地质学讲师，发表研究论文五篇。

光绪二十六年（1900）　三十岁，在哈佛大学得科学博士（S. D.）学位，升雷塞拉高等工业学校教授，同时供职密吉根[4]省地质调查所，发表研究论文四篇。

光绪二十七年（1901）　三十一岁，任哥伦比亚大学古生物学讲师，同时供职纽约地质调查所，被选为纽约科学院委员；与安丁·玛丽（Antin Marry）[5]女士结婚；发表研究论文六篇。

光绪二十八年（1902）　三十二岁，升哥伦比亚大学副教授，发表研究论

1　即麻省理工学院，下同。——编者注

2　应为一篇。——编者注

3　即伦斯勒理工学院（Rensselaer Polytechnic Institute）。——编者注

4　现一般译为密歇根。——编者注

5　应为玛丽·安丁（Mary Antin）。——编者注

文九篇。

光绪二十九年（1903）　三十三岁，威立士（B. Willis）及勃拉克维德（E. Blackwelder）由美来华地质调查。发表研究论文七篇。

光绪三十年（1904）　三十四岁，发表研究论文一篇。

光绪三十一年（1905）　三十五岁，升哥伦比亚大学古生物学正教授（至民国八年止），发表研究论文三篇。

光绪三十二年（1906）　三十六岁，发表研究论文二篇。

光绪三十三年（1907）　三十七岁，发表研究论文二十篇。

光绪三十四年（1908）　三十八岁，发表研究论文十二篇。

宣统元年（1909）　三十九岁，与辛默（Shimer）合著之《北美标准化石》第一卷在纽约出版；发表研究论文九篇。

宣统二年（1910）　四十岁，《北美标准化石》第二卷出版，赴欧洲考察，并赴瑞典京城 Stockholm 参加万国地质学会，发表研究论文四篇。

宣统三年（1911）　四十一岁，发表研究论文二篇。

民国元年（1912）　四十二岁，发表研究论文六篇。

南京临时政府设地质科，章鸿钊任科长。

民国二年（1913）　四十三岁，发表研究论文九篇。

《地层学原理》在纽约出版。

京师大学堂改称北京大学，章鸿钊创设地质研究所。

民国三年（1914）　四十四岁，发表研究论文五篇。

中国科学社成立，次年刊行《科学》。

民国四年（1915）　四十五岁，发表研究论文五篇。

民国五年（1916）　四十六岁，发表研究论文六篇。

六月，农商部设立地质调查局；十月，改称地质调查所，丁文江任所长。

民国六年（1917）　四十七岁，发表研究论文八篇。

民国七年（1918）　四十八岁，发表研究论文三篇。

北京大学恢复地质系。

民国八年（1919）　四十九岁，发表研究论文七篇。

《普通地质学》第一卷在纽约出版。

民国九年（1920）　五十岁，发表研究论文六篇。

非金属矿物质地质第一卷《食盐堆积之原理》在纽约出版。

应丁文江之聘来华，任北京大学古生物学教授，兼地质调查所古生物研究室主任，直至最近逝世。

恢复之北京大学地质系第一届学生毕业。

民国十年（1921） 五十一岁，发表研究论文二篇。

《普通地质学》第二卷出版。

民国十一年（1922） 五十二岁，发表研究论文五篇。

科学社在南京设立生物研究所。中国地质学会成立，《会志》[1]第一卷出版，先生被选为中国地质学会理事。

民国十二年（1923） 五十三岁，发表研究论文十二篇。

河南地质调查所成立，十六年停办，二十年恢复。

民国十三年（1924） 五十四岁，《中国地层学》第一卷由地质调查所出版。发表研究论文六篇，《地层学原理》再版。先生被选为中国科学社名誉会员。

翁文灏代理地质调查所长。

民国十四年（1925） 五十五岁，发表研究论文十三篇。

先生被选为德国皇家自然科学院会员。

中国地质学会授先生以第一次葛利普金质奖章。

北京博物学会成立。

民国十五年（1926） 五十六岁，发表研究论文八篇。

翁文灏继任地质调查所所长。

民国十六年（1927） 五十七岁，发表研究论文二篇。

先生被选为北京博物学会名誉会员及永久理事。

中央研究院地质研究所成立，两广地质调查所成立，中央大学地质系成立。

民国十七年（1928） 五十八岁，发表研究论文四篇。

与金叔初同著之《北戴河贝介小志》[2]再版（第一次在《中国杂志》刊印）。

《中国地层学》第二卷由地质调查所出版。

先生被聘为中亚考察队研究员，被选为费城自然科学院研究员。

北平研究院设地质研究所，静生生物调查所成立（十月）。中山大学

---

1 即《中国地质学会志》。——编者注

2 《北戴河的贝类》（*Shells of Peitaiho*）。——编者注

地质系成立。

民国十八年（1929）　五十九岁，发表研究论文七篇。

先生被聘为中央研究院研究员，协和医学院董事。

周口店发现中国猿人（*Sinanthropus pekinensis*）。

十一月十五日，赵亚曾在云南昭通遇害。

新生代研究室成立；中国古生物学会成立。

民国十九年（1930）　六十岁，发表研究论文十二篇。

静生生物调查所新厦立础。

民国二十年（1931）　六十一岁，发表研究论文八篇。

《蒙古之二叠纪》在纽约出版。

《中国地质学会志》"葛利普先生纪念册"（《会志》第十卷）出版；《地层学原理》三版。

民国二十一年（1932）　六十二岁，发表研究论文八篇。

赴纽约参加第十六次万国地质学会。[1]

清华大学地学系成立；西部科学院地质研究所成立；中国植物学会成立。

民国二十二年（1933）　六十三岁，发表研究论文一篇。

发表《脉动学说》和《两极管理学说》[2]。

中国动物学会成立。

民国二十三年（1934）　六十四岁，发表研究论文三篇，代理北大地质系主任。

步达生（D. Black）卒于北平（三月十五日）。

北京博物学会授先生以金氏金质奖章。

民国二十四年（1935）　六十五岁，发表研究论文三篇。

《腹足类之研究》由北京大学出版。

地质调查所由北平移至南京，仍在北平设置分所。

贵州地质调查所成立后停办。

民国二十五年（1936）　六十六岁，发表研究论文二篇。

《古生代地史之脉动观》第一卷及第二卷由北京大学出版。

1　葛利普参加华盛顿第十六次国际地质大会应为 1933 年。——编者注
2　1940 年，葛利普出版《年代的节律：从脉动理论和极控理论看地球的历史》（*The Rhythm of the Ages: Earth History in the Light of the Pulsation and Polar Control Theories*）一书，系统总结了他的脉动和极控理论，并用以解释地球的历史。——编者注

一月五日，丁文江卒于长沙。

重庆大学地质系成立。

民国二十六年（1937）　六十七岁，发表研究论文三篇。

《古生代地史之脉动观》第三卷由北京大学出版。

美国国立科学院授先生以汤默生[1]金质奖章（Thompson Gold Medal）。

七七事变，日军占北平，是时中央地质调查所由副所长黄汲清代所长职。

民国二十七年（1938）　六十八岁，发表研究论文六篇。

四川地质调查所成立，江西地质调查所改组成立。

《古生代地史之脉动观》第四卷由法文书店印行。

黄汲清继任中央地质调查所所长。

民国二十八年（1939）　六十九岁，发表研究论文一篇。

民国二十九年（1940）　七十岁，发表研究论文一篇。

《时代之律动》由法文书店出版。

美国康纳萨司（Coneseus）学院授先生以名誉科学博士学位。

一二八事变，新生代研究室及静生生物调查所被迫停顿。

黄汲清地质调查所所长职由副所长尹赞勋代理所长。

福建地质土壤调查所成立。

民国三十年（1941）　七十一岁，与德国吴兰芝（Volange）女士同居，集中英国大使馆。

《古生代地史之脉动观》第五卷由法文书店出版。

新疆地质调查所成立。

民国三十一年（1942）　七十二岁，发表研究论文三篇。

《古生代地史之脉动观》第六卷由法文书店出版，第七卷待刊。

李春昱继任地质调查所所长。

民国三十二年（1943）　七十三岁，发表研究论文一篇《西藏——人类之发源地》。

民国三十三年（1944）　七十四岁，写《我们居住的世界》待刊。

十二月，四川北碚中国西部博物馆成立。

民国三十四年（1945）　七十五岁。

1　现一般译为汤普森。——编者注

八月十五日，日军无条件投降。

民国三十五年（1946）　七十六岁。

一月十五日，移居丰盛胡同三号。

三月二十日，下午五时四十五分，因胃出血病故。

三月二十三日，下午二时遵照遗嘱在朝阳门外火化。

三月二十六日，昆明七学术团体开会追悼。

三月三十一日，下午三时，在北平协和礼堂举行宗教追悼仪式。

四月二十七日，中央研究院地质研究所、中国地质学会及中央地质调查所在四川北碚开追悼会。

六月二日，葬于北平北京大学地质馆前，并开追悼会。

——原载《科学》1948 年 30 卷第 3 期

In Memoriam: Amadeus William Grabau

(January 6, 1870 to March 20[1], 1946)

Dr. Grabau was the "Father of the Peking Society of Natural History", since it was his original idea and he remained an inspiration to all its members whenever he was able to attend its meetings. He was made a Life Councillor, and the Society came to count on him to be toastmaster at its annual dinners. His death during Japanese occupation is an irreparable loss to the Society.

Dr. Grabau was born in Wisconsin January 9[2], 1870. He came from Lutheran clergy stock. He received his training as a geologist and paleontologist in the Boston Society of Natural History, Massachusetts Institute of Technology and Harvard University where he took his Ph. D. in 1900. He taught at Rensselaer Polytechnic Institute, Tufts College and Columbia University from 1902 until 1919, and then in 1920, he came to China as Professor of Paleontology at the National University of Peking and Chief Paleontologist to the Geological Survey of China, to work with Dr. V. K. Ting and Dr. W. H. Wong. These positions he held until his death.

Before coming to China, Dr. Grabau had published 149 papers based on work on American material. From 1920 on papers on Chinese material began to appear and by his death the number of his publications had risen to 291. These were mostly published in *Paleontologica Sinica, Memoirs of Academia Sinica*[3], *Bulletin of Geological Society, Quarterly Journal of Sciences of Peking University*, as well as in other Chinese and foreign publications. In all of these his major objective remained the same: it was always to integrate his vast stores of knowledge in a comprehensive and generalized but accurate view of the geologic

---

1　原文为26，误。——编者注

2　原文为6，误。——编者注

3　应为 *Memoirs of the National Research Institute of Geology, Academia Sinica*。——编者注

succession, paleogeography and succession of life of the entire world.

In the Survey he began to issue the *Palaeontologia Sinica*, over one hundred numbers of which have been published. He started to train Chinese geologists and paleontologists. He gave lectures on paleontology, stratigraphy, and also on zoology for university students and research workers. He prepared Syllabuses in paleontology and historic geology. He was so successful that at least a dozen Chinese paleontologists have worked under his direction and now hold leading positions in geological institutions.

He published 2 volumes on the Stratigraphy of China and a dozen large Monographs on the paleontology of China. In 1930 there appeared a brief but significant article "Asia and the Evolution of Man", summarizing certain great geologic events in Asia, including the upheaval of the Himalayan Mountains. On the one hand, this upheaval had, he inferred, driven the forest-living anthropoid apes to the well-watered forests of the South, and on the other hand, it had forced the northern progenitors of man to give up the arboreal life and come out into the open plains of Central Asia.

Although painstaking in his records of fact, Grabau was always in search of generalizations of wider and wider scope and he vigorously defended his thesis against all opponents. Beginning with his lecture on the "Rhythm of the Ages" in 1933 which was later expanded into a work of 561 pages and 25 maps, he developed the theory of Oscillation or Pulsation, according to which the catastrophic rise and sinking of subcontinental areas was due to the rhythmic accumulation and release of pent-up forces of gravitation, which gradually brought about the shifting of the earth's poles. In order to pull all the lands into a great antarctic Pangea, with subsequent disruption and drift apart, Grabau invoked the aid of a distant passing star. These stupendous themes were developed in his 4-volume work *Paleozoic Formations in the Light of the Pulsation Theory*, totalling 3223 pages published in 1936–38, a work comprehensive and valuable for all stratigraphers and paleontologists, full of many new ideas.

Dr. Grabau was active in organizing the Geological Society of China and the Paleontological Society of China as well as the Peking Society of Natural History. He was the first recipient of the Grabau Medal of the Geological Society and also the recipient of the Thompson Medal of the Natural Academy of Sciences of the United States. He was a life member of the New York Academy of Sciences. He made one trip back to America when in 1933 he was invited to take part in the Geological Congress at Washington.

In his later years, Dr. Grabau was more and more hampered by arthritis, bug as long as the life flame burned on, he continued to discuss eagerly with his friends the great problems of geology. He suffered privation and grave illness during the Japanese occupation of Peking and was therefore prevented from finishing the remaining volumes of his pulsation theory and the revision of his *Textbook of Geology*. He finally succumbed peacefully on March 20[1], 1946. He is buried in the compound of the Geological Department of the National University of Peking according to his will. This Department has decided to name the departmental library the Grabau Library in his honor. The Chinese National Government has awarded him a testimonial, and the Geological Society has issued a Memorial volume in recognition of his great achievement for the geological science of China.

For all his vast and productive labors and for his vigorous genial and helpful nature we honor his memory.

(Compiled from several Biographical articles in the Grabau Memorial Volume of the *Bulletin of Geol. Soc. of China*, Nov. 1947).

——原载 *Peking Natural History Bulletin*, 1948, vol. 16 (3–4)

1　原文为 26。——编者注

科学家是怎样长成的？

——纪念葛利普先生逝世二周纪念作

杨钟健

（一）引论

说黄种人在人类进化程序上比白种人低，甚或举些似是而非的事实，说中国人脑量比外国人小，而断定中国人在科学成就上应该不如外国人。此等妄论，即明白的外国人亦不置信，中国人当然更不服气。此时代早已过去。黄种人，包括中国人在内，在智慧上，在能力上，一样的可与外国人（主要白种人）相同，一样的可胜任任何科学研究的工作。

我用不着远征古人，或外国非白种人的科学家，我只就我们这一代所亲自看见的许多地质学家、古生物学家，他们即能在很短期内，做出头等的科学贡献。活着的人，我不举例，以免有标榜之嫌。像死了的赵亚曾，在大学毕业后，短短五年之内，做出六七种古生物学专著，其中一部分，至现在还是在世界古生物学名著中不少逊色。又像死了的计荣森，也没有出过国，留过学，而所著的古生物学文章率多为标准作品。可惜其成名大著"泥盆珊瑚"未能及身完成。所以中国人能做科学的能力，绝对不容否认。地质学如此，其他科学亦然。在我们这一代中，都有不少的杰出人才。甚至于我们可说有许多中国科学家，其天分才具还在其外国同行同年之上。随便在任何科学门类中，都可找出例子。

但中国科学，虽然发展了三十多年，我们不能不承认两个重要缺点：

第一，我们虽然在各门科学中，出了不少杰出人才，但一般平均起来的水准，绝对不如人家。就拿地质来说，我国全国地质人才的总平均分数，无疑的远在他们之下。拿一个任何单独的机关来比，也是如此。机构不如人，组织不如人，合起来的成绩也不如人。虽然有少数杰出之士，只成了点缀太平的花瓶；谈起整个科学的进展来，还是一筹莫展。其他方面即如中国的官吏也如此。官场中诚不少一二竭诚奉公之士，然大多数甚不满人意。这当然不能说中国人不如外国人，这其间显然有其他原因。

第二，我国虽然在科学界出了不少杰出人才，但在每一科学部门中，养不出真正的权威，或者一代大师，像葛利普先生那样人才。我们需要头等的科学家，我们需要我们的每一门科学，均能像个样子进展，但我们尤需要每门科学，能出一二位大师。实在讲起来，我们现在的缺点，同我们缺乏真正的大师，有其因果关系。试想以一个像葛利普先生的外国大师，来到中国，不到数年立刻起了领导作用，使我国毫无根基的古生物学，有了根基。假使每一科学，真能有一二位大师，则其所起作用，必然可以弥补这里所说的缺点，至少可以减少这些缺点的严重性。何以我们养成不出若干科学家的大师，既不是中国人不如外国人，又不是这些有资格为大师的人不努力，当然尚有其他原因，可无庸置疑。

葛利普先生与其在中国科学界之功绩，知者甚多。光阴如流，葛先生逝世，忽届二年。而二年以来，中国之古生物学研究，因受环境影响始终不能步上按部就班，平庸推进之途，当然更谈不到发扬光大急起直追的善境。我因而联想到中国古生物学绝不能因葛先生之逝，而随之俱泯，又想到中国科学界之缺乏大师一类的领导人物，亦正如葛先生之在古生物学界。一个科学家是怎样养成的？一个科学界大师是怎样养成的？这些问题，需要合理的解答，从而可知我们的症结所在，才能予以革除，而步上真正的科学建设之途。今我所言，自然只能代表我一个人之意思。但刍荛之献，或者是国内科学界所乐于接受的。

(二) 论我们何以总平均不如人

为探本溯源计，我们先讨论上所提及的第一问题。这个至少又牵涉到几个基本问题。试一申述：

第一，是大环境当求改进，自然总平均分数可以提高。譬如中国四万万五千万人，虽然有不少杰出人物，而平均起来，距人家相差甚远。理由是我们的人大多数还是文盲，连字也不认识，更谈不上基本的国民教育。我们中国人，尽管有些像红楼梦上妙玉一样的清洁，但社会卫生环境，和大多数人的一年换不了几回衣服，一年洗不了几回澡，当然卫生的标准，十分低落。科学界也是一样。我们至少不能树立良好的科学环境。就地质学来讲，我们仅有的几个大学地质系，设备多很简陋，有的连教员也请不到，书籍很少，实习的标本也没有。在如此情形下，如何能希望有大量的好学生培养出来！其有一二杰出者，真是靠个人的天分与努力，绝不是大学校教育的成功。出了学校大门，进了地质机关。而这些地质机关，有的也不比大学的设备高明多少，而官场习气，则比大学更甚。弄得有志之士，吃不到羊肉，染了一身膻臭。这些人或者是缺乏人领导，或者是限于图书，或者是限于室内设备，或者外出工作限于经费，都弄得只能做些普通的工作，他们的才具，本来有十分，然在如此境况下，只能发展一二分。于是乎有些流于失望悲观，对现状不满意。有的人责备他们，其实他们很冤枉。他们应该责备别人，才较为合理。所有科学工作的大环境不能改进，是不能希望一般的提高的。

第二，新精神的缺乏。现在因对于胜利后的美丽梦境，为冷酷的现实所打破，心境都是悲观的。这影响，当然及于青年科学家。换言之，即一般人缺乏一种新精神的鼓励，因而或多或少表现出萎靡不振的气象，因而成了因循苟且，得过且过，混一天算一天的景况。造成如此悲惨情形的原因，实不能单责环境，个人亦应负一部分责任。胡适之先生谓青年须要有梦境。就是要有一个未来的理想目标，努力以赴，不能单拘囿于现在不满意的现状。说得再具体一点，此梦境在学科学的人尤为需要。科学研究，最要紧要有大胆的假设，也就是一个梦。此假设之成立与否，需要从各方面来证实或否认。这末以来，兴趣立至，虽在不良环境中，也可有作为，至少可不至全无作为。这一点看似不重要，但在一个人的心境上，至为重大。一个人做工作的主动或被动，全在此一点。如有兴趣的未来，大小梦境工作，可以发生兴趣，也就不难勇往直前，有居陋巷而不改其乐之乐。倘若无此梦境，则签到簿上签名，也为人家看；室内作文章，也为人家作；出外调查，也为了应付差事；全为被动，意境尽非，自然会兴致索然，也就不会有远大的成就。所以我所谓的精神，也就是梦境的精神，为我国科学界总平均分数提高的主要条件之一。

第三，乃是恒心的重要。天才的科学家，是很少的。即有，也只限于少数特殊之士，绝不能责于一般人。而且即就是天才科学家，聪明绝顶，也须要恒久的做去。如浅尝辄止，或半途改行，或分心他务，则至多也不过做一两篇好的文字，对科学有一些贡献，也还不能成一科学家，尤其不能成一够标准的科学家。科学工作，每一门类，浩如烟海，问题之多，无有止境。解决了一个问题，又有许多问题连序发生。科学家之工作为解决问题，而问题永远不完。科学家之工作兴趣，亦即在此。科学家而不能连续工作，只能算业余科学家，不能算真正科学家。真正科学之发达，自不能委之于业余工作的人。进一步言，虽资性平庸的人，如能继续不断的工作下去，人一己百，人十己千，积之既久，也可以成为大科学家。一人如此，集团亦然。譬如德国人的才具秉性不比其他国人士更高明，就因为德国人有一种锲而不舍的牛性，有打破砂锅问到底的奇癖，每遇一问题，必不怕麻烦，探求本源，悉心探讨，因而德国的科学水准，也显得特别高。所以聪明与天才，可以增加科学家的工作能力，但不是必须的资本。必须的资本，乃是日新又日新的持久精神。任何人有此决心，即可成为科学家。葛利普先生也具有德国人血液，也不是什么天才特殊的人，特殊的乃是他自出了学校四五十年，一直在地质古生物的问题上下工夫，直到临死也是如此。中国学地质学的人士，据粗略统计，自北京大学首有地质系毕业人才起，加上地质研究所毕业的人，在外国学地质的人，再加上以后各大学地质系毕业的人，总有一千之数。然目下在地质界实际做工作的人，至多不过上数的十之二。那十分之八的人那里去了？自然多数是改行，至少不能连续工作下去。倘使所有学习地质的人，均能连续工作二十年至三十年，中国地质界的发达，又当是如何的场面？地质如此，其他科门，更可想而知。因为一般人视地质是在中国比较发达早而进步速的一门科学，其现象尚且如此；其他科学部门之许多不能连续工作现象，当然至少相同。

无良好的环境，少梦境的精神，没有持久的历史，自然比起来不如人家。因为这三点在一般的上轨道而科学发达的国家，都是不成问题的。我们反其道而行，当然落后。

(三) 论科学大师的养成

上面所说的那三个主要条件，不但是养成科学家的必要条件，为科学要到基本水准的起码需要，也就是养成大师的先决因素。没有一个大师，不在科学工作优良条件滋养而成；也没有一个大师，没有他新奇的更伟大的梦境；也没有一个大师，不是皓首穷经，有假我数年，可以学易的精神。葛先生在美国时，于大学毕业后，即在其本国的良好的地质机关工作；到中国后，用地质调查所的材料，兼在北大教书，也有许多便利；可以说有良好的环境。他作学问，常有许多新奇理论，如关于人类起源，他主张系自中亚细亚而来，固然以后证其不可靠，然其大胆假定，则有助于理论与实际工作的推进，可以断言。他晚年倡脉动学说，也是一个美丽的梦境，直到他死，他还相信他的学说，过二三十年后，必可为世界地质家承认。他的持久精神，更不待说。我们在北平时，他常在家中请客。请客谈笑，饭后聊天，不到三句，就说地质或古生物方面的事，真是三句不离本行，平时可想而知。所有即以葛先生而言，乃是三者具备的人。

但是葛先生不止是一个杰出的科学家，而为一门科学的一位大师。大师与杰出的科学家不同。杰出的科学家，只可对其所习科学，有了不起的贡献，而一门科学的大师，则具有继往开来的作用。他可以对所习的科学，有结束以前人工作的能力；对未来青年，有循循善诱的雅量；对一门科学有双手推动的功能。此等我们目前最需要的科学大师，在中国尚不多睹，比杰出的科学家尤少。一代大师，对于一门科学之命运有了不得的关系，比科学的平均成绩尤为重要。因为有了科学大师，平均分数虽低，而不久可望增加。如无大师，则连低的分数，也可降而为零；少数杰出人才，也自有其山穷水尽的时候。

但是科学大师的养成，除了上述三点以外，还要靠许多优越条件。有些靠科学家本身的素养，有些赖客观的必要设备，今只能归纳起来，就下列数端，一为申论：

第一，科学大师需要高尚的素养，本人必须意志纯洁，以发展科学为己任，无有其他功利杂念，更不容"挂羊头卖狗肉"之行为。在道德方面，至少要修养到一种境地，够做一君子风度的贤人。要态度谦和，处事虚心，对科学研究，以真理为依归，不固执成见。对同事能和好相处，有容人之量。但对是非关头，不能像好好先生，随便迁就，而肯据理力争。对新进青年，能有热诚指导的耐心，循循善诱。凡是一个伟大的科学家，也就是一个伟大的做人模范，不仅其学问可为人尊仰，风度也为人尊仰，所以才能称得大师。这些人之使人佩服，不只是靠他们的学问，更不靠他们的地位或威权，而是有一种潜在的自然感召力，使人心悦诚服。大师地位的造成，不是勉强得来，而是自然获得。地位到了，不受也不由你。地位不到，勉强求之，也是枉然。回忆多年前北平豆芽菜胡同的葛先生寓所，无形中成了北平一个文化中心。每日来人不断，有他的上司，有他的学生，有他的同事，有他的同行的地质学家、古生物学家；也有虽不同行，而同为科学工作者的生物学家、人类学家、地理学家等。无论什么人去，他总是满面春风。无论什么人请教，他总是悉心教训，有一种说教的精神。谈天也好，论学也好，全在一种和蔼可亲的空气下。这里没有阶级，没有主属，甚至于没有老幼。他从来不拒绝人的请求，他以助人为乐。他替人改文章，而文章上不一定有他的名字。总而言之，大师要以德服人之成分相当之多。所以一个伟大科学家，也就是人生到最高境界的一个完人。完人成分越高，大师的光芒越大。

第二，大师的养成，固然靠他的道德方面的素养，而他本身对科学的工作，究竟最为重要。谈到工作，就不能不有一种方便，大凡上述的科学的杰出人才，无不有成为大师可能，可以说是大师的候选人。然而细察科学界杰出之士，有的不久蔚为一代大师，有的虽也努力不倦，而往往终其生只为一科学家。这其间就要看方便为何人获得，何人可以成为大师的希望就特别大。我这里所谓方便，与上述的大环境不同。大环境可以同时好的，或同时坏的，而方便与否，则有时不尽相同。譬如在美国纽约自然历史博物馆，其做科学工作的大环境，同时很好，而只出了沃士朋[1]，和葛雷高[2]两位大师（此只就古生物学言）。其理由就是因为这二位有了方便，而他人无之。我们知道一个大师的养成，需要许多优越条件。同时标本，一到实验室，他有首先研究的便利。有一个科学上很要紧的计划（譬如中国南方巨人化石的追寻），别人虽知道，而无法推进，他有方法推进，使成事实（如由沃士朋的哺乳类发见于亚洲的理论，而组成的中亚科学考查团）。在实验室中研究，须要许多人协助，别人无之，而他有。须要许多便利，别人无之，而他有。据说沃士朋在世时，研究化石的屋子有五六个。每一个有一特殊问题，和其标本，也有人在其领导下工作，所以能以一人之力，做许多人之事。葛先生在美国时候的情形不提，单就他到中国后讲，在中国科学环境那样不景气的情形下，他因外国人的关系，身体不良于行的关系，和他已有的成就的关系，他得了在那时无人可以比拟的便利。他要书，单子一开，有人替他找出，为他送去。化石的修理，标本的编号，文章的抄打，都有人替他帮忙，而他家中还有女书记替他办理一切。一个专家，须要许多人帮忙，古今中外，莫不如此。倘使葛先生在中国，也照现在中国大多数科学家，名为做科学工作，而实际上以公务员的姿态，应付一切，更谈不到要求一切。则葛先生之为葛先生，至多也不过一年出几篇文章的科学家罢了，岂能蔚为一代大师。

第三，科学大师之世界性，也十分重要。本来科学无有国界，中国大科学家，也就是世界的大科学家。科学上的问题，都是真理的追求。然而此等

1　即奥斯朋（Henry Fairfield Osborn，1857—1935）。——编者注

2　应为格雷戈里（William King Gregory，1876—1970）。——编者注

真理追求，往往需要世界性的研究。此在中国，一时实在办不到。中国的自然科学家，能把在中国发现的一些标本，合理的记述出来，已算上乘。如要把全世界相关种类，做彻头彻尾[1]的比较研究，能有新的理论与见解，实为数甚少。此非中国科学家不行，实在由于我们莫有充足的图书馆，没有随时可以借到的比较标本。就自然科学讲，世界上标本中心，不在欧洲，就在北美，他们每研究一门类，做必要的旅行，去考证他所要研究的材料，或借来研究，也容易办到。中国仅有的图书设备嫌不足，而标本的研究，尤为困难。在如此情形下，中国科学家有先天的困难，不能，至少不易养成为大师。葛先生来华时，其学力即已成熟，他自己就是一活的书库，又把他私人的收藏带来，又加上地质调查所的书，所以他能继续工作下去。我自己有一个切身经验，也表示在中国做工作之不易。抗战期间在云南找获的卞氏兽，经我鉴定，断定为和南非同年代的三瘤兽相近的化石。那时此兽在系统上归于哺乳动物。就我的标本所保存的样子来看，也当然以之归于哺乳动物。同类化石中，可与此标本相比者，在中国可以说一个也无有，我更不知道战事期间英国关于此类化石的新见。等到了美国，才知道有新的事实，证明这一类化石，应归于爬行类中最近似哺乳动物的一类。本来是一个惊人的大发见，经我们一弄，至少外国人分了一半的功勋，岂不可惜。所以在中国做工作，实有先天的限制，很不易养成头等科学家。

以上所说的几个条件，有的要反求诸己，然有的又非一己力量所可达到。我国从事科学工作三十余年，到现在不能有几个像样的大师出现，实在不是我们能力不够，乃受许多客观条件的限制。

（四）结论

在现在国家这样情形下，谈科学家的养成，乃至科学大师的养成，毋宁为一种白天做梦。纯粹科学研究，尽管有人力竭声嘶，认为重要，尽管政府也以提倡科学作为施政口号，然不幸一切完全为冷酷的现实所打破，成为一种可笑的讽刺。目下大学的教授，在研究机关工作的公务员，一月收入，不足其一家之温饱。大学教授所领的一百万元的所谓研究费，一些机关人员所领的三十万元的办公费，谓其可以贴补日用，多割几斤肉吃则可以，若说可用以研究或办公，恐发此钱者自己也不肯相信。人类什么都可以平等，但是

---

1 原文为"澈头澈尾"。——编者注

智力不能平等，造就也不能平等。而现在在中国苦学，成为科学界勇士也好，外国苦学成为科学家也好，他们半生辛辛苦苦，从事研究工作，为人类求真理，为社会增进文化，为国家争光荣的科学家，其待遇尚不及一银行之事务员，甚至不如一苦力之三轮车夫，如此情势，谈什么养成科学家！更希望什么科学大师！

就科学机构讲，现在任何研究机构，自中央研究院以至各省地质调查所，科学教育馆等都是苟且偷生尚可，发扬进展为难。仅有的费用，往往不够开销水电。以如此情形，要他们对研究科学的人，肯予以便利，肯帮忙，真是缘木求鱼。然而就研究的人言，个个感到不方便，进展为难，要求设备办不到，要求增加收入不可能。为了研究，不惜低声下气，向人叩头，然而所说的反响，仍是无有办法，自然谈不到进步，甚至消沉。就各研究机关的立场言，我开门营业，已是不易，还费了九牛二虎之力，汝等尚敢奢望其他！一方面觉得外界不但不帮忙，反有阻止发展之意；另一方面觉得我已尽最大努力，如何吃力不讨好，因而造成不协和的危险，当然也影响到科学整个的发达。此实今日中国科学界之大危机，凡稍留心者，类能道之。

旷观全世科学之能真正为政府为社会为人民所重视，实无过于苏联。即以美国科学在目下之根基，尚不免仍为资本主义社会之附庸。我曾在某处听美国人讲演及国家政治世界大势，他们说"我们利用科学云云"。夫科学家而被用作利用品，其地位可知。我国从政人士，其能真正了解科学家之重要者，百不获一，我尝也听见有人亦有"利用专家"之言。在他们看来，科学家正同他们的流线型汽车，长短波收音机，甚至于某机器上一个螺丝钉，必要时用用，以图便利或消消遣而已。然即就此点言，在中国尚不及外国。外国人要利用他们，尚知予他们以必要之扶助。在中国则只知利用，连扶助都没有，甚或不惜摧残之，破坏之，岂不可痛！

但我们亦不能完全悲观，而仍应鼓起勇气前进。纵观一部科学发达史，实可以说全在逆境中图生存，图发展。科学发达初期有不少冒性命危险，以身殉学之科学家。今日之情况，不见得比以前更坏。环境固可怕，但人亦可造环境。就连古生物学大师葛先生，也还是订书匠出身。中国科学家之出身，比之葛先生，均在其上。只要有勇气，不会无成就。以上所说的几个养成科学家的条件，和养成科学大师的条件，一方面固要靠环境方面之改良，一方面尤靠本身之努力。但尤所希望的，还是盼望今日能有几个具有远见的政治家与科学家，高瞩远瞻，能在社会培养出一个比较优良的环境，不要只利用他们。更不要把提倡科学，作为对科学的恩典或赏赐，而规规矩矩的导科学教育于正轨。使科学家均能尽力之所及，智之所能，在优良条件下工作。能如此，则不久可望整个科学的平均分数，可以提高，进而可与外国人相比，再进一步，也自然而然能产生科学大师资格的人。对这些人，也更要爱护，更要使之能尽其所长。这是国家百年大计，不只是建国，而是建国的根基。试想以一个外国科学大师，来在中国，不到二十年，尚可以发生如此伟大作用。如中国真能有良好的科学环境，产生出一群头等科学家，又养成了不少科学家的大师，则中国科学，岂能不独立，岂能不发扬！我想现在应该没有人迷信，只请几个外国顾问，即可把中国弄好。一切靠我们自己。中国地质界以后不会再有第二个美国葛利普，应该有许多中国的葛利普。地质界如此，他科亦然。然而科学家如何养成，倒成了大问题。今于葛先生二周纪念，特为一论，所以纪念以往，用励来兹，知我非我，非所计矣。

三十七年一月二十四日，南京

——原载《科学》1948 年第 30 卷第 3 期

Published Papers and Books of Amadeus William Grabau (1870–1946)

Compiled By A. W. Grabau

Amadeus William Grabau

Born: January 9, 1870, Cedarburg, Wisconsin, U.S.A.

Deceased: March 20, 1946, Peiping, China.

Degrees and Honors

1896   Received the degree of Bachelor of Science from the Massachusetts Institute of Technology, U.S.A.

1898   Received the degree of Master of Science from Harvard University, U.S.A.

1900   Received the degree of Doctor of Science from Harvard University, U.S.A.

1925   Awarded the first "GRABAU-MEDAL" founded by the Geological Society of China in his honor.

1934   Awarded "KING-MEDAL" by the Peking Society of Natural History, China.

1937   Awarded the "THOMPSON-MEDAL" by the National Academy of Science, U.S.A.

1940   Received the Honorary Degree of Doctor of Science from the Coneseus College, U.S.A.

Career in U. S. A.

1890   Appointed assistant in the Mineral Supply Establishment at the Boston Society of Natural History and concurrently Public lecturer Museum.

1891　Appointed student-assistant in Geological Department of Massachusetts Institute of Technology.

1896　Appointed assistant in Geological Department of Massachusetts Institute of Technology.

1899　Appointed Lecturer on Geology in Tufts College.

1900　Appointed Professor of Geology at the Rensselaer Polytechnic Institute. Appointed member of the Geological Survey of Michigan.

1901　Appointed Lecturer in Palaeontology at Columbia University. Appointed member of the Geological Survey of New York.

1902　Promoted to adjunct Professor at Columbia.

1905　Promoted to Professor of Palaeontology at Columbia.

Appointment in China

1920　Appointed Chief Palaeontologist of the National Geological Survey of China and Professor of the National University of Peking.

1928　Appointed Research Associate of the Central Asiatic Expedition.

1929　Appointed Research Associate of Academia Sinica.

Membership of Scientific Societies, etc.

Fellow, Geological Society of America.

Fellow, Palaeontological Society of America.

Fellow, American Association for the Advancement of Science.

Fellow, New York Academy of Science.

Fellow, Geological Society of China.

Honorary Member, Science Society of China.

Honorary Member and Life Councillor, Peking Society of Natural History.

Honorary Member, China Institute of Mining and Metallurgy.

Foreign Member, Kaiserlich Deutsche Akademie der Naturforscher zu Halle.

Correspondent, Philadelphia Academy of Science.

Dean, Peking Laboratory of Natural History.

Papers and Books

1890

1. On Historic Ground (A study of the Genesee River Region). *Buffalo Commercial*, June 14th, 1890. 2 pp.

2. Geological Observation (on the Genesee River). *Popular Science News*. Vol. 24, No. 11, Nov. 1890. 2 pp.

Total for 1890   4 pp.

1893

3. The Genesee River. *Wyoming County Times*, Sept. 7. 5 pp.

Total for 1893   5 pp.

1894

4. The Glacial Channel of the Genesee River. *Proc. Bost. Sec. Nat. Hist.* Vol. XXVI, pp. 259–269. 11 pp.

Abstract: *Am. Geol.* Vol. XIV, pp. 392–393. 2 pp.

Total for 1894   13 pp.

1895

5. The Emery Mines of Chester, Mass. *The Observer*, Jan. 1895 5 pp.

Total for 1895   5 pp.

1896

6. Three Hours in the Caverns of Luray. *The Observer*, Aug. 1896. 4 pp.

7. The Hamilton Group of Eighteen Mile Creek, Synoptic Table, published for the Buffalo meeting of the A. A. A. S. 1896. 1 pp.

Total for 1896   5 pp.

1897

8. The Sand Plains of Truro, Wellfleet and Eastham (Mass.) *Science*, New ser., Vol. 3, pp. 334–335, 361, 1897. 2 pp.

9. Physiographic notes. *Journal of School Geography*, 1898. 3 pp.

Total for 1897   5 pp.

1898

10. Palaeontology, Eastern Mass. *Am. Assoc. Adv. Sci.* 50 anniv. meeting. Guide to the localities illustrating the geology, marine

zoology and botany of the vicinity of Boston. Edited by A. W. Grabau and J. E. Woodman. pp. 37–62, Salem. Mass. 1898.　　26 pp.

11.　Marine Invertebrates (of Mass.) *Ibid*. pp. 67–96.　　30 pp.

12.　Palaeontology of the Cambrian Terranes of the Boston Basin, Mass. Abstracts: *American Assoc. Adv. Sci. Proc.* Vol. XLVII, pp. 303–306 ($^1/_2$ p.), 1898.　　$^1/_2$ p.

　　Science, new ser., Vol. VIII, p. 505 (4pp.);　　4 pp.

　　Am. Geol. Vol. XXIL, pp. 264–265 ($^1/_4$p) 1898.　　$^1/_4$ p.

13.　Siluro-Devonian contact in Western New York. *Science*, new ser., Vol. VIII, p. 800, $^1/_2$ p. 1898.　　$^1/_4$ p.

14.　Geology and Palaeontology of Eighteen Mile Creek and the lake shore sections of Erie County, New York, Part I. *Buffalo Soc. Nat. Hist. Bull.* Vol. VI, pp. I–XXIV, 1–91, pls. I–XXVII, 1898.　　105 pp.

　　　　　　　　　　　　　　　　　Total for 1898　166 pp.

1899

15.　The Palaeontology of Eighteen Mile Creek and the lake shore sections of Erie County, New York, Part II. *Buffalo Soc. Nat. Hist. Bull.* Vol. VI, pp. 93–403, 263 figs. 1899. Addenda & Corrigenda I page.　　311 pp.

16.　The Faunas of the Hamilton group of Eighteen Mile Creek and vicinity, in Western New York. *New York Geol. Surv.* 16th Ann. Rept. pp. 227–340, 5 pls. 6 figs. 1899.　　114 pp.

17.　Moniloporidae, new family of Palaeozoic Corals. *Bost. Soc. Nat. Hist. Bull.*, Vol. XXXVII, pp. 409–424, 4 pls, 1899.　　16 pp.

　　Abstract: *Science*, Vol. X, p. 65, ($^1/_2$p.), 1899.　　$^1/_2$ p

18.　Some modern stratigraphic problems. *Science*, N. S., Vol. X. Page 85.　　1 p.

19.　The relation of marine Bionomy to Stratigraphy. *Bull. Buff. Soc. Nat. Sciences*, Vol. VI, 4, pp. 319–367.　　49 pp.

　　　　　　　　　　　　　　　　　Total for 1899　491 pp.

1900

20.　Siluro-Devonic contact in Erie County, N. Y. *Geol. Soc. Amer.*

Bull. Vol. XI, pp. 347–376, pls. 21–22, figs. 1–8, 1900. Abstract: *Science*, new ser., Vol. XI, p. 105 (1/8 p.). 1900.　　　　30 pp.

21. Lake Bouve, an extinct Glacial lake in the southern part of the Boston Basin. *Boston Soc. Nat. Hist., Occ. Papers* IV, pt. III, pp. 564–600 map, 1900.　　　　37 pp.

22. Palaeontology of the Cambrian Terranes of the Boston Basin. *Boston Soc. Nat. Hist., Occ. Papers* IV pt. III, pp. 601–694, pls. XXXI–XXXIX, 1900.　　　　94 pp.

Total for 1900　161 pp.

1901

23. Guide to the geology and the Palaeontology of Niagara Falls and Vicinity. *Buffalo Soc. Nat. Sci. Ball.*, Vol. 7, pp. 1–280, 18 pls. 190 figs. and Geologic map. Also Reprint of Geologic part & map. State Comm. Reports on Niagara Falls.

24. *Ibid. N. Y. State Mus. Bull.* No. 45, pp. 1–284, 18 pls. 190 figs. and geologic map. 1901.　　　　284 p.

25. A preliminary geologic section in Alpena and Presque Isle Counties, Mich. *Am. Geol.,* Vol. 28, pp. 177–189, 1 pl. 1901.　　　　13 pp.

26. New species of Palaeozoic Corals. *Contribution to Indiana Palaeontology*, Pt. VII, 1901, 4 pp., 1 pl.　　　　4 pp.

27. Recent contributions to the problem of Niagara. Abstract: *Science*, N. S., Vol.4, p. 773, 1901; *N. Y. Acad. Sci. Annals*, Vol. 14, p. 139, 1901; *Amer. Geol.* Vol. 28, 329–330, 1901.　　　　3 pp.

28. Chronological classification of sedimentary rocks. *Columbia University publication*, 1901.　　　　2 pp.

Total for 1901　306 pp.

1902

29. Studies of Gastropoda, I. *Am. Nat.*, Vol. 36, pp. 917–945, 8 figs. 1902; *Columbia Univ. Geol. Dept. Contr.,* Vol. 10. No. 89, 1902.　　　　29 pp.

30. Stratigraphy of the Traverse group of Michigan. *Mich. Geol. Surv. Ann. Rept.* for 1901, pp. 163–210, 2 pls. 2 figs. 1902; *Columbia Univ. Dept. Geol. Contr.,* Vol. 10, no. 82, 1902.　　　　48 pp.

31. Hamilton group of Thedford, Ontario. Geol. Soc. *Am. Bull.,* Vol. 13, pp. 149–186, 5 figs. 1902; *Col. Univ. Geol. Dept. Contr.,* Vol. 10, No. 83, 1902 (Shimer and Grabau). 　　　38 pp.

32. A new species of Clavilitres from the Eocene of Texas. (Johnson and Grabau). *Phil. Acad. Sci. Proc.,* Vol. 53, pp. 602–603, 2 figs. 1902. 　　　2 pp.

33. The Geological Society of America (Proceedings and Abstracts of papers). *Science*, N. S., Vol. 15, pp. 81–91, 1902. 　　　11 pp.

34. Recent contributions to the problems of Niagara. Abstract: *N. Y. Acad. Sci. Ann.,* Vol. 14, p. 139, 1902. 　　　1 p.

35. Geology of Becraft Mt. N. Y. Abstract: *B. G. S. A.,* Vol. 14, 1902. 　　　1 p.

　　　　　　　　　　　　　　　　　　　　　Total for 1902　130 pp.

1903

36. The Washington meeting of the G. S. A. Dec. 30, 31, 1903. *Science*, N. S., Vol. 17, pp. 290–303, 1903. (Kemp and Grabau) 　　　14 pp.

37. Notes on the development of the biserial arm in certain Crinoids. *Am. Journal Sci.,* 4th ser., Vol. 16, pp. 289–300, 1903, 8 figs; *Col. Univ. Contr. from Geol. Dept.,* Vol. 11, No. 97, 1903. 　　　12 pp.

38. Stratigraphy of Becraft Mt. Columbia County, N. Y. *N. Y. State Mus. Bull.,* 69, pp. 1030–1074, 13 figs. 1903; *Col. Univ. Contr. from Geol. Dept.,* Vol. 11, No. 88, 1903. 　　　50 pp.

39. Palaeozoic Coral Reefs. *Geol. Soc. Am. Bull.,* Vol. 14, pp. 337–352, pls. 2, 1903; *Columbia Univ. Contr. from Geol. Dept.,* Vol. 11, No. 96, 1903. 　　　16 pp.

40. Studies of Gastropods. II. Fulgur and Sycotypus. *Am. Nat.,* Vol. 37, pp. 515–534, 19 figs. 1903; *Columbia Univ. Contr. from Geol. Dept.,* Vol. 11, No. 95, 1903. 　　　25 pp.

41. Limestone regions of Michigan. Abstract: *N. Y. Acad. Sci. Ann,* Vol. 15, p. 81, 1903. 　　　1 p

42. The phylogeny of the Fusidae. Abstract: *N. Y. Acad. Sci. Ann,* Vol. 15, pp. 86–87, 1903. 　　　1 p.

43. Traverse group of Michigan. Abstract: *Geol. Soc. Am. Bull.,* Vol.

13, p. 519, 1903.   1 p.

                                                    Total for 1903   120 pp.

1904

44.   On the classification of sedimentary rocks. *Am. Geol.*, Vol. 33, pp. 228–247, 1904; *Col. Univ. Geol. Contr.*, Vol. 12, No. 101, 1904.   20 pp.

45.   Relative ages of Oneida and Shawangunk Conglomerates. Abstract: *B. G. S. A.*, Vol. XVI, 1904.   1 p.

46.   Phylogeny of Fusus and its allies. *Smith. Misc. Coll.*, Vol. 44, pp. 1–157, 18 pls. 22 figs. 1904.   157 p.

                                                    Total for 1904   178 pp.

1905

47.   Physical characters and history of some New York formations. *Science*, N. S., Vol. 22, pp. 528–536, 1905.   8 pp.

48.   Evolution of some Devonic Spirifers. Abstract: *Am. Geol.,* Vol. 35, p. 195, 1905; *Science*, N. S., Vol. 21, pp. 426–427, 1905.   2 pp.

                                                    Total for 1905   10 pp.

1906

49.   Notes on the character and origin of the Pottsville formation of the Appalachian region. Abstract: *Science*, N. S., Vol. 24, p. 691, Nov. 30, 1906.   1 p.

50.   Guide to the geology and palaeontology of the Schoharie Valley in Eastern New York. *N. Y. State Mus. Bull.*, 92, (58th Ann. Rept. Vol. 3) pp. 77–386, 24 pls., 216 figs, 1906.   309 pp.

51.   Relative ages of the Oneida and Shawangunk Conglomerates. Abstract: *Geol. Soc. Am. Bull.*, Vol. 16, p. 382, 1906.   1 pp.

                                                    Total for 1906   311 pp.

1907

52.   Discovery of the Schoharie Fauna in Mich. *Bull. G. S. A.*, Vol. 17, pp. 718–719, 1907.   2 pp.

53.   Age and stratigraphic relations of the Chattanooga Black shale. Abstract: *Science*, N. S., Vol. 25, p. 771, May 17, 1907.   1 p.

54.   The Medina sandstone problem. Abstract: *Science*, N. S., Vol.

       25, pp. 771–772, May 17, 1907.   2 pp.

55. The Sylvania sandstone. A study in Palaeogeography. Abstract: *Science*, N. S., Vol. 26, p. 832, 1907.   1 p.

56. Evolution of some Devonic Spirifers. Abstract: *New York Acad. Science Annals*, Vol. 17, pt. 3, pp. 574–575, 1907.   2 pp.

57. North American Index Fossils. *School of Mines Quart.*, Vol. 27, No. 2, pp. 138–243, 175 figs. Jan. 1906. Vol. 28, No. 1, pp. 20–100, 46 figs. Nov. 1906. (Grabau and Shimer). Included in complete volume later.

58. New Upper Siluric Fauna from Southern Mich. *Bull. G. S. A.* Vol. 19, with Sherzer. pp. 540–553.   14 pp.

59. Nomenclature of subdivision of the Upper Siluric Strata of Michigan, Ohio and Western New York. (With Lane, Prosser and Sherzer). *Bull. G. S. A.,* Vol. 19, pp. 553–556.   4 pp.

60. North American Index Fossils, II. *School of Mines Quart.,* Vol. 28, No. 2, pp. 150–221, 89 figs. No. 3. pp. 251–352, 165 figs. 1907. Included in complete volume.

61. Types of sedimentary overlap. *Geol. Soc. Am. Bull.,* Vol 17, pp. 567–636, 17 figs. 1906. Abstract: *Science*, N. S., Vol. 21, pp. 991–992; *New York Acad. Sci. Annals*, Vol. 17, pt. 3, pp. 598–599, 1907.   70 pp.

62. Discovery of the Schoharie fauna in Mich. Abstract: *Science*, N. S., Vol. 23, p. 467, March 23rd, 1906.   1 p.

63. Studies of Gastropods, III. On orthogenetic variation in Gastropods. *Am. Nat.,* Vol. 41, pp. 607–646, 3 pls. Oct. 1907.   40 pp.

64. Subaerial erosion cliffs and talus in the lower Devonic of Michigan. Abstract: *Science*, N. S., 25, pp. 295–296, Feb. 22, 1907.   2 pp.

65. Type of cross-bedding and their stratigraphic significance. Abstract: *Science*, N. S., Vol. 25, pp. 296, Feb. 22, 1907.   2 pp.

66. Geology and scenery of the Upper Genesee Falls. *Science*, N. S. Vol. 25, pp. 538–539, April 5, 1907.   2 pp.

67. How Coney Island arose from the sea. *Brooklyn Daily Eagle*, N. S., Sunday, Jan. 13, 1907.   2 pp.

68. Seventh International Zoological Congress, Sect. of Palaeontology. *Science*, N. S., Vol. XXVI, No. 677, pp. 881–883. Dec. 20, 1907.   2 pp.

69. The geographical classification of marine life districts. *Science*, N. S., Vol. XXV, No. 631, pp. 184–185.   1 p.

70. Geology of Jamaica, *N. Y. Evening Post*, Jan. 13, 1907.   2 pp.
     Total for 1907   190 pp.

1908

71. Discovery of the Schoharine fauna in Michigan. Abstract: *N. Y. Acad. Sci. Annals*, Vol. 18, pt. 2, p. 294, 1908.   1 p.

72. Notes on the character and origin of the Potville formation of the Appaiachian region. Abstract: *N. Y. Acad. Sci. Ann.*, Vol. 18, pt. 2, p. 294, 1908.   1 p.

73. The scenery and geology of the gorges and falls (of central N. Y.) Abstract: *N. Y. Acad. Sci. Ann.*, Vol. 18, pt. 2, pp. 322–323, 1908.   1 p.

74. The Sylvania sandstone, a study in palaeogeography. Abstract: *N. Y. Acad. Sci. Ann.*, Vol. 18, pt. 2, p. 344, 1908.   1 p.

75. A revised classification of the North American Siluric system. Abstract: *Science*, N. S., Vol. 27, pp. 622–623, April 17, 1908.   1 p.

76. Notes on the Traverse group of Michigan. Abstract: *Science*, N. S., Vol. 27, p. 726, May 8, 1908.   1 p.

77. Preglacial Drainage in Central-Western New York. *Science*, N. S., Vol. 28, pp. 527–534, Oct. 16, 1908.   8 pp.

78. Continental formations of the North American Palaeozoic. Abstract: *Science*, N. S., Vol. 28, p. 936, Dec. 25, 1908.   1 p.

79. A new Siluric fauna from Michigan. Abstract: *Science*, N. S., Vol. 27, p. 408, March, 1908. (Sherzer & Grabau).   1 p.

80. Dovonic elements in the late Siluric fauna of southern Michigan. Abstract: *Science*, N. S., Vol. 27, p. 726, May 8, 1908.   1 p.
     Total for 1908   17 pp.

1909

81. A revised classification of the North American lower
Palaeozoic. *Science, N. S.*, Vol. 29, pp. 351−356, Feb. 26, 1909.　6 pp.

82. Physical and faunal evolution of North America during
Ordovicic, Siluric and early Devonic time.　42 pp.

83. Tertiary drainage problems of eastern North America.
Abstract: *Science*, N. S., Vol. 29, p. 632, April 16, 1909.　1 p

84. Some of the little-known geological terms and their application
in stratigraphic writing. Abstract: *Science*, N. S., Vol. 29, p. 750,
May 7, 1909.　1 p

85. Early development stages in recent and fossil corals. Abstract:
Science, N. S., Vol. 29, p. 917, June 4, 1909.　1 p

86. The Medina and Shawangunk problems in Pennsylvania.
Abstract: *Science*, N. S., Vol. 30, p. 415, Sept. 2, 1909.　1 p.

87. Nomenclature and subdivision of the Upper Siluric strata of
Michigan, Ohio and Western New York. *Geol. Soc. Am. Bull.*,
Vol. 19, pp. 553−556, 1909. (Lane and Others).　4 pp.

88. New upper Siluric fauna from southern Michigan. *Geol. Soc.
Am. Bull.*, Vol. 19, pp. 540−563, 1 fig. 1909. (Sherzer and Grabau).　14 pp.

89. North American Index Fossils. Invertebrates. (with Shimer)
1909. Vol. 1, 853 pp. 1210 figs.　853 pp.

90. Report on Examination of the Portage Dam site, New York.
*Amer. Scenic and Historic Preservation Soc. 14th Ann. Rept. to
the Legislature, N. Y.* pp. 45−51.　7 pp.

Total for 1909　930 pp.

1910

91. Palaeontology and Ontogeny. *Pop. Sci. Mo.* Sept. 1910, Vol. 77,
pp. 295−298.　4 pp.

92. Studies of Gastropods, IV. Value of the Protoconch & Early
Conch Stages in the classification of Gastropods. *Proc. 7th Int.
Zoological Congress*, 1910, pp. 753−766, 14pp. 10 figs.　14 pp.

93. Mutations of Spirifer mucronatus. (With Margaret Reed).

Seventh Inter. Zool. Cong. Proc., pp. 767–768.   2 pp.

94.   North American Index Fossil. Vol. II. (With H. W. Shimer). N.
       Y. (A. G. Seiler), 726 figs. 909 pp.   909 pp.

95.   The Monroe formation of Southern Michigan and adjoining
       regions. (With H. W. Sherzer). *Mich Geol. & Biol. Survey. Pub., 2.*   248 pp.

96.   Tertiary drainage problem of eastern North America. Abstract:
       Geol. Soc. Am. Bull., Vol. 20, p. 668.   1 p.

97.   Early developmental stages in recent and fossil corals. Abstract:
       New York Acad. Sci. Annals, Vol. 19, p. 299.   1 p.

98.   Intracolonial acceleration and retardation and its bearings on
       species. Abstract: *Science,* N. S., Vol. 32, p. 223.   1 p.

                                             Total for 1910   1180 pp.

1911

99.   On the classification of sand grains. *Science,* N. S., Vol. 33, pp.
       1005–1007.   3 pp.

100.   Palaeontology. (With H. F. Osborn). *Encyclopedia Britannica,*
       11th ed. pp. 579–592.   13 p.

                                             Total for 1911   16 pp.

1912

101.   Syllabus of Historical Geology. A. G. Seiler, N. Y.   51 pp.

102.   Stratigraphic features of ancient delta deposits. Abstract: *Science,*
       N. S., Vol. 35, p. 317. Abstract with discussion by J. M. Clarke,
       David White, G. W. Stose, Arthur Keith, E. T. Wherry, and H. B.
       Kummel. *Geol. Soc. Am. Bull.,* Vol. 23, No. 4, pp. 743–746.   4 pp.

103.   Structure of the Helderberg front. Abstract: *Science,* N. S, Vol.
       35, p. 319.   1 p.

104.   *Ibid.* Abstract (with discussion by J. B. Woodworth): *Bull. Geol.
       Soc. Am.,* Vol. 23, No. 4, pp. 746–747.   7 pp.

105.   *Ibid.* Abstract: *New York Acad. Sci. Annals,* Vol. 21, p. 210.   1 p.

                                             Total for 1912   64 pp.

1913

106.   Ueber die Einteilung des nordamerikanischen Silurs. Intern.

Geol. Congr. XI, Stockholm, 1910. *Compte Rendu*, pp. 979–995,
6 figs.   17 pp.

107.　Continental formations in the North American Palaeozoic.
Intern. Geol. Congr. XI, Stockholm, 1910. *Compte Rendu*, pp.
997–1003.   7 pp.

108.　Principles of Stratigraphy. A. G. Seiler, N. Y. 263 figs.   1185 pp.

109.　The origin of salt deposits with special reference to the Siluric
salt deposit of North America. *Min. & Met. Soc. America Bull.*,
No. 57 (Vol. 6, No. 2). pp. 33–44, Feb. 28.   12 pp.

110.　Early Palaeozoic delta deposits of North America. *Geol. Soc.
Am. Bull.*, Vol. 29, No. 3, pp. 399–528, 14 figs. 1 pl.   130 pp.

111.　Palaeontological notes (polyphyletic genera; 2. an illustration
of Waagen's theory of mutations. Abstract: *Geol. Soc. America
Bull.*, Vol. 24, p. 109.   1 p.

112.　Was there a former Goat Island at Niagara Falls? Abstract: *N. Y.
Acad. Sci. Annals*, Vol. 22, p. 378.   1 p.

113.　Irrational stratigraphy, the right and wrong of reconstructing
ancient continents and seas. Abstract: *Science*, N. S., Vol. 38, p.
282.   1 p.

114.　A classification of marine deposits. *Geol. Soc. America Bull.*,
Vol. 24, pp. 711–714.   4 pp.

115.　Glacial erosion in the Genesee Valley system and its bearing on
the Tertiary drainage problem of Eastern North America. *Geol.
Soc. Am. Bull.*, Vol. 24, No. 4, pp. 718–619.   2 pp.

116.　Preliminary report on the faunas of the Dundee limestone of
southern Michigan. *Mich. Geol. and Biol. Surv. Publ.*, 12, (Geol.
Ser. 9), pp. 327–378.   51 pp.

Total for 1913   1411 pp.

1914

117.　Irrational Stratigraphy; the right and wrong way of reconstructing
ancient continents and seas. (abstract, with discussion). *N. Y.
Acad. Sci. Annals*, Vol. 23, p. 288, April 30, 1914.   1 p.

118. Some new palaeogeographic maps of North America. Abstract: *Bull. G. S. A.,* Vol. XXV, pp. 136–7. 1 p.

119. Devonic black shale of Michigan, Ohio, Canada, and Western New York, interpreted as a Palaeonzoic delta deposit. *Bull. G. S. A.,* Vol. XXV, p. 137, (Title) paper presented.

Total for 1914 2 pp.

1915

120. Olentangy shale of Central Ohio, and its stratigraphic significance. Abstract: *Bull. G. S. A.,* Vol. 26, p. 112, 156. 1 p.

121. Hamilton Group of western New York. Abstract: *Bull. G. S. A.,* Vol. 26, pp. 113–158. 1 p.

122. North American Continent in Upper Devonic time. Abstract: *Bull. Geol. Soc. Amer.,* Vol. XXVI, pp. 88–89. $1^1/_4$ p.

123. The black shale problem, a study in Palaeozoic geography (Abstract). *New York Acad. Sci. Annals,* Vol. 24, pp. 378–379, May 14, 1915. 2 pp.

124. *Ibid. Science, N. S.,* Vol. 41, pp. 509–510, Apr. 2, 1915. 2 pp.

Total for 1915 7 pp.

1916

125. Comparison of American and European Lower Ordovicic Formation. *Bull. Geol. Soc. Am.,* Vol. 27, pp. 555–622, 10 text figs. Nov. 29, 1916. 68 pp.

126. Abstract. *Ibid.* p. 159, March 31, 1916. $^1/_2$ p.

127. Distribution and inferred migration of American middle and upper Devonic corals (Abstract). *Bull. Geol. Soc. Am.,* Vol. 27, p. 148, March 31, 1916. $^1/_4$ p.

128. Classification of the Tetraseptata (Tetracoralla) with some remarks on parallelism in development in this group; a study in orthogenesis (Abstract). *Bull. Geol. Soc. Am.,* Vol. 27, p. 148, March 31, 1916.

129. Sub-division of the Traverse Group of Michigan and its relation to the Mid-Devonic formations (Abstract). *Bull. Geol. Soc. Am.,*

Vol. 27, p. 159. $^1/_2$ p.

130. Systematic Rank of Mutations and sub-mutations in Orthogenetic series among the invertebrates. *Bull. Geol. Soc. Am.,* Vol. 27, p. 148. $^1/_2$ pp.

Total for 1916　70 pp.

1917

131. Stratigraphic relationships of the Tully limestone and the Genesee Shale in eastern North America. *Bull. Geol. Soc. Am.,* Vol. 28, pp. 945-958, 3 figs. Abstr. *Ibid.* pp. 207-208. $^1/_2$ p.

132. Were the Graptolite shales as a rule deep or shallow water deposits? (With Marjorie O'Connell). *Bull. Geol. Soc. Am.,* Vol. 28, pp. 959-964, 1 fig. 6 pp.

133. Problems of the Interpretation of Sedimentary rocks. *Ibid,* Vol. 28, pp. 735-744. 10 pp.

134. Age and stratigraphic relations of the Olentangy Shale of Central Ohio, with remarks on the Prout Limestone and so-called Olentangy shales of Northern Ohio. *Journal of Geology,* Vol. XXV, No. 4, May-June 1917. pp. 337-343, 1 text fig. 7 pp.

135. Comparison of the European and American Siluric (Abstr.) *Bull. Geol. Soc. Am.,* Vol. 28, pp. 129-130, March 31, 1917. 1 p.

136. New Genera of Corals of the Family Cyothophyllidae. Abstract: *Bull. Geol. Soc. Am.,* Vol. 28, p. 99. $^1/_2$ pp.

Total for 1917　40 pp.

1918

137. Isolation as a factor in the development of the Palaeozoic faunas (Abstract). *Bull. Geol. Soc. Am.,* Vol. XXIX, p. 143. $^1/_4$ p.

138. Relation of the oil-bearing to the oil-producing formations in the Palaeozoic of North America (Abstract). *Bull. Geol. Soc. Am.,* Vol. XXIX, p. 92. $^1/_4$ p.

139. Significance of the Sherburne bar in the Upper Devonic Stratigraphy (Abstract). *Bull. Geol. Soc. Am.* Vol. XXIX, p. 127. $^1/_2$ p.

Total for 1918　1 p.

1919

140.   Condition of deposition of some Tertiary patroliferous sediments (Abstract). *Bull. Geol. Soc. Amer.* Vol. XXX, p. 103. $^1/_2$ p.

141.   Inclusion of the Pelistocene period in the Psychozoic era (Abstract). *Bull. Geol. Soc. Amer.,* Vol. XXX, p. 104.   1 p.

142.   Migration of Geosyndines (Abstract). *Ibid.* p. 87. $^1/_4$ p.

143.   Prevailing stratigraphic relationship of the bedded phosphate deposits of Europe, North Africa and North America (Abstract). *Ibid.* p. 104. $^1/_4$ p.

143a.   Review of O'Connell's "The Schramnen Collection of Cretaceous Silicispongiae in the American Museum of Natural History". *Science*, N. S., Vol. L, No. 1288, pp. 231–233, Sept. 5, 1919.   3 pp.[1]

144.   Relation of the Holochoanites and the Orthochoanitesto the Protochoanites and the significance of the Bacteritidae (Abstract). *Ibid.*, pp. 148, 149. $1^1/_2$ p.

145.   Significance of the Sherburne Sandstone in Upper Devonic Stratigraphy. *Bull. Geol. Soc. Am.,* Vol. XXX, pp. 423–470, 2 text fig.   48 pp.

Total for 1919   55 pp.

1920

146.   Sixty Years of Darwinism. *Natural History*, Vol. XX, No. 1, pp. 58–72, 6 portraits.   15 pp.

147.   The Niagara Cuesta from a New Viewpoint. *The Geographical Review*, April, May, June 1920. pp. 264–276, 17 text figures.   13 pp.[2]

148.   Geology of the Non-Metallic Mineral Deposits other than Silicates, Vol. 1, Principles of Salt Deposition. pp. XVI & 435, text figs. 125, 1 plate. MacGraw Hill Book Co., N. Y. Sept. 24, 1920.   451 pp.

---

1 原文为4 pp，误。——编者注
2 原文为pp. 254–276，误。——编者注

149.   Text-book of Geology, Vol. 1, viii to xviii, and 1-864. pp. 734 text figs. and frontispiece. D. C. Heath & Co., Nov. 1920.   882 pp.

150.   A new species of *Eurypterus* from the Permian of China. *Bull. Geol. Surv. China*, No. 2, Oct. 1920. pp. 61-67, PI. IX.   7 pp.

151.   A Lower Permian Fauna from the Kaiping Coal Basin. *Ibid.*, pp. 69-79, pl. IX.   10 pp.

152.   Significance of the Middle Siluric in American and European Stratigraphy (Abstract). *Bull. G. S. A.,* Vol. 31, p. 138. $^1/_3$ p.

153.   Unicline, As term proposed for monoclinal ridges of erosion (Abstract). *Bull. G. S. A.,* Vol. 31, p. 153. $^1/_3$ p.

Total for 1920   1379 pp.

1921

154.   Text-Book of Geology, Vol. II, pp. 1-976, figs. 735-1980 (1245) and frontispiece, Jan. 1921.   976 pp.

155.   Topography and Geological Structure of North America. *Annual of the Geological Society of the University of Peking*, Vol. 1, 1921, pp. 1-7.   7 pp.

156.   Earthquakes. *Ibid.*, 3 pp.   3 pp.

Total for 1921   986 pp.

1922

157.   Age of the Coal Beds of the Kaiping Coal Basin in Northeastern China (Abstract). *Bull. Geol. Soc. Amer.,* Vol. 33, p. 201. $^3/_4$ p.

158.   Ordovicic Formations of North China (Abstract). *Bull. Geol. Soc. Am.,* Vol. 33, p. 202. $^3/_4$ p.

159.   First General Meeting of the Geological Society of China. *Peking Leader*, March 25, 1922.   2 pp.

160.   Ordovician Fossils of North China. *Palaeontologia Sinica*, Ser. B, Vol. 1, pp. 1-127, 9 pls. 20 figs., April 28, 1922.   127 pp.

161.   Palaeozoic Corals of China, Pt. 1. The Tetraseptata. *Palaeontologia Sinica*, Ser. B, Vol. II, Fascicle 1. Introduction and Petraidae Streptelamidae and Cyathaxonidae. pp. 1-76, pl. 1, 74 text figs. Sept. 1, 1922.   76 pp.

162. The Sinian System. *Bull. Geol. Soc. China*, Vol. 1, pp. 48–88. 45 pp.

163. The First Year of the Third Asiatic Expedition's Activities. *The Peking Leader*, Oct. 3, 1922. 7 pp. 7 pp.

Total for 1922 258 pp.

1923

164. Searching the Fossil Fields of China. *Peking Leader*, January 7th, 1923. 12 pp.

165. The Annual Meeting of the Geological Society of China. *Peking Leader*, January 14th, 1923. 11 pp.

166. Lectures on Evolution (in Chinese). The Commercial Press, Shanghai, Pt. 1, & Pt. 2. 164 pp.

167. Carboniferous Formations of China. (With Dr. W. H. Wong). *Transactions of the Science Society of China*, Vol. II, pp. 1–10, 2 text figs. 10 pp.

168. Report on Third Asiatic Expedition's Work. *The Peking Leader*, Sept. 29, 1923. 11 pp.

169. Cretaceous Fossils from Shantung. *Bull. Geol. Survey China*, No. 5, Pt. 2, Dec. 1923. 19 pp.

170. Cretaceous mollusca from North China. *Ibid.* 15 pp.

171. A Lower Cretaceous Ammonite from Hongkong. *Ibid.* 10 pp.

172. Contribution to the Fauna of the Kweichow Formation of Central China. *Ibid.* 9 pp.

173. What is a Shantung? *Bull. Geol. Soc. China*, Vol. 2, No. 1–2. (Title).

174. The Devonian in China. *Bull. Geol. Soc. China*, Vol. 2, No. 1–2. $1^1/_2$ p.

175. Carboniferous in China. *Ibid.* $2^3/_4$ pp.

176. Carboniferous Formations of China. (With W. H. Wong). Congress Geologique Interhational, XIIIème Sess., 1922, Belgique. *Compte Rendu*, Liège, 1923, pp. 657–689, 4 text figs. 33 pp.

Total for 1923 298 pp.

1924

177. The Annual Meeting of the Geological Society of China. *Peking Leader*, Jan. 6. 12 pp.

178.	Geological Conditions Bearing upon potash prospecting in China. *The China Institute of Mining and Metallurgy Bull.* Vol. III, pp. 1–28, 2 figs.	28 pp.
179.	*Stratigraphy of China*, Pt. 1. 1–528 pp. 306 text figs, 6 pls.	546 pp.
180.	The Oldest Scarab in the World. *Peking Leader*, Sept. 9. 1924.	4 pp.
181.	Celebrate Publications of First Important Chinese Contributions to Palaeontology. *The Peking Leader*, Dec. 30, 1924.	3 pp.
182.	Migration of Geosynclines. *Bull. Geol. Soc. China*, Vol. III, Nos. 3–4, pp. 207–349, 12 Text figs. *Ibid. Contributions from the Geological Institute of the National University*, No. 6, Peking, pp. 141–283.	143 pp.
	Total for 1924	736 pp.

1925

183.	Annual Meeting of the Geological Society of China. Proceedings and summary of papers. *Peking Leader*, Jan. 4, 7, 8, 10, 1925.	22 pp.
184.	Misunderstood Factors of Organic Evolution. Annual Address, Wen Yu Wei. *Peking Leader Reprints*, No. 8, 1925.	15 pp.
185.	Report of Meeting of Geological Society of China on April 7. *Peking Leader*, April 13.	5 pp.
186.	New Light on Ancient Chinese History, Part I. *Peking Leader*, May 5. 1925.	2 pp.
187.	Adventures in Antarctic Exploration. *Peking Leader*, May 10, 1925.	5 pp.
188.	New Light on Ancient Chinese History, Part II, *Peking Leader*, May 13.	4 pp.
189.	Report of Meeting of Geological Society of China. Sept. 23, 1925. *Peking Leader*, Sept. 25, 1925.	13 pp.
190.	Science and Superstition. Address before the Rotary Club of Peking, Dec. 17. *Peking Leader*, Dec. 22, 1925.	6 pp.
191.	Palaeogeographic Maps of Asia. 36 maps and Explanations.	7 pp.
192a.	The Peking Laboratory of Natural History. Leaflet I.	8 pp.
	Total for 1925	87 pp.

1926

192b. The Peking Laboratory of Natural History. Leaflet No. 2, pp. 1–8.   8 pp.

192c. *Ibid.* Leaflet No. 3, pp. 1–12.   12 pp.

193. Silurian Faunas of Eastern Yunnan. *Palaeontologia Sinica*, Series B, Vol. 3, Fasc. 2, pp. 1–100, plates I–IV, 2 text figs., March 25, 1926.   100 pp.

194. Religious teaching must be sane, or China must reject it, Published in *The People's Tribune*, June 1st, 1926. (Newspaper columns 3).   6 pp.

195. Discovery of tooth of Peking Man-most ancient fossil Man known, *Peking Leader*, Oct. 27, 1926. 2 cols.   4 pp

196. China in the making: The Ancient Rock Floor of China. *The Oriental Engineer*, July 1926, pages 3–6.   4 pp.

197. China in the Making: The Geosynclines. *The Oriental Engineer*, September 1926, pages 27–31.   5 pp.

198. China in the Making: Cathaysia. *The Oriental Engineer*, November 1926, pages 12–17.

199. Also *Peking Leader* Dec. 5 & 7th, 1926.   6 pp.

200. Reply to response on presentation of the Grabau Gold Medal. May 3. 1926.   1 p.

201. Elements of Petroleum Stratigraphy. pages 1–11.   11 pp.

    Total for 1926   157 pp.

1927

202. China in the making. *The Oriental Engineer*, February. 1927, pages 21–26.   6 pp.

203. A summary of the Cenozoic and Psychozoic Deposits, with special reference to Asia. *Bull. Geol. Soc. China*, Vol. VI, No. 2, 3, 1927, pp. 151–264.   114 pp.

    Total for 1927   120 pp.

1928

204. China in the Ordovician Period. *Bull. Geol. Soc. Nat. University*,

Peking, Vol. III, July. 1928. pages 7–22.   14 pp.

205. *Shells of Peitaiho*. (Grabau & King) 1st. edition. Peking Society of Natural History. Reprinted from *The China Journal*, Peking, 1928, pages 11–64.   64 pp.

206. *Shells of Peitaiho*. (Grabau & King) 2nd. ed. Published by the Peking Laboratory of Natural History. pages 1–279 with plates. 1928.   279 pp.

207. Studies of Gastropoda V. The significance of the so-called ornamental characters in the Mulluscan shell. *Bull. Peking Soc. Nat. Hist.,* Vol. II, Part 4. pp. 27–36, February 1928.   10 pp.

208. Second Contribution to our knowledge of the Streptelasmoid corals of China & adjacent territories. *Palaeontologia Sinica*, Series B, Vol. 2, Fascicle 2, 1928, pages 1–175, Plates 1–6, pages 3.   175 pp.

209. *Stratigraphy of China*, Part II: Mesozoic with four hundred and forty-nine Text Figures and four plates, pages 1–774. Published by The Geological Survey of China, 1928.   774 pp.

210. Syllabus in Palaeontology. Department of Geology, National University of Peking, 1924 & 1928. pp. 1–291.   291 pp.

211. Laboratory syllabus in Historical Geology. National University of Peking. pp. 1–53.   53 pp.

212. Glimpses of the Earth's Past. A lecture delivered before the "Things Chinese Society." *Peking Leader Reprints*, No. 41, 1928, pages 1–12.   12 pp.

Total for 1928   1672 pp.

1929

213. Origin, distribution and mode of preservation of the Graptolites. *Memoir Research Institute of Geology*, Academia Sinica, August 1929, pages 1–52.   52 pp.

214. Geological Studies in China. Notable New Information, The "Peking Man" & other Early men. *Peking Leader Press*, 1929, pages 1–20.   20 pp.

215. The outlook for science in China. Address delivered at the Annual Dinner of the Science Society of China, Aug. 25, 1929.

	Peking Leader Reprint, No. 49, pp. 1–15.	15 pp.
216.	Y. Y. Chao. *Peking Leader*, Nov. 30, 1929. $1^1/_2$ Column.	2 pp.
217.	Problems in Chinese Stratigraphy, Part I. *Science Quarterly of the National University of Peking*. Vol. I, No. I, pp. 67–98. Oct. 1929.	31 pp.
218.	Terms for the shell Elements in the Holochoarites. *Bull. Geol. Soc. China*, Vol. 8, No. pp. 115–123.	9 pp.
219.	Palaeontology in China. *Journ. Science*[1], Vol. 8, pp. 679–697, 1929.	18 pp.
	Total for 1929	147 pp.

1930

220.	Lecture on *Sinanthropus pekinensis. Peking Leader*, Feb. 2, 1930.	2 pp.
221.	Problems in Chinese Stratigraphy, Part 2, *Science Quarterly of the National University of Peking*, Vol. I, part 2, pp. 33–64.	31 pp.
222.	Geological World Problem solved in China. *Contributions of the Dept. Geography and Geology, Yenching University*, No. 27, 1929–1930.	3 pp.
223.	Asia and the Evolution of Man. *The China Journal*, Vol. XII, No. 3, March 1930, pp. 152–163.	11 pp.
224.	An Outline of the Geological History of North China. *Peking Nat. Hist. Bull.* Vol. 5, 1930–31, Part 1, pp. 1–13.	13 pp.
225.	A Decade of Research in Chinese Geology. *Bull. Geol. Soc. Nat. Univ.*, Peking, Vol. 4, pp. 1–8, April, 1930.	8 pp.
226.	Contributions to Geology by N. U. Graduates. *The Sciences Quarterly of the National Univ. of Peking*, 1930, pp. 235–250.	15 pp.
227.	Ancient Geography of China. *Geographical Journal*, Tsinghua University, June 1930, pp. 1–7.	7 pp.
228.	One Big Garden of Eden. *Peking Leader Reprints*, pp. 25–27.	2 pp.
229.	Corals of the Upper Silurian *Spirifer Tingii* Beds of Kweichow. *Bull. Geol. Soc. China*, Vol. 9, No. 3, pp. 223–240, Plates I–III, 1930.	17 pp.

1　应为：*Lingnan Science Journal*。——编者注

230. Problems in Chinese Stratigraphy, Part 3. *Science Quarterly of the Nat. Univ. Peking*, Vol. 1, No. 4, pp. 303–340.　　　37 pp.
231. Problems in Chinese Stratigraphy, Part 4. *Science Quarterly of the Nat. University Peking*, Vol. 2, No. 1, pp. 47–90.　　43 pp.
232. Published papers of A. W. Grabau, 1890–1930. pp. 1–26.　26 pp.

　　　　　　　　　　　　　　　　Total for 1930　241 pp.

1931

233. Problems in Chinese Stratigraphy, Part 5. *Science Quarterly of the Nat. Univ. Peking*; Vol. 2, No. 2, pp. 91–162.　71 pp.
234. Studies for Students; The Brachiopoda, Part 1. (pp. 1–20). *Science Quarterly of the Nat. Univ. Peking*, Vol. 2, Part 2, pp. 235–254.　19 pp.
235. Why we study Geology. *Bull. Geol. Soc. Nat. Univ. Peking*; No. 5, pp. 1–9, 1931.　9 pp.
236. *The Permian of Mongolia*; —A Report on the Permian Fauna of the Kisu Honguer Limestone of Mongolia and its Relations to the Permian of other Parts of the World. *Nat. Hist. of Central Asia*, Vol. IV, pp. I–XLII and pp. 1–665, Text Figs. 1–72, Plates 1–35.　665 pp.
237. Devonian Brachiopoda of China; Part 1, Devonian Brachiopoda from Yunnan & other Districts in South China. *Palaeontologia Sinica*, Ser. B, Vol. III, Fascicle 3, pp. 1–538, Text Figs. 61.　538 pp.
237a. *Ibid*. Plates 1933, I–LIV.
238. Studies for Students; The Brachiopoda, Part 2. *Science Quarterly of the Nat. Univ. of Peking*, Vol. II, pp. 397–422.　25 pp.
239. Problems in Chinese Stratigraphy, Part 6. *Science Quarterly of the Nat. Univ. Peking*, Vol. II, pp. 367–396.　29 pp.
240. The Significance of Sinal Formula Devonian and Post Devonian *Spirifers. Bull. Geol. Soc. China*, Vol. XI, No. 1, pp. 93–96. 2 Plates.　3 pp.
241. Palaeontology; Preliminary Paper Prepared for the Fourth Biennial Conference of the Institute of Pacific Relations; 1931.

China Institute of Pacific Relations. pp. 1–16.   16 pp.

242. Palaeozoic Centers of Faunal Evolutions and Dispersal. *Bull. Geol. Soc. China*, Vol. XI, No. 3, 1931, pp. 227–239.   13 pp.

             Total for 1931 1,388 pp.

1932

243. Problems in Chinese Stratigraphy, Pt. 7 (Middle & Upper Mississippian). *Science Quarterly of the Nat. Univ. Peking*, Vol. II, No. 4, pp. 423–479.   57 pp.

244. Geological Investigations in the Western Kwenlun and the Karakarum-Himalaya, by H. de Terra. *Bull. Geol. Soc. China*, Vol. XI, No. 4, pp. 471–482.   13 pp.

245. The Availability of the Bar Theory in the Elucidation of Ancient Potash Deposits. *Bull. Geol. Soc. Nat. Univ. Peking*, No. 4, pp. 1–26.   26 pp.

246. Studies for Students; The Brachiopoda, Part 3. *Science Quarterly of the Nat. Univ. Peking*, Vol. 3, No. 2, pp. 75–112, 1930. 38 pp.

247. Studies for Students; The Brachiopoda, Part 4. *Science Quart. of the Nat. Univ. of Peking*, Vol. 3, No. 4, pp. 85–117. 32 pp.

248. Problems in Chinese Stratigraphy, Pt. 8, The Dinantian of Europe. *Science Quarterly of the Nat. Univ. Peking*, Vol. III, No. 3, pp. 149–217, 1932.   69 pp.

249. The Silurian of China and its Relationship to that of other Parts of the World. *Bull. Geol. Soc. China*, Vol. XII, Abstract, 1 pp. 1 p.

             Total for 1932 145 pp.

1933

250. The Rhythm of the Ages—The Pulsation Theory, A New Aspect of Earth History. *The Peking Chronicle*, May 1933. 8 pp.

1934

251. Early Permian Fossils of China, Pt. 1; Brachiopods, Pelecypods and Gastropods of the Lower Permian Beds of Kweichow. *Palaeontologia Sinica*, Ser. B, Vol. VIII, Fasc. 3, pp. 214, 11 plates (Pt. II, issued in 1936).   241 pp.

252.　Syllabus in Historical Geology; 3rd. ed. 665 pp. The Nat. Univ.
　　　of Peking.　　　　　　　　　　　　　　　　　　　　　　665 pp.

253.　Syllabus in advanced Palaeontology; 2nd. ed. 226 pp. The Nat.
　　　Univ. of Peking.　　　　　　　　　　　　　　　　　　226 pp.

254.　Laboratory Guide and Syllabus in Palaeontology; pp. 1–306.
　　　The Univ. Press, Nat. Univ. of Peking.　　　　　　　　306 pp.

255.　Oscillation or Pulsation. *Report of the XVIth Int. Geol. Cong.,*
　　　Washington, July-Aug. 1933, Vol. 1, pp. 539–553; Washington
　　　1936; (Reprint issued August 1934).　　　　　　　　　15 pp.

256.　The Carboniferous of China and its Bearing on the
　　　Classification of the Mississippian and Pensylvanian (with V. K.
　　　Ting). *Report XVI Intern. Geol. Cong.,* Washington, 1933. Vol. 1,
　　　pp. 555–571, 1 plate (Reprint issued Aug. 1934).　　　17 pp.

257.　The Permian of China and its Bearing on Permian
　　　Classification (with V. K. Ting). *Ibid.* pp. 663–677. (Reprint
　　　issued Aug. 1934).　　　　　　　　　　　　　　　　　15 pp.

258.　Palaeozoic Formations in the Light of the Pulsation Theory; Pt.
　　　1, Lower Cambrian Pulsation. *Science Quarterly of the Nat.
　　　Univ. Peking,* Vol. IV, No. 1, pp. 27–184. Reprint pp. 1–158.
　　　Map & 3 Text-Figs. 1934.　　　　　　　　　　　　　158 pp.

259.　Palaeozoic Formations in the Light of the Pulsation Theory; Pt.
　　　2, The Middle Cambrian or Albertian Pulsation. *Science
　　　Quarterly of the Nat. Univ. Peking,* Vol. IV, No. 4, pp. 355–832,
　　　1934, Reprint pp. 159–636.　　　　　　　　　　　　448 pp.

　　　　　　　　　　　　　　　Total for 1933–1934　2064 pp.

　　　　　　　　　　　　　　1935

258a.　Indexes and Table of Contents for Vol. 1 (Palaeozoic
　　　　Formations in the Light of Pulsation Theory); 1st.

259a.　and 2nd. pt; 1st edition; pp. I–XXI and 637–679.　　64 pp.

260.　Palaeozoic Formations in the Light of the Pulsation Theory,
　　　Part 3 (Vol. II, Part 1), The Cambrovician Pulsation. *Science
　　　Quarterly of the Nat. Univ. Peking,* Vol. V; No. 1; pp. 1–120;

1935; Reprint pp. 1-120.   120 pp.

261. Palaeozoic Formations in the Light of the Pulsation Theory; Part 4 (Vol. II, Part 2); The Cambrovician Pulsation. *Science Quarterly of the Nat. Univ. Peking*, Vol. V; No. 2; pp. 121-316; 1935; Reprint pp. 121-316.   296 pp.

262. The Beginnings of the Human Race. Lecture, Shanghai, Sept. 29, 1933. *Journal of the North China Branch of the Royal Asiatic Society*, Vol. LXV, 1934, pp. 1-20 (Separates printed in 1935).   20 pp.

263. Did Man Originate in Asia? *Asia Magazine*, Jan. 1935, pp. 24-27.   4 pp.

264. Tibet and the Origin of Man. *Sven Hedin 70th Anniversary Publ., Geografiska Annaler*, 1935, pp. 317-325.   9 pp.

265. Studies of Gastropoda, (reprint in 1 Vol., 5 Parts). Nat. Univ. Peking, The Univ. Press, 1935, pp. 159, 56 Text Figs., 3 Plates.   159 pp.

266. Studies of Brachiopoda (Reprint in 1 Vol., 4 Parts). pp. 1-117, 57 Figs. The Univ. Press, Nat. Univ. Peking, 1935.   117 pp.

    Total for 1935   789 pp.

1936

267. Early Permian Fossils of China, Pt. II; Fauna of the Maping Limestone of Kwangsi and Kweichow. *Pal. Sinica*, Series B, Vol. 8, Fasc. 4, 31 Plates, 1 Text-fig, pp. 1-411.   412 pp.

268. Revised Classification of the Palaeozoic System in the Light of the Pulsation Theory. *Bull. Geol. Soc. China*, Vol. XV, No. 1, 1936, pp. 23-51.   29 pp.

269. The Great Huangho Plain of China. *Journal of the Association of Chinese and American Engineers*, Vol. XVII, No. 5, pp. 247-266, 1936.   20 pp.

270. Palaeozoic Formations in the Light of the Pulsation Theory, Vol. 1, Taconian and Cambrian Pulsation Systems, 2nd. Edition (revised) pt. I-XXI, 1-680, 17 Text-figs, 5 Palaeogeographic Maps. The Univ. Press, National University of Peking, 1936.   680 pp.

271. Palaeozoic Formations in the Light of Pulsation Theory, Vol. II, The Cambrovician Pulsation System. Part I Caledonian and St.

Lawrence Geosynclines, pp. I–XXIV, 1–751, 42 Text Figures, 1 Plate (Pate III-a). The University Press, National University of Peking, 1936.   751 pp.

Total for 1936   1892 pp.

1937

272. Fundamental Concepts in Geology and Their Bearing on Chinese Stratigraphy. *Bull. Geol. Soc. China*, Vol. 16, 1936–37, pp. 127–176, 1 map, 2 figs.   49 pp.

273. The Polar-Control Theory of Earth Development. *Journal of the Association of Chinese and American Engineers*, Vol. XVIII, pp. 202–223, Plates 3.   21 pp.

274. *Palaeozoic Formations in the Light of the Pulsation Theory*, Vol. III, Pt. II, Cambrovician Pulsation. 4 plates, 6 correlation charts, 58 text-figs., 850 pages.   850 pp.

275. The Evaluation and Dating of Palaeozoic Orogenies. *Journal of the Association of Chinese and American Engineers*, Vol. XVIII, pp. 371–379, 1937.   9 pp.

Total for 1937   929 pp.

1938

276. Dr. Nahaniel Gist Gee (Biographical Sketch). *Peking Natural History Bull.,* Vol. 12, Part 3, March 1938.   2 pp.

277. Sea-Salts and Desert Salts: Part I. *Journal Assoc. Chinese and American Engineers*, Vol. XIX, No. 3, May-June, 1938.   24 pp.

278. Sea Salts and Desert Salts: Part II. *Journal of Assoc. Chinese and American Engineers*, Vol. XIX, No. 4, July-August, 1938.   15 pp.

279. Palaeozoic Formations in the Light of the Pulsation Theory, Vol. IV, Ordovician Pulsation System: Part I, pp. 970, Text-figs. 67, 1938.   970 pp.

280. Ice-Ages or Polar Glaciation; Abstract of paper read at the Fifty-first Annual Meeting of G. S. A., *Bull. Geol. Soc. America*; Vol. 49, No. 12, Pt. 2, pp. 1883–1884. $^1/_2$ pp.

281. The Significance of the Interpulsation Periods Chinese Stratigraphy. *Bull. Geol. Soc. China*, Vol. XVIII, No. 2, pp. 115–120, 2 tables, Changsha, 1938.　　　　　　　　　　　　　　　　　　9 pp.

282. Classification of Palaeozoic Systems in the Light of the Pulsation Theory. Abstract: *Bull. Geol. Soc. America*, Vol. 49, No. 12, Pt. 2, pp. 1932–1933.　　　　　　　　　　$1\,^1/_2$ pp.

　　　　　　　　　　　　　　　　Total for 1938　1022 pp.

1939

283. Present Status of the Polar Control Theory of Earth Development. *Bull. Geol. Soc. China*, Vol. XIX, No. 2, pp. 189–205, 1 fig, July 1939.　17 pp.

1940

284. Revision of some Lower Palaeozoic Sections of Central China. *Bull. Geol. Soc. China*, Vol. XIX, No. 4, 1940, pp. 455–478, 1 folded plate.　　　　　　　　　　　　　　　　　　27 pp.

285. The Rhythm of the Ages: Earth History in the Light of the Pulsation and Polar Control Theories. Jan. 1940, Vetch, Peiping. pp. IX 561, 126 Plates, Maps I–XXV, and Tables 1–32.　570 pp.

　　　　　　　　　　　　　　　　Total for 1940　597 pp.

1941

286. A New Interpretation of Earth History. Address at 15th. Ann. Meeting Peking Soc. Nat. History, Peking *Nat. History Bull.* 1940–41, Vol. 15, Pt. 1, pp. 1–12, Text-fig. 1.　　　　　12 pp.

287. Was there ever a great Ice-Age? Address before the Sigma club of Peking at the 9th. Annual Meeting, May 2, 1941. *The China Journal*, Shanghai, Vol. XXXV, No. 2, Aug. 1941, pp. 61–66, Plate 1 & 2.　　　　　　　　　　　　　　　　6 pp.

　　　　　　　　　　　　　　　　Total for 1941　18 pp.

1942

288. A new view of the "Ice-Age". *Collectanea Sinodalis*, Peking. Vol. XV, No. 8 & 9, Aug-Sept. 1942, pp. 307–311.　　　5 pp.

289. The Geological Dismemberment of Ancient Cathaysia. *Ibid.* Vol. XV, No. 10/11, Oct. -Nov., 1942, pp. 496–498.　　　3 pp.

290. The Pulsation Theory. *Marco-Polo Magazine*, Vol. IV, No. 13
Oct. 1942, pp. 1–10, 1 Text-figure.   10 pp.

            Total for 1942   18 pp.

1943

291. Tibet, The Cradle of the Human Race. *Collect. Comm. Sinodalis*, Vol. XV, Jan. -Feb., pp. 62–78, 1 plate.   17 pp.

Total pages printed

1890	2 titles	4 pp.
1893	1 title	5 pp.
1894	1 title	13pp.
1895	1 title	5 pp.
1896	2 titles	5 pp.
1897	2 titles	5 pp.
1898	5 titles	166 pp.
1899	5 titles	491 pp.
1900	3 titles	161 pp.
1901	7 titles	306 pp.
1902	7 titles	130 pp.
1903	8 titles	120 pp.
1904	3 titles	178 pp.
1905	2 titles	10 pp.
1906	3 titles	311 pp.
1907	20 titles	190 pp.
1908	10 titles	17 pp.
1909	10 titles	930 pp.
1910	8 titles	1180 pp.
1911	2 titles	16 pp.
1912	5 titles	64 pp.
1913	11 titles	1411 pp.
1914	3 titles	2 pp.

1915	5 titles	5 pp.
1916	6 titles	70 pp.
1917	6 titles	40 pp,
1918	3 titles	1 pp.
1919	7 titles	55 pp.
1920	8 titles	1379 pp.
1921	3 titles	986 pp.
1922	7 titles	258 pp.
1923	13 titles	298 pp.
1924	6 titles	736 pp.
1925	10 titles	87 pp.
1926	11 titles	157 pp.
1927	2 titles	120 pp.
1928	9 titles	1672 pp.
1929	7 titles	99 pp.
1930	13 titles	245 pp.
1931	10 titles	1388 pp.
1932	8 titles	145 pp.
1933	1 title	8 pp.
1934	10 titles	2064 pp.
1935	8 titles	789 pp.
1936	5 titles	1892 pp.
1937	4 titles	959 pp.
1938	7 titles	1022 pp.
1939	1 title	17 pp.
1940	2 titles	597 pp.
1941	2 titles	18 pp.
1942	3 titles	18 pp.
1943	1 title	17 pp.
Total 1890–1943	299 titles	20, 862 pp. [1]

1 原文统计为 305 titles，10892 pp.，疑误。——编者注

Books in manuscript or partly printed

Palaeozoic Formations in the Light of the Pulsation Theory, Vol. V., Ordovician (Part II). In course of publication. (400 pages printed, 6 colors maps etc. in 1941)

Palaeozoic Formations in the Light of the Pulsation Theory, Vol. VI, Ordovician (Part III). Publication begun, 120 pages printed in 1942.

Palaeozoic Formations in the Light of the Pulsation Theory, Vol. VII, Silurian. In manuscript.

The World We Live in: a Popular Book on the Earth History as Interpreted by the Pulsation and Polar Control Theories. In Manuscript.

——原载 *Quarterly Bulletin of Chinese Bibliography,* 1946, vol. VI, nos. 1–4

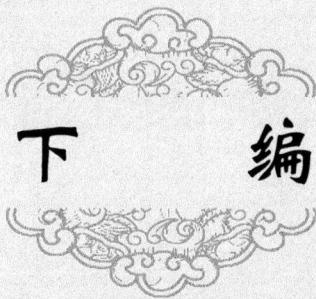

下　编

葛利普论中国和科学

生物进化的误解
（Misunderstood Factors of Organic Evolution）

北大古生物学教授葛拉包[1]博士（Dr. A. W. Grabau）原著
斯行健译述

　　生物进化的事实，到现在已无辩论的余地了。一般智识阶级的人，只要他们对于这个问题，有深足的研究，俱已承认现代生物或古代生物俱是他们以先的生物经历过一种发展的历程进化而来的，不过进化的原因与其相互的价值，仍为各派争论的焦点，而对于进化论各种的要素，一般人仍不免有普遍的误解，兹分别释之。

达尔文学说（The Darwinian Theory or Darwinism）

　　达尔文学说，已成为一般人的口头禅，虽然他们不知道达氏学说为何物，而自以为对于他有特别研究的！假使问他们，何谓达氏学说？百人中定有四十九人回对说："人类最初是由猴子生出来的。"还有四十九人却回对说："达尔文学说就是进化论。"这两派的人都没有真正读过达尔文的创著《种源论》[2]（The Origin of Species）。他们对于种（Species）的一字并没有明了而坚持说："猴子就是人。"一个教士带了一只猴子在讲台上对着听众说："看！达尔文这样告诉我们——你和我，我和你的祖先，是这样的一个生物！"自然这个猴子装一装脸，现出不肯承认的样子！但是很多智识的人们，对于达氏学说，仍不免有所误解。不过因为达尔文是最先指导人们生物进化是一个科学上的问题，不仅是哲学上的理论。达尔文以先的时候，物种原始，犹为一般人的疑问，虽然他们已经否认生物是由万能的上帝的谕旨所造功的。所以达尔文以先的时候，仍可称为黑暗的时期。因为他们没有知道只要苦心孤诣的收集事实，就无异在黑暗中举起烽火，结果可使火焰冲天，照耀全球。

1　葛利普初来华时，还有葛拉普、葛赍普、葛拉包、葛拉伯等中文名字，后统一为葛利普。——编者注
2　现在一般译为《物种起源》。——编者注

有些人紧闭双目 (Some Refuse to See)

达尔文经过二十余年的苦心研究，将一束一束的薪柴，一点一点地收集起来，当初他自己曾没有想到这一束束零零碎碎的薪柴，可以举起猛烈的烽火，但是他总是一年又一年的努力下去，等到他自信他所收集的材料，与真理渐渐接近。至一八五八年火光爆发，全球震惊。至一八五九年，他的种源论发表以后，火光的照耀，竟可使其同时的人类，因其闪射出来的余光，而求得进化的真理。虽然那时候仍有许多人曾犹抱疑豫不决的态度，而谓尚须进一步的考察和研究。而一般宗教观念极深的人，仍紧闭双目，坚不承认，直到如今，这般人的后裔，犹有存在！(《种源论》的原名为 *The Origin of Species by Means of Natural Selection, or Preservation of Favored Races in the Struggle for Life*)。

天择律 (Natural Selection)

达尔文贡献天择律于世人，为进化论中最重要的要素。佛莱士（A. R. Wallace）同时发明此定律，而承认天择律即可名为达尔文学说。佛莱士和达尔文自己都没有说天择律可以完全解释进化，不过说是天择是进化论中最重要的要素罢了！现在一般人都以为达氏学说与进化论是一个名词，没有分别，这固然是现代一般人普遍的误解。就是他们明白达氏学说或天择律不过是进化论中的要素或原因，而非进化论的本身的人们，也常常对于此种重要的天择律不能完全明了，或竟完全误解。我们常常可以听见他们说："天择为生存竞争而保存适者，而破坏不适者。"这就大误而特误了！天择不是一种天然的力量，天择不是一种实体（Entity），天择不能做什么，天择不过是一种方法或是历程（Process），在天然选择的方法或历程中，最适宜于生存竞争者，就可以保存，而可以逃出一种破坏的力量，而他们不适于抵挡此种破坏的力量者，就不能生存了！这就叫做天择律。譬如一个耳聋的人，和一个耳聪的人在火车轨道上，相偕平行。火车从后飞至，耳聪的人听见其危险而跳出轨外，同时大声疾呼其同伴使其出险，然而这个耳聋的人，因为太聋了，仍旧没有觉察火车之将至，到最后一刻钟，方才觉察，然而太迟了。这就是这个耳聋者，在其不适的环境中吃了大亏了！又如两个洋车夫同向哈德门疾驰，互相争先，在前面一个胡同里，忽然听到汽车的呼声，谨慎的车夫立刻停止，不再前进。愚笨的车夫仍向前驰，结果被汽车辗坏了，这就是表示这个愚笨的车夫不适于生存在一个不慎的汽车夫的环境中。又如一个城子，被围甚久，四门俱闭，粮食无从运进，居民预防此困而预备粮食者得生存，没有预备的，只有坐以待毙。除非将适者所有而分于不适者，则方可共存。但这已经是干涉天择了！又如两只鹿在草野中食草，忽然来了一只饿狼，两鹿俱大惊而思逃避，一鹿因为腿略长，筋肉之力较强，得幸免于难。而其余的一鹿，遂不免被狼吞食了。腿长，筋肉强，而能疾驰，就是天择的根本条件！

物种性质未受天择以前的进化 (Preselection Evolution)

天然选择，仅可能于物种的个体，选择或除绝某种特性以后。这是大家已经明白了。但是这种特性，怎样发生的呢？达尔文的门徒告诉我们说，这是骤然或偶然的变化；而德维里（De Vries）的门徒却告诉我们说，这是因为物种主要或根本的性质，经过一种忽然的重行铺排，好像万花镜（Kaleidoscope）移转的时候，其颜色玻璃的分子重行安排无异。经过一种跳变（Saltation）的历程，忽然发生一种新的布置（Pattern）。德维里学说，称为突变论（Mutation）。不过我们应该明白的，就是物种许多细小而不可约束的变化，我们处处可以见到，我们名之曰偶然，曰骤然，因为我们没有知道这些细微性质的由来，但是这些细微性质的变化，其本身并无选择的价值。性质而有选择的价值，那必定是主要的性质，直接可以影响到物种的个体，使其有益或有害。不过物种的多数性质，是无关紧要的，没有选择的价值的。如果某种性质的变化，到了一定的时期，直接影响到个体的有益或是有害，这就是选择的根基。

鹦鹉螺（Niutilus）的贝壳，其壳片与膜片相接合之处，叫做缝合线（Suture），凹处叫做腰，凸处叫做鞍，此种缝合线甚为简单，并不比其同时生存之他种螺类其膜片仅如碟状者，适于生存。因为这两种螺类，经过极长的时期，犹能同时并存。假使其缝合线之腰与鞍，生长极速，由单简而繁复，则其贝壳片结合之面必同时扩大。

譬如电灯罩两个，其直径相等，不过其一的外缘系平直的，其一的外缘系曲折的。我们用一直线靠住灯罩的外缘而量其长短，我们知道屈边的灯罩，总比平边的要长得多。膜片与壳片相接之处就是外面的边，假使此外边（缝合线）是简单的，则膜片与壳片，俱须一定的厚度，方可保存。假使此外边是曲折的而繁复的，则有较薄的壳片与膜片，即可保存。但较厚而沉重的贝壳如鹦鹉螺，仅可生存于海底，除非受了特别的情形，方可偶然浮到海面。而现代之舡舟螺（Argonauta），其贝壳片较薄而轻，故可飘浮于海面而分布甚广。如一旦海中的环境受了一种特别的物理变化，则生存于海底的生物，必将绝灭而不克幸免，只有飘浮于水面的生物，因为受影响较轻，且分布较广，仍可继续生存与发育。古生代的末期，海洋中受了极激烈的变化，结果惟飘浮的生物，仍可继续生存，就可以证明了。因此我们可以知道飘浮的贝壳，其生存的原因，是因为飘浮的缘故，亦即因为其贝壳较薄而轻的缘故，而贝壳之所以较薄是因为其缝合线的曲折繁复。但是最先的鹦鹉螺类，这些缝合线是很简单的，不完全的，随后在一个方向上逐渐进化，由简单而繁复，所以最初的鹦鹉螺类，其缝合线并无选择的意义，与鹦鹉螺本身的生存，并无多大关系。后来缝合线逐渐繁复，逐渐进化，遂直接影响于其种类的保存，而为中生代侏罗纪时最繁殖的菊石螺类（Ammonites）的祖先了！

古生物学的重要 (Importance of Palaeontology)

自古生物学发明以后，于是生物的各种性质在一个方向上，逐渐增加，逐渐进化，直接或间接关系于生物的生存或绝灭的事实，乃大显著。现代生物学家研究此种性质，好像在黑暗中照玻璃仅可得其闪光而已。只有古生物学家可以看得很明白，可以真正懂得生物进化的方法。现代生物学家研究生物进化，好比一个新到北京的外国人，要想在西城的许多胡同中寻觅他的道路，不迷路者几希！只有古生物学家，可以用其过去的经验，在许多错综的道路中，找到其要想达到的目的地的最近的几条捷径，如有经验的北京人一样，假使有一张指导的地图，那末，更容易了。如果要想编一部中国历史，仅仅观察现代中国人的性质与习惯，或仅观察其极小部份，这样的历史，是何等的靠不住呢！真正史学家，是要收集许多过去的记录，或流传的事实，所以只有他可以值得一般渴望研究真实历史的人的注意。生物进化，是地球上生命的历史，如果要想研究此种历史，而不注意他的记载，研究亿万年以前以迄于今日生存于此地球上的生物，那是一定无效的！

个体进化为种族进化的先导 (Ontogeny as a Guide to Phylogeny)

现代生物学家研究进化论，惟一之道，是根据胚胎学。胎子在母体未生出以前，是从简单的蛋，或最初的原生细胞，经过很多复杂的历程，如从单细胞分身，成功多数而繁复的细胞，又经过很多的时期方才生成。繁复的生物，是由于最初和最简单的单细胞的蛋，经过很多的历程和时期，所成功的。这是无怀疑的余地了。但是这样过去的历史，是何等的残缺，何等的简略呢！在数周或数月一转瞬间，此个体在母体内，即将在地球上经过亿千万年的历史复演一次，这又是何等的速率呢！（行健按，黑格尔氏 Ernst Haeckel 创明复演津 Biogenetic Law，他说地球上的生物，是由单细胞的原生动物逐渐进化到无脊椎动物，又逐渐到鱼类、两栖类、爬虫类、哺乳类以至人类，在地层中找到的化石，证明的确如此，而人类在母体的十个月当中，即将这悠久的历史，复演一次。譬如精虫与卵子成功的蛋，就是等于单细胞的原生动物，后来这个蛋逐渐分身，成功多细胞，又逐渐成功腔肠时代，这时候就是等于下等的无脊椎动物，后来又逐渐发育，又好似鱼类的胚胎，又渐渐像两栖类、爬虫类、哺乳类的胚胎，最后方才像人，所以人类在母体的十个月的短期间中，已经将他祖宗的形状，亿千万年的历史复演一次了！）

所以胚胎学家根据个体的胚胎时期，而想得到生物进化的详细历史，其失败是不足为奇的！不过生物自离母体后，自幼而长而老而死，其性质的变化，比较迟缓，易于观察和记录。此种变化生理方面与心理方面，俱影响于生物的全体，与其形态。所以此种性质，是易于观察的。但人类和其他脊椎动物，须长久而坚忍的观察其自幼而长而老而死的变态，这差不多是很不容易的事了！同时观察一群生物个体的各种时期的变化，或是[1]比较容易的事，但有时又不免将各个体各时期变化的性质，平均的概括和缩小，这样又不免失却少许进化的意义了。

天然的记载（Natural record）　许多无脊椎动物，尤其是软体动物，自幼而长，其个体形态的变化，俱可以永远保存，而便于观察。就是软体动物的贝壳，其形态变化的时期与形状，载得非常明了，此种贝壳，如果变成化石则其壳上的记录，便可永远保存。这种被海水冲[2]流到海岸的空壳，乃是生命发育时中最奇妙的记录，而可以启示一种种族发展的历史，并且可为地球上各种生物进化的公证，这是无论何人所梦想不到的！

（行健按，软体动物的贝壳，在初生时，条纹甚少或竟无有，至幼年初期，则发生原生条纹，至幼年中期，则发生次生条纹，至幼年末期，则每一原生条纹与每一次生条纹之中间，又发生新生条纹，其逐渐发育的情形，俱记载

1　原文为"者"。——编者注
2　原文为"充"。——编者注

在其贝壳上面历历可数，瞭如指掌。故软体动物的贝壳为进化论中最重要最有价值的证据，兹作两图于下以证明之）。

古生物学家可以将此种贝壳的化石——在地球上亿千万年以迄于今的过去许多海洋中的生物——陈列在一处，观察其形态变化的记录，编成一部完全的种族历史，而解释此种普通的进化公理。在最近的将来，如果古生物学不仅为几个专家所研究，而成为一种普通的学问，成为一般相争欲得生物原始与其进化的真理的人们，必须研究的学问时，则我们可以强有力的态度，而试看各种力量限制各种下级生物与人类各方面的进化，可以明了各种生物是主体为天然的公理，并且可以知道我们人类及人类以外的各种生物，不仅在其身体上物质上的发展，即其精神上心灵上亦向前发展不已！

直系学说（Orthogenesis[1]）

> 行健按，希腊字 Ortho 直字意，Genesis 发育意，即在一方向上或一直线上向前发育之意。有人译为正统发生学说，也有人译为直线进化学说，现在姑译之为直系学说，不知对否？望科学家有以教之。

古代软体动物的贝壳化石，与现代软体动物的贝壳的天然的记载，不仅是动物个体形态上变化的记载，乃有更深的意义。贝壳形态的变化，实在说起来，不过是壳内动物的软体的变化，也就是动物分布及其生长平均速率的变更，简而言之，不过是动物生理上环境变化的表示，这才是根本的意义。为明了起见，可以举几个例于后。

现代海扇蚌（Scallop）或帆立贝（Pecton）的贝壳，其特性因为有放射状的坚粗纹线，但当其幼年时期，此种纹线并未发生，而仅为一光滑的平面。现在我们可以看：在同样大小的两贝，其一贝壳上有粗纹，其一则平滑而无纹线，则此两贝之横切面之直径，应何等的差异！虽然两贝之直径大小相等，但是粗纹贝的外缘，总比平滑贝的外缘要长得多。这是表示粗纹的贝壳特别澎涨的缘故。但是贝壳的成因，是由于壳内动物的软体外面所包裹的网织或薄膜所成功的。如故薄膜生长太速，因为保护或包围软体使其安适起见，则此种薄膜，因为大阔，非曲皱不可，迨至后来，薄膜渐渐变硬，造成贝壳，此种曲皱，遂变成曲折的纹线。

在一个方向上，贝壳上此种曲皱生长不已，继续增加。虽然经过若干时

---

1 原文为 Orthogensis，疑误。——编者注

后，进行的大道，可以分出许多支路。如有些贝类，经过很多的时间，其原生条纹与两纹中间凹下之处，并不发生次生条纹，不过其原生条纹中间凹下之处特别发展而已。此种贝壳生长的方法，其条纹与凹处很一致的。有些贝类，则其原生条纹增大，甚为迟缓；而两纹中间凹下之处，却增大甚速。此种贝类生长的方法，是特别集中于两纹中间凹下之处的，因此次生条纹，即插生于其间，如次生条纹与原生条纹中间凹下之处，又特别发育，则新生条纹即插生于其间。如此继续不已，则贝壳表面，可以发生很多细微的纹线。有些贝类，其生长的方法，却专注于条纹的，此种条纹，发育增大极速，则每一原生条纹上面可中分而为两次生条纹。如故次生条纹又复发育极速，则每一次生条纹又可中分而为两新生条纹，如此继续不已，贝壳上亦可发生不少的纹线。因此我们可以明白，不过是因为薄膜发育的集中的不同，而可为贝壳形态分成三路进化的发轫。

假使时间充足，这些例子，是举不尽的。简而言之，则凡生物的性质，其进化在一方向上发轫以后，其进化与变形的性质，和其分出的支路，俱可继续前进，为一无穷尽的旅行。不过其变形的成功，是由于生长力的集中在许多方向中的一方向，而且须有遗传的前定（Hereditary Predisposition）。命运（Predestination）一字，旧的意义，是根据于一种特别的外力的。照新的意义，应该说，非由于外力，乃由于生物自身的遗传线上。——此种直线不至中断，约束或限制我们，不仅传至第三代或第四代，须一直约束到此直线上的最后一代。这就叫做直系学说（Orthogenesis）在一个方向上继续发展的。进化的道路，可高可低！进化的最后，或生或灭。假使此直线的最后一代，不适于环境就不免于绝灭。否则就可以生存。此即谓之天择。但是此直线到了末代时天择亦随之而停止了。

讲到这里，一般人一定要失望了。假使我们是被遗传所限制的，我们的良否，俱是先天预定的，我们是命运的生物。否！否！我们不要忘记，生物是一种复杂的实体，而环境仅能限制性质的个体。无论无数形质中之每一种性质，形态上的、生理上的、心理上的，造成我们这样复杂的实体，是不受遗传或直线发育的限制的。遗传的限制或约束，虽然不能取消，但是因为特别加重于某种性质的发展，而压制他种性质的进化，则其结果，可使遗传的效力减弱或消灭。讲到此处，我们又遇到一种新的限制或约束的要素，这就是环境（Environment）。

先进与后进（Acceleration and Retardation）

在我们未研究环境与适应的先，我们可以简单的讨论一点，也是古生物学家所发明，在直系学说上面一种新的要素，就是先进与后进。生物之每一性质的发展，不仅不受遗传线上的约束，即其每一性质的发展的程度的差异，也是独立的。每一性质，可以发展特别快，可以进行特别缓，甚至可以完全停止发展。我们可以将他来比一群定好的旅行队，其出发点与目的地俱同，虽然此目的地的意义，是各人不同的。在出发点与目的地两者之间，有许多车站，每一旅行者，达到或经过每一车站时，其进行的速度，俱用时间记录。一人步行，一人坐洋车，一人骑自行车，一人坐汽车，一人坐特别快车，一人乘飞机，同时同处出发，这是没有疑问的，当步行者达到第一站时，其余的早已过去了。骑自行车的，经过此站，定比坐洋车的为先！坐汽车的定比坐自行车的先；坐特别快车的，定比坐汽车的为先；而坐飞机的却都比他们为先了！

```
甲（出发点）              乙（第一站）              丙（目的地）
步　行_____
洋　车_____
自行车_____
汽　车_____
快　车_____
飞　机_____
```

（斯行健作表）

坐飞机的，到了目的地时，步行者或仅至第一站，而其余的，俱在其中间，以其进行的速度为比例。到一定时期的终了，而观察各旅行者进行的先后，或在一天以内，或无论何时均可。我们现在可以知道物种性质的发育，或形态的发展，亦犹如是。我们试看每一生物的个体生命的历史，到某一时期（Certain Period）的时候，各种性质发展的程度，也可以分成各种不同的阶级，或则甚速，或则甚缓，或则完全没有发展。

新种的发生（a New Species is Produced）　如故某一时期适在生物的壮年时，则各种性质的发展，亦须视生物的个体而不同。因为虽然有许多生物的个体，其各种性质发展的情形，可以同一群旅行者相似，尤其是发展的程度，可以分成各种不同的阶级，而每一阶级可以保持其相互的距离，或仅具极微的差异。但许多别的个体，其各种性质的发展，可为别样的程度。譬如步行者，或者善跑而能疾驰；骑自行车者，或者没有纯熟；乘汽车者，或者可以超过特别快车；而飞机师或者是没有技能的，永远不能达到其目的地。生物个体各种性质形态上的变化，也是如此。故到一定时期的终了，这些发展程度各异的性质，可以变成新的物种。

环境可以限制遗传的力量，仍与限制旅行者的进行一样，如同道路的性质，天气的境况，食物的丰否，等等；俱可限制旅行者的速度。假使在另外的一种环境，则依照此群旅行者的天然技能，而可得另外的结果。

最后，如果此群旅行者的几人，志气很弱，或体格不强，则在无论何种时期，都可以淘汰，有些简直不能出发。生物个体的各种性质，也是如此，有些性质，当初即不能发育。有些性质，发育到了中途，即被淘汰。有些性质，则一直向前发展不已。故结果遂造成各种不同的新种。

直线进化的倾向，是遗传的而发展的程度，和进化的迟速，则被环境所限制。许多有天才的人，和许多卑恶的人，当他们初离母体时即可死亡，被一种不适于他们的环境扼杀[1]。

环境与适应 (Environment and Adaptation)

我们已经知道天然选择，无论何时，无论何处，俱是一种普遍的历程，普遍的方法，也就是普遍的公理。无论在莽野树林之中、在市场、在会计室、在法庭，甚至在迷信的教堂中，无论何时何处，总是优胜劣败，弱肉强食。这就是天择。很简单的！很明了的！我们更知道，天择仅可能于物种的性质，发展到了一定的时期，而有选择的价值时。而此种性质，是直线进化的，也就是遗传所预定的。因此我们应该明白，这是环境可以限制此种性质的发展，和发展的程度，天择就是环境的代表。

这样说起来，环境乃是极重要的要素，适应于环境是要造成适者，使其生存的意思。在一定的环境中，总是强者或适者生存，弱者或不适者屈伏。如故改变其环境，则适与不适的意义，亦随之而不同。譬如一个教士到了野蛮的地方，野人将教士吃了，天择适于野人，而不适于教士，是根据于这一种特别的环境而使行的。假使改换一种环境，则教士为强者，为适者，为社

1　为使句意通顺，"扼杀"二字为编者所加。——编者注

会的领袖，而野人反不能不屈伏而做奴隶了！

但是环境是和缓的，并不需要完全的适应。有些适者，其本身有否适存的价值，很可怀疑，但是环境总是宽纵他们。只有实在不能适存者，方始被淘汰了。

一条可怜的鱼罢了（Was a Poor Fish） 如故环境是永远不能变的，那么一种适者的简单标准可以定出来了。但是这些乌托邦的梦，是永远不会做成功的。环境好像天然间的各种事物，是常常改变的。因此适与不适的意义，亦常常随之而改变。一个哲学家，没有熟练游泳的技能，假使沉入海中，那末，只有灭顶，只有溺死。虽然他的学问是可以惊人的，道德是很完全的，但是他总不过是一条可怜的鱼罢了！

反嚼兽（Ruminant）是完全适应于食草的习惯，如果森林占据了牧场，沙漠侵入了草野，那末，除非他迁移到别处，可以得到他的食物，否则只有饿死了！刀齿虎（Sabre-tooth tiger）的牙齿，虽然尖利，但是与厚皮而缓行的大懒兽（Megatherium）争斗，大懒兽屈伏以后，仍不免受饥饿的痛苦，而较大之猫族，具有疾驰之足者，比较可以生存。

假使环境改变甚缓，则适应的改变，大半与之并驾齐驱。要是环境改变甚快，则惟有最适者，可以生存。但是我们不要忘记，所谓适与不适俱是相互的名词，专根据于一时的环境的。迷信者可以优胜在左道邪说的时代，而不适于正理和文明的世界。要想造成人类都适于环境是一种极可赞扬的奢望；要想造成环境使其特别适于最善良的人类，那不过是一般诗人和哲学家的梦想罢了！

只有绝灭，不能进化（Not Progress but Death） 环境既然可以时时改变，则我们可以将他分成各种不同的环境。假使一个生物到了他的直线进化上面的某一时期，需要适于一种较新的环境，他可以遇到此种环境。要是此种环境的势力扩大，则适应的生物，必占优胜。古生代泥盆纪时有些鱼类，并不发展其呼吸空气的力量，使较适于未来的环境。同时有些鱼类，在旧时的潮湿空气的环境中，呼吸空气的力，已较为发育。所以只有这些鱼类，可以迁移到干燥的地方。要是后来潮湿地方的热量开始向上加高，那末，不能呼吸空气的鱼类，不适于生存而相继绝灭，只有呼吸空气的鱼类可以继续生存，而一部份且渐渐进化而为两栖类，渐渐进化而为陆地脊椎动物。又如第四纪时喜马拉雅山加高以后，于是南部潮湿的海洋空气与西藏及亚洲北部的人猿（Anthropoids）隔绝，森林因之而勃发，而繁盛，此时惟已经发展能生存于新的环境的人猿，可以生存。因为受了热度继续加高的刺激，他们可以发展他们潜伏的智力，逐渐进化而为用器与用火的人类。要是此种智力，不是先前潜伏于其遗传上面的，那末，一旦环境突然改变，只有绝灭，不能进化！

结论（Conclusions）

生物进化的事实，此篇讨论，甚为简单，但详细研究的练习，是根据于此种简单的讨论的。此种练习，并不甚难，就是研究古生物学——生物进化的历史的记载——可以替我们翻开此天然的书籍（The Book of Nature），而使我们懂得过去的故事，并希冀未来的进化。

蒲德莱氏（President Butler）解释"教育"的意义，为求得适应于环境的势力，这是初步的教育。芬莱氏（President Finlay）解释"教育"的意义，为求得限制环境的势力，这是高等的教育。余则以为教育的最高目的，是要造成人类可以创造身体、智力，和灵魂各方面进化的环境，这些人方才不愧生长于地球上面了。

这篇文章，是葛先生讲演时，我用英文笔记下来的，现在又把他译成中文，自然有许多谬误。希望阅者加以原谅！又葛先生的原文，并没有图画，此篇所有图画，都是我自己加进去的，不过先前曾得葛先生的同意。

斯行健附识

一九二八年四月十日译于东山

——原载《现代青年》（广州）1920 年第 31—35 期；
又载《自然科学》1928 年第 2 期（本书以后者录之）

六十年的达尔文学说

葛利普著　江锦梁译

　　达尔文学说怀胎了二十年，才在六十年以前产生出来。当时的科学界，对于这婴孩的生活力，意见分歧。在宗教界方面，对于这"新来者"是敌视的，看他做为一种虚伪的救主，是破坏的，不是救世的。而宣言无偏见的智识界，他的态度亦不一致。有些人看他的降生，为一种新的"神约"；有许多人是大声嘲笑他为假冒。听嘲笑他的话："诸君勿笑，有一只鹿，其头长于他同类者一倍；一伸再伸，变成一只麒麟，无人能否认。"

　　然此现在不关涉达尔文学说，而说到更老的大博物家拉马克（Lamarck）。然而这是嘲笑者所不能领会的。现在一般人的错处，是将达尔文学说同进化论混乱。续说我们的暗比法：进化不是产生的，是像 Topsy，"忽然发生的"。亚历斯多德（Aristotle）亦知这件事，或者他自以为知道，以及他以前的、他以后的哲学家，曾经默认他的存在，并悬想他的性质。Goethe, Buffon, Saint-Hilaire, Lamarck 和达尔文的祖父诸人，对于进化论，或讲或著作过。但是他们关于他的真正品质，不过是一个朦胧的概念罢了。

达尔文学说是进化的方法

　　达尔文学说不是进化论，是一种进化论的方法，拉马克学说也是如此。达尔文是设法去解释进化论，去找出天然的原因，进化论的自身，则是视为已经成立。

　　但是达尔文虽没有给世界以进化论的观念，然他乃是第一个搜集一群势力雄厚的事实，这些事显然不容人有别的解释，并且将这种事实用一种法则，提出于智者之前，使他们不得不诚切的视之为有哲学上的重要。此外达尔文亦是第一个贡献进化论的方法到博物界者，并且建设天然选择的历程，唯有最适应于环境者能生存之定律。赫胥黎（Huxley）说："直到达尔文的'种族起源'出现时代……那赞助变化说者之证据，是完全不充足的，关于假定的'变化'之原因，并没有提议，能适于解释此现象。"

赫胥黎续说："这个建议，以为新物种可由外面的情况及于个体所呈现之特殊形体之选择作用所生的变异而出，我们称之为'自发的'，因为我们不知道他们的原由，此事在一千八百五十八年前的生物学专家亦完全不知道，科学思想的历史家亦是同样的。但是这种建议是'种族起源'的主要观念，包括达尔文学说的精华处，我们所寻求而未得者，乃是一种关于已知的有机形体之起源，这假定假设为除却可证明其实际上有所动作者外，没有原因之作用。我们不是要固定我们的信心，于此或彼，明白与确定的理论，而乃得着一个可以和事实相印证，并可将他们确实的试验一下之明显一定的概念。这'起源'的概念，给我们以所寻求的应用的原理。"

什么是达尔文学说的精华，这个天然淘汰的主义，引伸到种类变化，受智识界一大部分之一般的承认？

为环境所薙除者即是天然淘汰

第一件事情，我们必定要完全解除我们心上一切极普通的意见，以为天然淘汰是一种能作某事之实体或是一种势力。有许多比方说法，像"天然淘汰是薙除不适宜者""天然淘汰是保存适宜者"和别种类于这样的话，这都是引人入邪道的。天然淘汰不是一种势力的，是一种历程。由自然方法而薙除不适宜者，即是天然淘汰；在生物竞争中，适者的能够保存，是为生存之天然界里的淘汰。你可以问，是什么势力或主体做这些事呢？这一点我可以回答：这是一种环境。将此用在他们很广大的意义里边，比方：一只满载人的船颠覆了，倘然各人只顾自己，那末，只有能够游泳并且是力量充足者，可以达到岸上而生存；有不能游泳的，必定溺死；还有能够游泳但是没有力量足以抵御那个震撼和危难，必定同是死亡的。由此那些在特别境遇里面，能够去适应暂时的生存者，就可以生存的。这才是天然淘汰。倘然其中有人想起他亲爱的伴侣不会游泳，于是就去救他的伴侣，而不愿去救别人，那末，他就执行了人择的法则了。倘然这只船颠覆在狂洋之中，近边没有救，那末，一定都是死亡的，因为他们没有充足的力量去适应他。

另一例证：一群行商在沙漠中，缺少了水的供给，在得着第二个井前，只够供给他们所需要的一半，于是强悍者和狡猾者就掠夺这个水，去让其余得不到水的商人死亡。那就是天然淘汰。他们都是人，他们都共缺少水的供给，他们全部或都死亡。或者达到第二水井时，已是在孱弱的境况。天然淘汰对于全部行商身体上之势力，是被妨阻了，于此所得获者，实在在天然淘汰之中的动作，所能及者以外。

再者：有两只鹿被一只饿狼所袭击，其中的一只是较为机敏，他就迅速的逃去了，还有一只是给狼吞食了。于此天然淘汰的进行，是以机敏及足的敏捷为基础，于这个方面最能适应者，就能够生存。英国麻雀子，将本地的鸣鸟驱逐，这种雀子的身体上，是很适宜于生存的，但是在美术上，他是低劣的。所以从别方面观察起来，这适应者亦不必常常是最好的。

在天然界中，最强健和最狡猾的，能够得到食料。那些最能抵御来攻击之敌人，用逃跑或是用欺骗，或是别种的智力，去逃避他的仇敌。那些能以不受别种的疾病攻击之害，那些能抵御天气大的变化者，或是迁徙到温和的地方，或是在地上掘穴，或是用旁的智力去免除这势力。那些能以抵抗干旱的，如预先贮藏好水，像植物里的仙人掌和动物里的骆驼。一总而言之，那些最能抵抗天然界里的一切破坏的势力，并且能以得到他们所需要食物的供给，那些就是最适者，他们就可以生存了。在别一方面讲起来，文明援助弱的，给东西与饿的人吃，给衣服与没有衣服的人穿，所以天然淘汰不能进行。

天然淘汰的历程之其他方法

然天然淘汰也依别的方法而进行，两个妇人在那边转磨，取其一个而遗其一个，要怎样去判定？在人类社会中去选择一个配偶，不是每每根据在心理的，而不根据在身体的品质；根据在精神上的美，而不根据身体上的美；虽是身体上的品质，其中如财产所有，和由财产而来的社会之身分，亦常造成选择的基础。身体上的品质，在一定范围里，亦成为吾人所云"雌雄淘汰"的基础。动物中此事的进行，是在自然状态中，像鸟的有悦目的羽毛，或是悦耳的声音，以及别种动物之有各种的华采或才艺，这种属性，为异性方面所有吸引力的。乃是由性的选择而供给候补者发展出来。然而有很好理由，可以信仰，如若不是最多的品质，普通看作于有性淘汰有意义的，其实仅仅为有性的成熟，在雄性方面的表示。如牡鹿的角，在配合时期，虽常常用来和其他雄的相斗，然而很难看作他本来是为响应他的斗争冲动而发的，因为此事的本体，乃性成而表示。倘然是这样的，那末，我们要怎样去解释这种事实，当他们生长完全，这种角变为战斗的武器，并且当这容易损伤棉绒状的外皮的时候，供他的用处时，立即脱去，并且我们要怎样去解释当有性的兴奋过了以后，而角自己脱落的时期？

形式上颜色上特别性质的发展，在雄鱼中有许多例，但是他从来不和雌性相匹偶，不过单独将雌鱼所产的卵浸润之，使他受胎。在动物中如同人，"亲近"在影响选择配偶上有重大的势力，但是这是无疑的，其实[1]，这个苗裔能够为表示于父母之性的成熟上之高度之生殖力所影响，发达的悦目的羽毛，乃受有很复杂并且很强壮的角，很错杂的跳舞或是别种奇怪之形状的影响；在人类方面，最是有闪动的眼睛，愉快的颜色，勇敢的行为，灵敏的才能，及鼓起的情绪，发为歌唱，或是不联贯的曲子。然这些品质，是表明一种情况，而非对于目的之方法，虽然，大致真可以有将心身调和人之个人，弄到一起的势力，由此去做改良种族的工作。

当有意的为那些在自己操制之下的，去决定他的伴侣时，那就不然了。人的畜养动植物，为的是对于他有一定用途的性质，或是他以此目光而选择其亲，这种叫做"人为淘汰"。各种奇异的，所想象的，家种动植物，就是出于此法。实在是由于研究养育者的实施人为淘汰，才引达尔文到辨别天然界淘汰的进行上，并且在考究这个朦胧问题变化之方法和相互的适应！中间，确然找出这层，在家养下之变化的知识，虽然不很完全，但确能给吾人以很好很安全的踪迹，并且他在他的《物种由来》[2]一书的开首一章，就是"豢养变异"，这即是人为淘汰的原理。

但是人不单是对于家种动植物上行配偶选择，他亦常常也要为自己的子孙而行之，于此他跟一种原理，为养育好动物者所不取的。一统时的犹太人，当他不能替他女儿选得一个学法律的新郎时，他就替他女儿选一个富商，这种压制治理者，是为政治的理由而配偶他的子女。对于社会上有野心的母亲，把女儿卖给出价最高的人，也不顾他身体上和智慧上的怎样，以及精神上的适宜与否。有时对于人种上很侥幸，前列之父母的计划，被勇敢的女儿所阻，而自己从较高的平面上去选择，而不管眼前的境况。在近来的"优生学"科学中，才有考察选择的标准，而这程序是提到高的水平线上，确保将来最适宜者之永续不断，和时时改进生存和永久的改良，到真正的适宜。

变态或均衡的缺乏是天然淘汰的开端

对于此事，无需许多深思，就可以看出淘汰的发动，只有当在有机体里边，有几种不同的特质时，然后才能从这上面行其选择，若是一群的动物各

1　原文为"真实"，疑误。——编者注
2　即达尔文《物种起源》。——编者注

方面都是相似的，倘然许多人都是真正同等的，那末，"生存"将仅为偶然的事了。所以差异乃是选择的开端所必需者。但是变态要到足够的程度，而有选择的价值。腿之少微的增加，能以保其充分的轻快，足以使具有此腿能跑过他的伴侣，并且逃避追赶的仇敌，而别个就要被这种仇敌所克服了。但是为什么蝶翅的四周有一种分外微小的斑点，或甲虫的翅上，有两点之联合？一只饿的鸟，要不要在选择他们的食物里，去停在那边，计算这个斑点，或是注意他们的整列？显然有一种变态，这种变态，有选择的价值，而那些是没有的。这是确实的，这种特别颜色的斑点，或条纹的品质，也许是一种别的变异相伴而生，或者与之有相互的关系。类如较大的躲避能力，和其他相似的事，这乃是保障生存的实在之品质。犹之在一黑制服肩上的徽章上，这种细微不同的标识，不能去保护他，使不受敌人的弹丸。像一个军官，他是受命令要极谨慎小心，并且求较好的保护，这肩章不过是他和兵卒的中间，很有相关的变态，随他品位阶级所生的互相关联之差异。

但是依多数生物学家的信仰，较小的差异，不必是常常的，亦不是通常的，为属于与别有选择价值的差异相关的。他们的存在，固没有疑惑，实在，这种族的建设，就在微小的差异之基础上。我们是决不能一致，对于要有何种数量的差异，为产一个新种类所必需者。一在事实上，没有两个生物学家曾经对于此点完全一致；其实我们可以相信这个"自然"，不知种类，只知个体，但是我们都同认这种变异在天然界中是存在的，并且在事实上，没有两个个体是真确的相似处。进者，我们承认多数变异是先天的，就是说有机体产生时之天赋的两部分，是他们的生的权利，或是生来的患害，皆随情形而定。并且我们另外又知道这种品质，出此种变异的结果，可以遗传到子孙，直到后裔的三四代。或者若是他们能服役于自然天命之下，则甚至可以递传到千代。当各个体生活时期，所获得的品质，能否遗传到子孙，这是直到现在的一个争论的问题，但是关于这种品质，不是我们现在所关心的。

虽然，我们在天然界的先天变异上，决没有一致的意思；他们是一定的还是不一定的呢？继续的还是不继续的呢？逐渐的还是骤然的呢？达尔文相信他系由"中心"而生微小，波动的变异，这种由选择作用，那些个体恰适巧在幸运的方向而变异的，并且由此而递传他的幸运的变异。就是，在最与环境谐和的方向上，就能够生存。由新的"中心"而更向前的变异，和合宜的品质之继续的选择，久后就能生出一种特别和善能适应的型体来，这种少数的得天独厚者，在群众死亡绝灭中，独合于生存。这是一种偶发事故，这是"偶然主义"，并且是过激达尔文派（Ultra-Darwinians）的主要主义，那种达尔文的学生，不像他的先生观察天然淘汰，不根据在幸运的，而根据于偶然的变异，可为有机的世界之进步的惟一的方法。

新达尔文派的学说根据在细微的偶然变异

达尔文学说，照近来达尔文派——即新达尔文派（Neo-Darwinians），他的主义环境是破坏不适宜者而保存适宜者——天然淘汰的起始；适宜者的精义，即在适应于生存的境遇——最能适应者能生存，并且将适宜生存的特性，传给他的子孙；因为偶然变异的结果，就发生这种特性，即显然不受任何已知规律所节制的变异。

但是环境是常常变的；上帝乐园，能变成荒芜，历史以前的沙漠，现在能开美丽之花，沧海能变成桑田，桑田亦可变为沧海。适应是在一种环境底下，保障生存，在变化的境遇底下，即促进死亡。大懒兽（Megatherium）者，是生存最久之懒兽类，很适应他的天然来的食料，但是他因为天气变化的结果，食料减少之故，变成孱弱，以至成为利齿虎的食物。这种利齿虎是很适宜的，用齿和爪去适应到猎捕，刺杀那些迟笨、厚皮的草食兽；当大懒兽食尽了，他就要受饿，因为这猫类过于向一方面发达，以致不能适应捕获捷足的草食兽，这草食兽就成为较敏捷同类——猫类——的食物；即虎、狮、豹，和肉食猫，因主要仇敌的人类，听其生存，所以亦仍能存在。

生存竞争，不许不适宜者以机会；到处是适宜者能生存的，到处永久是弱者失败。这是达尔文主义的纯粹而简单者。今天能适宜，亦许明天不适宜，是环境的变化。森林法律不能比森林常久，军法不能存于非武器时代；资本家不能存于经济均等时代，军阀不能存于世界和平时代。当常识环境再建立的时候，工作者和资本家各得其所，对于全体的利，可于那些使今日不适合者，能适于在争"人道"之联军中，占一相当地位之品质为能力之发展中见之。这又是达尔文主义。

偶然变异淘汰价值的疑问

我重复申明，达尔文学说不是进化的惟一要素，至少不是多数博物学者之信仰。以偶然变异为淘汰的基础之价值，为那班人不肯承认形态或性质的细微变化，足以决定生存。与四周住处色的调和，可以欺其仇敌而保护动物；伪假的色，可以逃避潜水艇的机会，没有这种颜色，必定被毁，但是这种适应，并不是微小的，与平常的大不同。一单行的斑纹，是船的保护色的第一步，不足以欺他的仇敌；正如一个绿色斑点，不足于青草上的褐色螳螂；正如美味的蝴蝶，近似恶味的蝴蝶之花纹发展第一步，不能蛊惑饥饿的鸟。但是偶然的变异的第一步，乃是全体变异的起始，不是当为完全的产物。要有选择的价值。变异一定要适当；要使成为合宜的，必须忽然发生，或在一定的方向渐渐增加，但是并没有直接的淘汰性。

晚近生物学家注意，渐趋向于那些大而明显的猝然变异，这种变异，间或发现于自然界，这种变异，像熟知的"顿变"（Sport）。这种现象，可以用短腿的羯羊来解明，一千七百九十一年，忽然发现于 Massachusetts，从前所宝贵的 Ancon 羊，因为他不能跳篱笆，即从这种所遗传下来的。达尔文起初非常重视这顿变的价值，后来归到次要的位置。

骤变选择

但是当荷兰的生物家佛利氏（Hugo de Vries）在离 Amsterdam 不远的田里，发现晚月见草（Anothera lamarckiana）的顿变，并且用试验寻出所生的各种形态，非变动不定的，而乃系有常久的新性质者，于是科学世界的注意，重新倾注到这种现象上，并且经心的培殖试验成为当日的习尚。佛利氏和他的多数学生，于是相信进步不是由从中心渐出来一种细微的偶然变异，而乃猝然的、躐等的。动物或植物的单纯特质，由于一种还在未知的原因，有时忽然经一种重新安置，正像万花镜里的许多颜色玻璃的微片之忽然重排置，而形成新花样一般，而此新有机的形态，即构成一种新物种。佛利氏叫他做"变异"（Mutation），亦有人叫他做"跳变"（Saltation），并且好多人信为一种不受已知律所制之进程。在佛利氏派看来，由此产出之新物种，乃是唯一可注意之变化。唯他们受制于天然淘汰；唯他们表出有机世界进步中的历程。

"幼稚时期"和"古生物学"之研究示人以变态不是偶然的而乃受一种定律之支配者

达尔文派和佛利氏学徒有一个共通的出发点——他们两者皆注意成熟的

个体。老 Agassiz[1]的声浪，前半世纪在生物学的荒野中，大声疾呼，忠告博物学者去注意动植物的未成熟时期——即居于胚胎和成年之中间的时期——没有达到他们的耳中，倘然他们已听见了，他们亦不曾注意。但是有一个门徒随从其师漫游地下世界，此世界为数世纪来死并且忘掉动物生命的，写他们自己的墓志铭在石片上之迅法，得了一新的信心，指示出生物学的救星之真途。黑德（Alpheus Hyatt）[2]证明 Agassiz 的说法，以为这种动物未成熟时期的生命历史，可以供给研究他的种族历史，这种幼期动物再现他祖先的成长性质，这种性质在最近的前一期之地质学时期里表示的，并且这个祖先依然也有他自己的幼年时期，去再现他最早的一种时期的祖先时性。Agassiz 不能领受进化的学理，但是黑德因他的研究，承认这种事实有哲学的态度。每个个体形像的历史，渐次回溯种族的历史，直至回溯到最初情形而止。他所经过的，详细研究，随时告诉他的学生。热心考察每个新得的记载，且每个学生选择他们最有望的，于是开始去解释祖先生长的记载。时代无论如何长，研究的精神有时衰颓，而新的通景可以开出来，新的向前研究之心又可兴起。在这个探索里，可以变成很明白的证明，那种变态不是偶然的，亦不是无规则的，但是支配于一定的方向。这种支配势力尚未显露，但是变态是很有定向是无疑的。

研究小孩及青年生活，自有一种奇趣，但是借此研究以说明人类进化历史者，直至现在始有人提出的。我们必定要在他们发育的不同时期里，去研究和比较各个个体，但是我们必定要在单纯的一个人的各个生长时期，去研究各时期的生长情形。从生后到老或死，都不可间断。在这种方法里，我们可以除去发育里变态，如加速或退化等，使记录不能明了的事实。凡是一个时代的记录，必定要去和前一时代的记录，用比较的方法去研究的，而和后来的一个时代的记录，也要有比较的研究，如此则某种族的进化的趋向，始得而明。这样的研究是能够办到的吗？如把小孩自从呱呱坠地之后，以至于老死，年年用照相、测量，并记录等法，以记其特性，而这个小孩又继续着这种大事业去记录他的子孙，如此子孙相续，以每于千年之久，那么研究人

---

1 即路易·阿加西（Louis Agassiz，1807—1873），1807 年生于瑞士，1829 年获慕尼黑大学和埃尔朗根大学哲学博士学位，1830 年获慕尼黑大学医学博士学位。1846 年应邀前往美国，长期担任哈佛大学教授，直至 1873 年逝世，在美国生活了 27 年，培养了美国第一、二代博物学家，对美国科学贡献巨大。——编者注

2 路易·阿加西的学生。——编者注

类进化学者，自然可以有许多极有价值的材料了。但是我们大概没有忍耐心去担任这种职务的一部，或者倘然是我们有的，我们的小孩，或是小孩的小孩，必定也有不能继续这个工作的；虽然，可以使他们信服他的价值，而事实上这种伟业很有望的。倘然记录是无心的，一世又一世的保存下去，以供后人的比较，则后人必定要相信人类生存的秘密，从那些从前的记录上，可以推想而知，于是要想寻出钥匙，去解决人生的问题，和人生命运的支配。

这个是否是一个无望的梦想？我相信不全然的，虽然几许人类进化的记录，人决做不到的，即是别种有机物的进化记录，也不能完全的，因为无论那种有机体，他们都已经生存到不知几许万年，他们的性质形体，都是由更下等更单简的生物进化而来，所以任何动物的已往的进化历史，是不可得而知的。

软体动物的贝壳在几许万年的历史上可以推知是原来的软体动物进化而来的

软体动物贝壳的样子，珊瑚虫的石灰分泌物，珊瑚，和许多别种动物生命的下等形状，都是含有他们生存之初以至于今日的长久历史，虽然只涉及具体的形状和匀配，可是对于研究进化学者，至少也能贡献几个基本定律，以说明支配全生命的发育。没有一个古生物学家或是搜索者，在地球的古生命记录里，不去研究奇怪的空壳，而不注意的闲游者，在海岸上，倒能用闲散的好奇心去注意。只有古生物学家，能遨游已往的海岸，在已消灭的海里，收集贝壳；并且只有他能在那些贝壳的形状、组织里，研究记录的共通之点；并且能以定律窥探他的意义。除去以前所有的记载，我们现在尚须上千的学者研究贝壳的化石——我们须要男的女的，很留心的搜索已往的生命记录，预备做将来工程的基础。下等的软体动物，可以为执政者和哲学者的师，就是立法者和改革者，也要奉之为师，因为软体动物能保存既往的历史，欲得学问上的原理者，亦不得不在软体动物上用功夫。倘然你以为这个思想太奇，而把我的说话，当作狂言，也不足为怪，但是我可以担任使你信服，假如你变成对于软体动物的贝壳，有诚心去研究，而缺乏充足的能力者，我一定能设法使你明白其进化的路迹。暂时脱离达尔文而为黑德的学生，然后你再回到达尔文，乃可以领会他的实在的意义了。

软体动物在他很早的生命里，已起始构有贝壳。这个贝壳，不仅是保护的器具，但是由化石上的痕迹和外皮的情形看起来，这个贝壳，确是一种进化而来的成绩。软体动物的外皮，由渐渐生长复杂起，新式外皮已生，而旧者未去，于是化合而为永久的硬壳。发育里的每个步骤，自始至终，都记录得很详。但是贝壳化石，他的样子，于长时间之中，也没有重大的变化，并且从天然的软体动物的习惯，贝壳又能保存软体动物，变成埋置于海底的沉渣之生物，以后产生的，亦如此经过他的生存。连续贝壳皮层如斯日积月累，永久记录起来，可以排成编年的次序，最老的在下，最幼的在顶上。

从贝壳的性质上，研究单一个体的生命历史，我们发现那些记录的变化，不是偶然的，是有次序的；不是都在各个方面里，是属于一定的方向；不是躐等的，是有精密的次第；每一个小的进步，必在同一方向的。比较上千个体的同层的同一片，我们注意都遭遇相似的发育，在同样普通的方向，从普通的出发点，不过有许多进行很快，有的是很慢。加速度或延缓，只更改进行的速率，无关于进行的方向。像贝壳性质增加数目，全群中的某某几个动物，在某种性质上特别发达，其结果就有了种差。但是每一群，起初发育在一定的方向里，很慢的继续在这线上进行而扩大性质，起初很微小，不注意的，但是达到成长时，有几许性质变成很分明，并且做变化特性的基础。观察许多成长的，有几许加速的形状，可以表明，可有千变万化的现象。但是用各个单一的个别生命历史的情形看起来，发育是很有次序的。佛利氏派所举骤变的几种例，是从这一点观察试验，他们可显出他们自己也是一种常态，不过大多数的个体所演进的程度，都相同的，无十分迟速之差而已。

有次序的渐进发育，并且在一定的方向里连续更改，是个体发育之定律，延长去研究连续的地质上的各时期的贝壳情形，也可以表明他的种族发育的定律。在一定方向里开始发育，从无论如何的原因，有机体继续在一条线上发育，直等到生出一种将来的能胜任天然淘汰的基础而止。倘然循序发育的结果，是很猛烈的，而不与环境相调和；倘然在一定方向里，锐意进化，一时超过寻常的进化之度，而天然淘汰的作用，就要乘机以起。所以尚武的进行发育，最后必能超过寻常的进化限度，所以往往自招祸害。所以时式的装束，和装饰的流行，常在一定的线上进行，直至超过世间一般人所能容忍的限度之时，就有生存和绝灭的结果，这是淘汰作用做成的。

在一定方向里所有的进化作用，不顾及最后的结果者，Shaler[1]和黑德熟思

---

[1]　谢勒（Nathaniel Southgate Shaler，1841—1906），路易·阿加西的学生。——编者注

这种现象。而叫他"惯性进化"（Inertia of Evolution）。Tuebingen 大学的动物学家欧姆（Theodore Eimer）他曾经有一个宣言，他做一个真实的研究，生存动物在个体发育时期里的生命，采用名字叫"Orthogenesis"，因为一定的直接的发育底原则，给在很少的方向里的微小连接之更改作用。但是他的工作没有研究到古生物学上历史上的记录，虽然他发明了这个学理的公式，而证明这种学理的功绩，必须归之于黑德。黑德虽然他未尝用过那 Orthogenesis 的名词，而说明其事实者，确是黑德一人之功。

变态是一定的，在一定的方向里进行，是必有进步的；他往往有微小的变化前进，这微小的变化，他们自己没有淘汰的价值，但是因为他们连续和累积下去，就产生那些异态。这种异态，在生存竞争里，有时有用，有时或可有害。淘汰在进化里，不是第一步，乃是第二步的要素。进化诚然要受淘汰作用的支配，但是不是绝对的由淘汰作用去决定进化的运命的。以上所说，乃由于研究软体动物的贝壳而得的经验。

Orthogenesis 是什么？什么是可决定变态在一个方向里，或是在别的方向里？有那些人，像欧姆相信是环境的作用。黑德不是他们中之一，他相信环境是一种刺激物，不是创造者。在一定的方向里，除非他们有发育的可能性，否则环境是无力的。有机体的从事于进化，研究在何方向是可以预定的，但是有力的几种进化的可能性，必先预有。

Jean Valjean[1]在争战的环境之中，就变成残酷不仁，后来入于 M. Bienvenu[2]所设的环境之中，竟变为圣徒。古生物学证明 Orthogenesis 的倾向是天然的，新达尔文派必当注意此点，并且应将淘汰视为环境的结果。适者生存的定律，是永远不能改变的，因为天理是永恒不变的。然而选择适宜的环境，也可以轻减进化活动的范围。知道他将走何路而预先供给适当之道，是便利得多了，环境和选择，都不能使母猪的耳生一个丝钱袋，但是有好环境的刺激，母猪的耳能发达到很高的能力，虽然这耳的发达终有限度，而在不适宜的环境中，则发达之度就要极低，永不能达于极限了。

——原载《博物杂志》1921 年第 4 期、1922 年第 5 期

1　雨果著名小说《悲惨世界》的主角冉·阿让。——编者注
2　雨果著名小说《悲惨世界》中的米里埃主教（Charles-François-Bienvenu Myriel）。——编者注

Science and Superstition: An Address Delivered before the Rotary Club of Peking on Dec. 17, 1925

By Amadeus W. Grabau

When we were young, we were superstitious. We Westerners believed in ghosts and fairies, in Santa Claus, in angels and in a devil. We believed in heaven and in hell. We believed that the world was created in six days, and that the Creator rested on the seventh day—that's why we keep holy the first day of the week. We believed that Eve was made from Adam's rib - that the whale swallowed Jonah and that we all were eternally damned because Eve made Adam eat an apple. You Easterners had other beliefs equally superstitious.

As we grew older we questioned some of these beliefs and we frankly discarded others. We began to observe, to relate cause and effect, to experiment-we became scientific.

Like Race Experience

That is only a recapitulation of the history of man. In a primitive state he is superstitious-believes in evil and in good spirits, in signs, omens, feng-shui and oracles. And there are always priests and medicine men who try to keep him in that belief. As the race grows older it becomes more skeptical. Why? The old ideas do not harmonize with experience and the facts observed seem to contradict the old dogmas. We laugh at a belief in witches—but what about the devil? Hell may be difficult to locate, but we sometimes think there ought to be one if there isn't, to accommodate some of our friends. As the race progresses some men begin to observe the world about them in detail and to reason from the observed facts. Science has been born, and the battle with superstition is on. But science is still a youngster, a David ready to fling his pebbles. But the Goliath of superstition is a monster-it takes more than one pebble from David's sling to down him.

Science Will Win

I take it no one here has any doubt as to the ultimate outcome of the battle, nor any question that a world enlightened by science is a better world than one steeped in superstition. But how many of us are prepared to join David against Goliath? How many of our friends and acquaintances are on the fence, doubtful as to the relative merits of the two combatants, and how many bow to the rule of Goliath. Let us respect the adherents of Goliath, in so far as they are sincere. If they pray for rain in the churches, or even as I have heard them do, at the commencement exercises of a leading American University—if they bring the iron tablet to Peking to invoke the rain—why laugh? Sometimes the rain comes, and comes in torrent, and then the people rejoice in the vindication of their belief and the efficacy of their prayer. To believe is human, to know is divine.

Faith the Last Resort of Feeble Wills

But how much easier it is to be merely human, to place our burdens somewhere else than on our own shoulders. Surely it is easier to pray for rain than study dry farming, or construct expensive irrigation systems. Faith may be called the last resort of feeble wills, a reliance on authority—and how we worship authority! It saves us the trouble of thinking for ourselves—and most men hate to think. But the authority that most men look to, is the authority of the majority—the minority is necessarily in the wrong. So superstition sways the earth and science battles on, gaining ground, but oh how slowly!

I said we can sympathise with the adherents of Goliath, because we pity them, we can even forgive them for their attacks on us—for they know not what they do. But what about the multitude of onlookers? When a crowd of ignorant clergymen, under the leadership of a born notoriety seeker, starts an anti-evolution campaign, and brings it to the disgraceful climax of the Scopes trial, why are they permitted to run amuck? Any other group of atavistic morons would be rigidly confined and controlled, if they proved a menace to society?

Inadequate Education

The clergymen can't be blamed, they, at least the older generation, are boys from the farms who never had any education worth mentioning except what they got in the theological school. I know—for I am well acquainted with the older theological schools and their class of students. My paternal grandfather founded and conducted one of these schools—and my maternal grandfather another, of a different denomination. The younger generations got a better education, they even had a smattering of science, but how pitifully inadequate. For that matter how pitifully inadequate is the teaching of science in most of our educational institutions. You all had some of it. What do you remember, and what did it give you, except in your own particular branch, if you specialized in science. What did it give you to fight superstition whit? What has the average business man, and in America everybody is a business man—what has he to oppose to the claims of infallibility of the orthodox believer? Perhaps the fundamentalist "is" right, perhaps the world "was" created in six days, with all the species of animals and plants, perhaps men "was" made out of mud, perhaps Noah "did" build an arc, which was stranded on Mt. Ararat, releasing all those that begat the modern types of beasts and men, which haven't changed a whit since then, except to grow from bad to worse, because of the Edenic curse. Perhaps why not? —The scientists may be all wrong, all their ideas may be pure speculations, how do we know?

Developing the Scientific Interest

Some of my impatient colleagues say, why doesn't the business man read and become familiar with the facts of science? I know why. I used to work from 7 A.M. to 6 P.M. and after supper I tried to read Darwin, but, disgraceful as it may seem, I went to sleep over the book. But if there was a good show in town, or a lively party, I could easily keep awake half the night. Finally I made a discovery. I began to collect the wild flowers of my district, to preserve them, study and classify them and I founded a Natural History Society among my friends. And each Sunday we got up at 2 A.M. and walked out of the city ten miles into the country—that was before the days of trolleys—and spent the day searching for wild flowers. And the joy of finding a new species! I never had any difficulty in getting up at one or two o'clock in the night—even though I worked overtime the day before until ten o'clock—to change the dryers on my specimens-much, let me say, to the disgust of my brother, an embryonic clergyman, whose rest I disturbed.

That was the secret—to study things, not books, to get a first hand knowledge of nature, to speak to the earth that it may teach thee.

Nature's Endless Variety

No one can take up the study of natural objects and fail to find a new world, and to get a rational view point of the old. Collecting postage stamps is not in the same class—they, like all the things of man-made origin, are limited—but Nature's children present an endless variety, no two are exactly alike, and even our species groups are numerous beyond conception. Among insects alone over 380,000 species have been determined and no one knows how many other thousands are awaiting discovery. Among mollusks, already 62,000 species are known.

Surely when we begin to realize the wealth of life around us, we can not fail to gain a new conception of the world we live in. And when we take another step and look into the profundity of the past, where period after period had its own no less voluminous life record, we lose all respect for the time-honored homocentric attitude of the man who holds that all revelation is confined between the two covers of a book. The man who questions nature gets a truer answer to the riddle of the universe than the man who juggles words until their perfectly obvious primitive meanings are distorted into a grotesque imagery of things they never were meant to portray.

We Are Creatures of Custom

We are creatures of custom. Some things are right and proper because they have always been considered so, while others to which we are unaccustomed seem trivial or wrong. The man who would commit the intellectual absurdity of praying that the laws of nature may be altered for his personal benefit, scoffs at the collector of shells or insects as a man occupied with childish things.

"Isn't it nice that he can amuse himself so" was patronizingly said by a disciple of the old faith on seeing me collect shells at Peitaiho. That's the usual attitude of the ignorant-amusing—perhaps, fit for children but not for a grown up person who should be making money or studying the mystery of the records left by the fathers...

The Danger of Specialization

That they are the children, specimens of arrested development, who have failed to grow up, never occurs to them. That is because the mass of humanity has not yet grown up, is still in the childhood stage of superstition and primitive occupations, is still swapping knives and pitching pennies. Of course, the swapping of knives and pitching of pennies has been carried to a high degree of perfection, that is specialization, always a one-sided business. We are all specialists, and the more intensely we specialize the more complete is the atrophy of the uncultivated part. Darwin lost his taste for music, art and poetry which in his youth he loved, by too continued dwelling on problems of heredity and adaptation. Henry lord thinks history bunk and science foolish because his vision is limited by motor cars.

Scores vs. Species

No one can deny that the world needs specialists, we have too few of them now. But there is something wrong with our educational system that makes a man a genius in one field and keeps him a child in all others. Perhaps that is the price of "geniosity." If so, let us accept it and be thankful for the genius. But that need not apply to the near-genius, the semi-genius and the non-genius, that is the mass of mankind. Such a man has time for golf, which is an exercise I envy him for. He might find time for natural history if his early tendencies had been directed that way, and might find that even more absorbing. Suppose the most frequent question asked on Saturday night was not "what is your score?" but "how many new species did you get?" What would that signify. Let me tell you. It would be an indication that the mass of mankind was beginning to awake intellectually that the day of the supremacy of knowledge over ignorance and superstition had begun to dawn.

Observation, Then Deduction

That may sound extravagant, but it isn't. The first step in knowledge is observation—the next deduction. Every one of you appreciates a keen observer and a sound reasoner from observed facts. Enlarge the field of your observations, and you broaden the scope, and increase the force of your deductions. The study of the shells of the sea-shore or that of any other group of natural objects, will give you justification for the faith that is in you. You are convinced that the great American doctrine, that all men are free and equal is a superstition and fallacy. The study of natural history objects will show why men are not and never will be equal, and that the freedom of one is unattainable by another.

"You Can't Keep a Good Man Down"

You believe in competition. Nature teaches you that the struggle for existence is the most fundamental of all natural laws, that always and everywhere the strong will win, and the weak go to the wall. That is natural selection and you can't get away from it or ignore it. But it doesn't stop there. As your vision clarifies you see that natural selection is wedded to environment. The fittest survive—but the fittest is the one most in harmony with the environment. Then you turn your attention to environment and find that it is not a fixed quantity. And as the environment changes, so does the standard of fitness. The average individual becomes welded to his environment, and when that changes he cannot change with it, but falls by the wayside. The superior man has potentialities of adaptation denied to others. Change the environment, prevent him by statute from continuing his leadership, and he will come to the top again in some other capacity. The leader of a band of highwaymen is leader by virtue of some superior quality. Under other conditions, that might elevate him to the generalship of a great army, in still another environment he might become a captain of industry, a

Napoleon of finance, a builder of empires, a preeminent statesman, or a man of science. In the well-known phrase said to have been invented by the whale as he spat up Jonah—"you can't keep a good man down."

Environment Also Struggles

The student of nature soon learns that environment is not a fixed quantity. Interpreting the results of his studies in terms of human relationships, he sees the age of militarism replaced by the age of commercialism, but though he may not know what comes next, he knows that that is not the end. And in every age he sees the superman—the man endowed with potentialities, which make it possible for him to rise again to the top in the new environment. And he sees the struggle of environment and realizes that natural selection is active there too, and that in the end better environment prevails.

I should like to sweep the present type of science teaching from all our schools, except the universities, and substitute for it the collecting and classifying of natural objects, shell, insects, plants, etc. Thus I would have the study of natural science begin at the bottom, for such work is the foundation on which we must build. You cannot erect a scientific edifice by beginning with the roof, even in China.

And for heaven's sake, get away from the notion that such study is fit for children only, that grown men have no concern with it. Thank your stars if you have been admitted into this fascinating field in your youth—but be especially thankful if you have leisure to carry on such studies throughout life—for such occupation makes men.

The Records All About Us

If I were a clergyman I should quote to you the words with which the Great Teacher answered the tempter who would make him believe in the all-sufficiency of the materialistic attitude: "Man does not live by bread alone, but by every word that comes from the mouth of the Father", and I should refer you to my favorite volume, where in all these words are recorded: But being only a clergyman's son I dare give you better advice. I would direct you to the original records and bid you study them, the records all about you, open to all who wish to read.

The Great Reward

"The language is too hard to learn" you say? Not at all—not as hard as to learn Chinese. Observation and common-sense reasoning, that is all there is to it. And the reward! A new heaven and a new earth. A new conception of man's place in nature, a new understanding of the forces that rule the universe. New energy to attack the problems of life that confront you, be they big or little, and new power to adjust yourself to the demands of your immediate environment and the requirements of that which may follow. And though you will never disregard the material returns—who of us, that is wholly alive, would you will gain a new understanding of the words of Kipling[1], when he describes the true attitude of the big man towards his work, the attitude which recks not gain, nor praise, nor blame, as the chief reward, but works for work's sake, for the sake of doing that which is worthwhile, for the sake of accomplishment. You know the words:

"Then none but the Master shall praise us.

And none but the Master shall blame.

And no one shall work for money.

And no one shall work for fame.

But each for the joy of working,

And each in his own native star,

Shall draw the thing as he sees it.

For the God of things as they are."

——原载 *The Peking Leader*, December 22, 1925

---

1 拉迪亚德·吉卜林（Rudyard Kipling，1865—1936），印裔英国记者、文学家、诗人，1907 年获诺贝尔文学奖。本文最后引用的诗即出自他的诗 "When Earth's Last Picture Is Painted"。——编者注

The Peking Laboratory of Natural History

By Amadeus W. Grabau, Dean

The following account of the above institution has been received, and we make no apologies for publishing it *in toto*.

Organization

The Peking Laboratory of Natural History was founded in 1925 with headquarters at 11 Kaka Hutung, Peking. It is a voluntary association of scientific men engaged in the study of the fauna and flora of China, and has for its purpose the systematic survey, and monographic publication of the animals and plants of China. In this work the Laboratory stands ready to cooperate with, and if it seems advisable, divide the field with other organizations. Thus it is already affiliated with the Chinese Geological Survey, and supplements the work of the Palaeontological division of that Survey, by taking over the study of the Pleistocene Mollusca—to be published, however, in the *Palaeontologia Sinica,* and by continuing the work in the modern fauna and flora.

The work of the staff of the laboratory is entirely voluntary, no salaries or remuneration of any kind being paid. The buildings and equipment of the central laboratory on Kaka Hutung were provided entirely by Mr. Sohtsu G. King, who is at the same time the Custodian of the Laboratory. Branch laboratories may be established, if it is deemed desirable, at other places and centers, the central laboratory acting as a clearing house for correlation of the work of all the branches. The laboratory aims to associate with itself active workers in the systematic zoology and botany of China, in any part of the country, whose work is in the line of that aimed at by the laboratory. This work is the systematic description and illustration of the plants and animals of China and their publication in a uniform series of monographs, according to a broad general plan.

At present partial lists of species, and descriptions of some of the Chinese animals and plants are found scattered in scientific periodicals all over the world. These are for the most part unavailable to the average student in China, not only because they are so widely scattered, but also because they are in a multitude of foreign tongues. English is the language to be employed in the monographs of the laboratory and with it for each volume, as comprehensive a summary in Chinese as appears desirable. The aim is to present in series of uniform character the complete fauna and flora of China. This is of course an undertaking of many years, even decades, and the plan of publication has sufficient elasticity to enable the publication of any complete unit whenever it is available. Though each field is under the direction of a specialist connected with the laboratory staff, who at the same time is the editor of the particular series of monographs pertinent to his field, contributions from others are accepted provided they are complete systematic descriptions of a unit group, whether that unit represents a genus, a family, an order or some other unit, biologic, geographic, etc.

Staff

The Scientific Staff of the laboratory at present is as follows:

Administrative Officers

A. W. Grabau, S. D., Dean.

Sohtsu G. King, Custodian and Honorary Secretary.

Heads of Department

Sohtsu G. King, Mollusca.

F. N. Kolarova, Cand, Ph. D., Assistant in Pelecypoda.

N. Gist Gee, M. A., Fresh-water and terrestrial invertebrata.

C. Ping, Ph. D., Fish, Amphibians and Reptiles.

G. D. Wilder, D. D., Birds.

Bernard E. Read, Ph. C., Ph. D., Plants.

Kungpah T. King, Illustrations.

Work Under Way—Mollusca

The work so far begun by the staff includes extensive collection of the marine

Mollusca from Pechili Bay by Prof. Grabau, Mr. Sohtsu King and Miss Kolarova, and other marine invertebrates by Mr. R. J. Dobson. A marine biological laboratory was established at Peitaiho, and several months devoted to the collection of the marine fauna with almost daily dredging expeditions during part of the time under the direction and personal charge of Mr. Dobson. As a result a very large amount of material has been brought together, including well over a hundred species of Mollusca represented by thousands of specimens. In addition, Curator King of the department of Mollusca, has arranged with local collectors all along the Chinese coast, to supply him with Mollusca, the shells as well as the animals, and much material has already been obtained in this way. Arrangements have also been made with a number of individuals to supply fresh-water and land Mollusca from different parts of China. A very respectable beginning has been made to bring together the literature on Mollusca, and the curator is now in correspondence with librarians and booksellers all over the world in an endeavour to bring together in time a complete library on this subject.

The first fascicle of Volume I, Series A, of the Fauna Sinensis devoted to the *Rapanae* of the Chinese waters is now in preparation and will be published during the winter.

Other Marine Invertebrates

At present there is no one on the Laboratory staff in charge of the other marine invertebrates, but extensive collecting has been begun and will be carried on in subsequent years.

Land and Fresh Water Invertebrates of China
(Exclusive of Mollusca and Insecta)

The land and fresh water invertebrates, exclusive of the Mollusca and Insecta, are in charge of Curator N. Gist Gee who has for some time been collecting specimens representing several of these groups, notably the Rotifers, the Fresh Water Sponges, and the Leeches of China. Curator Gee has also brought together lists of many of the animals and plants which occur in the vicinity of Soochow where he formerly taught. Some of these lists, especially the ones of Ants, Leeches, Rotifers, and Fresh Water Sponges, represent a wider range than Soochow. These lists have been arranged from intensive search into the scattered literature and from extensive correspondence with specialists in various parts of the world and from sending large quantities of material abroad for study and identification by specialists. Some of these lists have been published in various periodicals but they will be made available by their publication in the series of Bulletins of the Peking Society of Natural History, a larger organization devoted to all aspects of Natural History and to the spread of interest in it, but one with which the Peking Laboratory of Natural History works in close harmony.

A monograph on the Fresh Water Sponges of China is being prepared by Curator Gee and collections of sponges from the various parts of China are being made. This work covers a virgin field and will be a definite contribution to the knowledge of this group.

Preliminary lists of a number of the groups will be issued in order to furnish a basis for those who wish to work in these groups.

Insecta

At present there is no one on the staff of the Laboratory in charge of Insecta and related groups, but it is hoped in the near future to establish, probably in connection with other Peking institutions, an Institute for Entomological research.

Lower Vertebrata

The lower vertebrates are in charge of Dr. C. Ping at present resident in Nanking. Dr. Ping has already under way extensive researches on the fishes of China, and some of his associates are engaged in the study of amphibians and reptiles as well as insects. It is hoped that in the near future this department may develop into a larger institute because of its great importance in connection with the question of food supply in China.

Birds

Curator G. D. Wilder has for many years studied the birds of North China and in co-operation with Curator Gee, Lacy I. Moffett, and other ornithologists in this and foreign countries has brought together an extensive list of Chinese birds which will be published shortly in the Bulletin of the Peking Society of Natural History.

A monograph of the Chinese *Corvidae* is now being prepared by him, the coloured illustrations for which are being made by Mr. Kungpah T. King. This will be published as the first fascicle of Series I, of the Fauna Sinensis.

Material for other monographs has been brought together and these will be issued as soon as completed.

Botany

Curator Read of the Department of Botany has for many years collected and studied the plants of medicinal value in China. This work is carried on partly under the auspices of the Peking Union Medical College with which Dr. Read is connected, but will be brought out by the Laboratory. A complete list of medicinal plants of China, with their original and romanized Chinese names, the parts in use, their constituents, habitat and references to botanical and pharmacopoeialliterature, Chinese as well as foreign, arranged in botanical classification, has been compiled by Curator Read and his assistant, Dr. J. C. Liu, and will be published as Volume I, Series A of the Flora Sinensis.

Other phases of botanical work will be arranged for as soon as specialists in these branches are ready to undertake the work. Two volumes of Series B, containing the systematic description of species are nearing completion. These will comprise about 40 plates of colored illustrations and from 600-800 pages of text for the 2 volumes.

Illustrations

Mr. Kungpah T. King, the well-known Peking artist, has undertaken the charge of the illustrations. He will prepare the paintings for the Chinese birds himself, this being purely a labor of love on his part. Furthermore, he has undertaken to defray the cost of publication on the monographs of the Chinese birds, as his brother Sohtsu G. King, has undertaken to defray the cost of publication of the monographs on the Mollusca. This truly patriotic attitude towards the development of Science within their own country, cannot receive high enough commendation, and, it is hoped, may lead others to come forward to the support of this and similar scientific undertakings.

Finally, Mr. Kungpah King is giving gratuitous training to a number of the younger Chinese artists, to enable them to become experts in the accurate and scientific delineation of the animals and plants of China, and these men will be employed in the production of the illustrations for the monographs on the Fauna- and Flora Sinensis.

Publications

The publications of the Laboratory will be confined to the monographs on the fauna and flora of China. All smaller papers, such as lists of species, preliminary descriptions of new forms, natural history notes on Chinese animals and plants, etc., will be referred to the Bulletin of the Peking Society of Natural History. The monographs to be issued by the Laboratory will be of quarto form, identical in size and general character of text and plates with the volumes of the *Palaeontologia Sinica*, issued by the Parent Institution, the Geological Survey of China. Whenever possible, however, the plates will be colored. The monographs will be in two groups: I *Fauna Sinensis* and II *Flora Sinensis*. Under each group there will be a number of series, each of which is devoted to a special field, this division being in some cases biologic, in others geographic and in still others determined by other reasons. Each series will be in charge of the Curator of the respective department, who is at the same time the Editor of the series.

Under each series there will be a greater or less number of volumes. Each volume will comprise from 300 to 400 pages of text and about 30 plates. The volumes will be issued in fascicles, each fascicle a complete treatise of a related group (genus, family, etc), but the larger fascicles may be issued in parts. As each fascicle is paged separately (as in the *Palaeontologia Sinica*) there need be no rigid sequence in the issuance of the fascicles, any fascicle of any volume appearing whenever the text and plates have been completed.

The treatment is to be systematic in the first place—there being a complete description of the species, according to modern systematic methods, with the aid of full illustrations. A complete synonymy and bibliographic reference will be given whenever it seems desirable. The description is to be complete—whether the species has been described before or not, though part of the description may be a quotation of the original description, the object being the presentation of a complete monograph without requiring the student to refer to the original scattered literature, except for purposes of more extended research. The description and illustration shall be so complete that any one using them can readily identify his material.

Following the systematic description there will be an account of the habits and habitats, conditions, and all other purposes, and all other natural history notes of interest together with reference to Chinese folk-lore and literature and such other facts as will be of interest to the general reader desirous of acquiring knowledge of Chinese animals and plants.

It is proposed to distribute a limited number of monographs to museums and institutions of learning in all parts of the world, including the Universities of China. Also to specialists in the respective branches, though this generally is taken care of by the author, who receives a liberal supply of copies of his monographs. The remainder are to be put on sale, the receipts to be added to the publication fund. A preliminary analysis of the several series of the *Fauna Sinensis* and the *Flora Sinensis* has been prepared, this being, of courses subject to modification as the subjects are developed.

Peking, November 1925.

——原载 *The China Journal of Science and Arts*, 1926, vol. IV, no. 4

Memorial of Yatseng T. Chao

By Amadeus W. Grabau

The death of Mr. Y. T. Chao at the hands of bandits in Yunnan is a tragedy, the magnitude of which few can realize, and a loss to China which is absolutely irreparable. Modern science in China is so young and able men of science are so few, that the ruthless cutting off, at the very threshold of his career, of the ablest and most promising young scientist that China has ever had, is a national calamity.

Mr. Chao was one of my first students in China. Almost from the beginning I recognized that he was one of those exceptional types destined to become a leader in his chosen field of science. I have been engaged for more than 30 years in training young men for a scientific career, and I can truthfully say that in all that time I have never had one of whom I have had greater expectations, and a more confident belief in the certainty of future achievement than I have had of Mr. Chao. He was a brilliant refutation of the charge sometimes made by superficial western observers, that the Chinese mind is not adapted to critical scientific work. Mr. Chao was as critical, as accurate, as painstaking and as unsparing of himself in his work as any young man whom it has ever been my lot to know. Of robust health, great physical vigour, possessed of a driving ambition, which was nevertheless held in check by a keen mind, and determined to become the master of his subject, he was of that type of student that delights the teacher engaged in training specialists for the exploration of new fields, and makes him feel that his work has not been in vain and that it will be carried forward when he himself relinquishes it.

Even before graduation Mr. Chao began his research work and after his four student years with me, at the National University, he became my assistant and at the same time a junior member of the Geological Survey where he soon rose to become a valued member of its staff of geologists and palaeontologists. He might have taken for its own the motto adopted by the oldest scientific academy of the world, *Nunquam Otiosus*, for he was ceaselessly active at the work which he loved, and which meant more to him than opportunities for advancement and pecuniary gain. Mr. Chao had many offers of positions which would have gained him a comfortable livelihood and much prestige among his fellow men, but he preferred to remain quietly in the laboratory of the Survey and carry on his research work. And that work was thorough. Realizing the importance of foreign literature, he set about, after having mastered English, so that he spoke and wrote it with remarkable fluency, to become thoroughly conversant with German and French, and he had even begun the difficult task of acquiring a knowledge of the Russian language. With Japanese, he was perfectly familiar. He was a critical student, but he also was an extremely modest one, and never undervalued the work done by others. Indeed, he had a young man's reverence for the great names in his science, and always thought of himself as a beginner and hoped some day to achieve and gain a position where he might be a worthy representative of his country and his nation among the brotherhood of scientific men.

For a man only six years out of college, Mr. Chao's contributions to his science are exceptional. He has published four monographs in the *Palaeontologia Sinica* and they are absolutely his own work, although he was always ready and indeed eager to accept advice and criticism. He has also published many shorter papers. I consider his work thorough, critical, scientific, and in every way equal to the best type of work done in this field elsewhere. Not content with mere laboratory work, he again and again took the field to investigate the stratigraphic relations and make careful collections of fossils from the formations in which he was interested. He was engaged in field-work when he met his end.

Mr. Chao was a critical observer of details in the field. Professor Arnold Heim, who spent some time in the field with him, has testified to Mr. Chao's exceptional qualifications as a stratigrapher and during our many discussions of stratigraphic relationships I learned thoroughly to appreciate his ability as an observer, and his critical analysis of observations. That stratigraphic and palaeontological science has gone forward by leaps and bounds in China, is in no small measure due to Mr. Chao, whose own restless activity was a constant spur to his colleagues and his younger associates.

And with it all he was a man of exceptional likeability. His modesty, and his appreciation of the work of others, endeared him to his teachers and older colleagues as well as to visiting men of science, with whom he came in contact. He had many correspondents among scientific men in other countries who helped his work with advice and with material loaned or exchanged. With his young colleagues he was on exceptional friendly terms and he was a good teacher and eager exponent of the science to the young men of the succeeding classes in college and of those picked few of these who were accepted as junior members on the Survey.

Science has lost one of its most earnest and most promising devotees. China has lost one of its future leaders, and we, his teachers and colleagues, have lost a valued friend and co-worker. But China's loss is the greatest. It has urgent need for every one of its builders on the slowly rising temple of science, and of its master builders before all.

Let the Nation mourn, for the loss is National.

Yatseng T. Chao was born in 1898 in Li Hsien, Hopei province, North China, graduated in 1923 from the Geological Department of the National University of Peking. Entered to the Geological Survey service in the same year. Promoted to geologist of senior class in 1926 and in charge of the Palaeontological Laboratory of the Survey in 1927. Recipient of First Scientific Price awarded by the China Foundation for the promotion and education and culture. Killed by bandits during his field work at Cha Hsin Chang of Chao Tung district, North Yunnan, on Nov. 16, 1929.

Peiping, November 28, 1929.

Bibliography

1. The Structure of the Nankou district. *Bull. Geol. Soc. China*, Vol. II, No. 1–2, pp. 111–115, June, 1923.

2. Geology of the gorge district of the Yangtze (from I-Chang to Tze-kuei) with

special reference to the development of the gorges. (with J. S. Lee) Ibid. Vol. III, No. 3–4, pp. 351–392, Dec. 1924.

3. On the Stratigraphy of Tze Chow and Liu Ho Kou Coal Fields, S. Chihli, and N. Honan. (with C. C. Tien) *Bull. Geol. Surv. China*, No. 6, pp. 67–86, Dec. 1924.

4. Stratigraphy of Lin Cheng Coal Field, Chihli Province. (with C. C. Wang and C. C. Tien) Ibid. No. 6, pp. 27–36, Dec. 1925.

5. Geology of I Chang, Hsing Shan, Tze Kuei and Pa Tung districts, W. Hupei. (with C. Y. Hsieh) Ibid. No. 7, pp. 13–76, Dec. 1925.

6. The Mesozoic Stratigraphy of Yangtze gorges. (with C. Y. Hsieh) *Bull. Geol. Soc. China*, Vol. IV. No. 1, pp. 45–51, April, 1925.

7. A Study of the Silurian section at Lo Jo Ping, I Chang district, W. Hupei. (with C. Y. Hsieh) Ibid. Vol. IV, No. 1, pp. 39–44, April, 1925.

8. On the age of the Taiyuan Series of N. China. Ibid. Vol. IV, No. 3–4, pp. 221–249, Dec. 1925.

9. Succession of the marine beds in the Chang Chiu Coal Field of Shantung. *Bull. Geol. Surv. China*, No. 8, pp. 1–5, Dec. 1926.

10. Carboniferous Stratigraphy of South Manchuria. Ibid. No. 8, pp. 6–9, Dec. 1926.

11. Classification and correlation of Palaeozoic coal bearing formations in N. China (with J. S. Lee) *Bull. Geol. Soc. China*, Vol. V, No. 2, pp. 107–134, Dec. 1926.

12. Geology of Western Chekiang. (with C. C. Liu) *Bull. Geol. Surv. China*, No. 9, pp. 11–28, Oct. 1927.

13. Productidae of China. Part I (pp. 244 English, 23 Chinese, 16 plates and 7 text-figures) *Palaeont. Sinica*, Geol. Surv. China, Series B., Vol. V, Fasc. 2. Sept. 1927.

14. Fauna of the Taiyuan Formation of North China, Pelecypoda. (pp. 64 English, 10 Chinese, 4 plates) Ibid. Vol. IX, Fasc. 3. Dec. 1927.

15. Brachiopod Fauna of the Chihsia limestone. *Bull. Geol. Soc. China*, Vol. VI, No. 2, pp. 83–120, Augt. 1927.

16. Productidae of China. Part II (pp. 81 English, 23 Chinese, 6 plates and 3 text-figures) *Palaeont. Sinica*, Geol. Surv. China, Series B, Vol. V, Fasc. 3, Oct.

1928.

17. The geological age of the limestone of Chihsia Shan, near Nanking. (in Chinese) *Science*, Science Society of China, Vol. XII, No. 9, pp. 1161–1179, 1928.

18. Origin of the Hot Spring at Wen Chüan Se, Western Hills, Peking. (in Chinese) *L'Education Franco-Chinoise*, No. 17, pp. 18–21, Augt. 1928.

19. Carboniferous and Permian Spiriferids of China. (pp. 133 English, 6 Chinese, 11 plates and 20 text-figures). *Palaeont. Sinica*, Geol. Surv. China, Series B. Vol. XI, Fasc. 3, June, 1929.

20. Geological Notes in Szechuan. *Bull. Geol. Soc. China*, Vol. VIII, No. 2, pp.137–150, Augt. 1929.

——原载 *Bulletin of Geological Society of China*, 1929, (3)

悼赵君亚曾文

葛利普作　胡伯素译

　　赵亚曾君在滇之死于匪手，为中国之一种饮恨终天无可补救之损失，抑亦空前之惨剧，于此甚少有人知其重要性。夫今日中国科学如此幼稚，科学人才又如此稀少，而赵君实为最能干最有希望之青年科学家，前程万里，未可限量，今不幸遭此惨厄，不可谓非国家所受之一种莫大灾殃；一种莫大打击。

　　赵君为余抵中国后最早一班学生中之一员，自始余即认定赵君，对于所学，必有成就，必有建树，而将不后于人，余教学凡三十余年，固无日不在领导青年使从事科学作业之中。实言之，在此时期中余从未对于余之任何一弟子，深信其将来之成就，或予以较大之希望有甚于赵君者。外人不察。每谓中国人无科学头脑，不足与谈精细之科学，此种谬见，观赵君则可不攻自破。赵君明敏耐苦，无稍逊让于人之所能者，对于学问，再接再厉，素不放松，要为青年科学家中所不多觏。赵君更体格健强，气魄雄厚，极富进取之雄心，常兢兢自惕，有非于所学能登堂入室，窥其玄奥，显其英华不止之势。赵君既为如此优异之学生，实足使其以训练专门家为学术界开疆辟土为终身事业之师，心中得到无上安慰，深觉一番心血并未虚掷，并觉其老而无用时，其未竟工作，将亦不虞无人起而肩以行之也。

　　赵君于未毕业前即已开始研究工作，在完毕其与余相处四年之学生生活后，仍留校为余之助教，同时亦为地质调查所之一少年技术员，未几即已一跃而占重要之位置，该所地质学家古生物学家虽跻跻称盛，而彼亦渐能于其中崭然露头角，赵君平素垂帷攻读，自强不息，或已取世界上最古科学研究会（Scientific academy）之遗风余教如 Nauquam Otiosus（即永不休息之意）[1]者为其座右铭。彼之奋勉所学，自有其乐，金钱之求得，盖在其次，亦非所计也。人之于赵君盖多力为推毂，赵君苟允就之，未始不可飞黄腾达，度其

---

[1]　此处"Nauquam Otiosus"（永不停息）是指 1652 年成立于施韦因富特的自然奇趣学院（Academia Naturae Curiosorum）的箴言。该学院 1687 年以德国皇帝利奥波德命名，2007 年起成为德国国家科学院，总部在哈雷。——编者注

安乐之生活，而为侪辈所艳羡，但赵君宁愿埋头所中之实验室进行其研究工作，固视富贵荣华如草芥，外界之利诱势导不足稍动其心也。赵君鉴于外国文之重要，故于英文说写两皆通顺后，更从事学习法文，于德文亦深具根底，对于深玄难解之俄文亦曾开始研讨，至于日文则完全娴熟，已得到其精髓矣。若赵君者固为杰出之才，但其谦虚亦有非常人所能企及者，对于他人意见著作，总是悉心研究，从未妄加菲薄，以自尊大；若夫地质界前贤更彼所景仰无已者，因而无时不视所学为在发蒙之时期，故日惟孜孜不倦，冀其有一日能在学术界得到相当地位，而不愧对友邦学者，藉为祖国吐气。

赵君出校才及六年，而对于科学上之贡献，殆为异外之成绩，不能仅求之于人。彼在《古生物志》[1]曾刊行专门论著四卷，为其呕心之作，类皆创见，固然其中亦有从人之处，因彼尝以热诚预备接受他人之意见与批评，若觉其是，总不惜舍己从人，牺牲成见也。至其所发表之短篇论文亦甚多，微言名论，完全从科学上着眼，实为最有价值之文品，其重要无论从何方面言，皆不在已往任何巨著之下。顾赵君并不以实习室工作为已足，亦且进而取大自然界为其研究场所，如考察地层层序之关系，细心化石之采集，已一再努力，视为至乐，直至赵君之末日，固仍未脱离其野外工作也。

赵君出外调查，无论巨细，必尽力推敲，穷其究竟，广东中山大学教授韩墨[2]先生，固曾一度与彼共同考察者，亦甚称赞赵君之能，认为上驷之材，不愧为一完善之地层学家，当余等讨论地层相互关系时，盖亦已令余认识其鉴别能力之伟大，而引为畏友，今日地层学古生物学在中国骎骎乎日有进境，赵君之力，盖不在少，因其踏步直进发愤忘食之精神，实足以鼓舞其同僚进取，使不敢稍懈，故有美满结果之可言耳。

赵君固又不特身负异才已也，亦且虚怀若谷，和蔼可亲，故其师长友朋及外来科学家，莫不乐与之交游，其与国外之名科学家，尤多书札往还，互相参证，裨益于彼者不浅。赵君与其青年同事，亦切磋观摩，弥觉友好，而亦为贤明教师热心科学之指导者，奖掖后进，不遗余力，故在校及毕业而在所中任事之学生，莫不敬服之也。

今赵君死矣，科学界顿失去一最诚恳最有望之同志，中国丧失其一未来之领导者，吾辈——其友若师——失去一益友而少他山之助，而尤以中国之损失为最大；良以此方在慢慢繁荣之科学之宫，正需各个大匠维持，而以此

---

1 即《中国古生物志》。——编者注

2 韩墨（Arnold Heim，1882—1965，亦译为哈安姆），曾在中山大学地质学执教。——编者注

种类领袖人才需要为尤迫切也。

噫！赵君死矣。赵君之死，为国家之损失，吾人应一致哀悼，以志不忘！

一九二九年十一月二十八日葛利普于北平

——原载《国立北京大学地质学会会刊》1930 年第 4 卷

Palaeontology in China

By Amadeus W. Grabau
Chief Palaeontologist of the Chinese Geological Survey and Professor of Palaeontology, National University, Peping

Fossils have been known to the Chinese as objects of curiosity for thousands of years. Who it was that first attributed medicinal potency to these objects that were found in the soil and weathered rock material, and at what period of Chinese history such belief led to active collecting and barter of these fossils, is unknown. It is a very ancient practice—so much is certain; and the exploitation of deposits of "dragon bones" especially for the teeth, is a recognized and legitimate pursuit, and the secret of the location of such deposits is handed down from generation to generation.

Most of the "dragon bones" and "dragon teeth", in North China at least, are either of late Tertiary (early Pliocene *i.e.* Pontian) age or belong to the early part of the Quaternary. The most abundant types of the larger vertebrates, whose teeth are especially valued, are horses (especially *Hipparion*) rhinoceroses of various genera, deer, and some others. These bone-fragments and teeth are even now purchasable at Chinese medicine shops in villages as well as in the larger cities, such as Shanghai, Tientsin, and Peking. It is from such sources, that the first collection of vertebrate fossils was obtained which was scientifically studied and described by the Munich Palaeontologist Dr. Max Schlosser.

But it is not only the remains of vertebrates that can thus be obtained. Fossil Brachiopoda, especially those of Devonian age, which are the best preserved and most abundant fossil in South China, are eagerly collected and sold to the native drug-stores as stone swallows or Shih-yen.[1]

1　A considerable series of such brachiopods (*Spirifer, Cyrtiopsis, Atrypa, Schizophoria, Yunnanellina, etc.*) has recently been purchased by my students in Peking, Shanghai and elsewhere, to supplement the collections made in the field, for a monograph on these fossils which is now passing through the press.

Weathered outcrops of Devonian and, sometimes, other rocks, from which fossils may easily be collected are well known to the natives of those districts in which they occur. They have indeed been considered as of sufficient interest to be mentioned in the histories of the districts, and the wise collector will first of all consult these local histories for any record of outcrops of fossiliferous rocks.

But we must not suppose that fossils have been regarded by all Chinese as miraculous or wonder-working objects of unknown origin. Chinese philosophers have not overlooked them and as early as the 12th century, correct notions of their origin and significance had been arrived at. Chu-Hsi wrote in 1200 A. D, "In high mountains there are shells. They probably occur in the rocks which are the soils of older days, and the shells once lived in the water. The low places became high, and the soft mud turned into hard rock." This, it must be noted, was two and a half centuries before the birth of Leonardo da Vinci to whom is commonly credited the first correct interpretation of the origin and meaning of fossil shells.

Although the ancients knew about fossils, and had correct notions regarding them, no attempt at a scientific study of these objects was made. These men were philosophers, not scientists. In China, philosophy is old, but science is young. Half a century covers the period during which Chinese fossils were studied scientifically, and less than a decade measures the period when Chinese themselves began the systematic study of their fossil faunas and floras.

Almost all of the Chinese fossils brought to Europe or America in the early days were collected by missionaries and other travellers from Chinese medicine shops. Some of these were described in Belgian, English, Italian and German literature by such men as de Koninck[1], Davidson[2], Woodward[3] and Crick[4], and the Italians, Martelli and Pellizzari, all of whom confined themselves to invertebrates. Smith Woodward studied some of the fishes and Schlosser described the first mammals. Systematic collecting did not begin until the year

1　Laurent-Guillaume De Koninck（1809—1887），比利时古生物学家。——编者注
2　Thomas Davidson（1817—1885），英国古生物学家。——编者注
3　Arthur Smith Woodward（1864—1944），英国古生物学家。——编者注
4　George Charles Crick（1856—1917），英国古生物学家。——编者注

1860 when Raphael Pumpelly, the American geologist made the first geological explorations in China. His collections, however, comprised chiefly plant remains, which were described by Dr. John Strong Newberry.

The epoch making systematic investigation in Chinese geology was that of Freiherr Ferdinand von Richthofen, during the years 1869 to 1871, followed by the publication of that great work, *China—Ergebnisse eigener Reisen und darauf gegründeter Studien* in 5 quarto volumes, several of which were issued after the death of the great explorer, together with two large atlasses of maps. This work has become a classic of Chinese geology and an inexhaustible mine of information. Two of these volumes are devoted to Palaeontology: Volume 4, chiefly prepared by the late Professor Emanuel Kayser of Marburg, with contributions by G. Lindström, Wilhelm Damas, Conrad Schwager, A. Schenk, and later emendations and revisions of parts by Fliegel; and Vol. V, the work of the late Prof. Fritz Frech of Breslau.

Following Richthofen came the expedition of the Hungarian count Belalt Szechenyü[1] in 1877–1880[2], the geological and palaeontological work of which was done by Ludwig von Loczy, who himself described many fossils in Vol. III of the report of that expedition. Here too, other collaborators contributed descriptions of selected groups, among them Lorenthy on the Foraminifera, Frech on the corals,—Schenk on the plants, Neumayer on the Quaternary and recent mollusks, and Koken on the vertebrates (Palaeontographica). Other expeditions which brought back fossils and found space for their description in their own or other publications, were those of Futterer[3], Obruschew[4], Sven Hedin from Tibet, and the Merzbacher[5] Tienshan expedition of 1907 to 1908[6]. Among the collaborators in the descriptions of the fossils obtained by

---

1 现在一般写为 Széchenyi Béla，匈牙利人名字姓在前，名在后。——编者注

2 *Reise des Grafen Bela Szechenyii in Ost-Asien.*（全称为 *Wissenschaftliche Ergebnisse der Reise des Grafen Béla Széchenyi in Ostasien*, 1877–1880。——编者注）

3 Karl Josef Futterer（1866—1906），德国地质学家、探险家。——编者注

4 Vladimir Afanasyevich Obruchev（1863—1956），俄国地质学家，曾于 1892 年在华北和西北地区考察。——编者注

5 Gottfried Merzbacher（1843—1926），德国地质学家，曾于 1902—1903 年在新疆天山地区考察。——编者注

6 Merzbacher 在天山地区的考察应为 1902—1903 年，并于 1905 年出版 *The Central Tian-Shan Mountains 1902—1903* 一书。——编者注

these expeditions were: Potonie, Schellwien, Gröber, Keidel, Krasser, Krenkel, and Leuchs.

Independent collections made by themselves or others were also described by Bergeron, Lorenz, and Monke, and by the Japanese palaeontologists, Yabe and Hayasaka. Finally Mansuy[1] and Cowper Reed described many fossils from Yunnan, which were collected by Deprat[2] and others, and by Coggin Brown, while Zeiller[3] described many coal plants from South China. It must not be forgotten that many of the fossils described in the *Palaeontologia Indica* came from the Tibetan border, if not actually from that territory.

The last two important foreign expeditions designed chiefly for the collection of fossils, were those of Bailey Willis and Eliot Blackwelder under the auspices of the Carnegie Institute of Washington, and the Central Asiatic expedition of the American Museum of Natural History, and Asia Magazine under Dr. Roy Chapman Andrews. The former devoted itself exclusively to invertebrates, which were subsequently described by Walcott, Weller and Girty, while the latter paid chief attention to vertebrates, though a rich invertebrate fauna of Permian age was collected, which has furnished the material for one of the final volumes now in press, the preparation of which has been entrusted to me. Unlike the collections of all previous expeditions, the invertebrate material on which this volume is based has been deposited in the Museum of the National Geological Survey in Peking by the Andrew's expedition.

The Chinese Geological Survey was organized in 1913, but active work did not start until 1916, as the first three years were required for the proper training of a staff of field geologists. From the beginning, palaeontology was regarded as an important branch of study that should be undertaken in connection with the work of this Survey. When the Survey was organized there were no adequate courses in geology at the higher institutions of learning in China and none in palaeontology. Three Chinese geologists, Dr. V. K.Ting, Dr. W. H. Wong, and Mr. H. T. Chang took charge of the new Survey. They had received their chief training, the first in

---

1 Henri Mansuy（1857—1937），法国古生物学家、考古学家。——编者注

2 Jacques Deprat（1880—1935），法国地质学家。——编者注

3 René Charles Zeiller（1847—1915），法国古生物学家。——编者注

Great Britain, the second in Belgium and France, and the third in Japan, but all had traveled more or less extensively. To train an adequate staff of field geologists was the first step to be taken, and these men, in 1913, organized the school of geology, in which for three years graduates of Chinese colleges and middle schools received an intensive training in the necessary subjects. The graduates of this school were appointed on the staff of the Survey. In this school Dr. V. K. Ting taught, among other subjects, the first course in palaeontology given in China.

In 1914 Dr. J. G. Andersson, the noted Swedish geologist, director of the Geological Survey of Sweden and geologist of Swedish Arctic and Antarctic Expeditions, was called to China as adviser to the Ministry of Mining. From the first Dr. Andersson kept up a close relation with the geological Survey though he was not an official member of it. With his aid and under his supervision, the Museum of the Survey was organized.

Dr. Andersson early recognized the importance of the rich deposits of Pliocene and early Quaternary vertebrates, which for so many years had been exploited for "dragon bones" and "dragon teeth". With the co-operation of the director of the Survey, Dr. V. K. Ting, and aided by a grant from a Swedish Committee organized for the purpose under the patronage of H. R. H the crown Prince of Sweden, Dr. Andersson began to collect these vertebrate remains, which were sent to Sweden for study and description by specialists under the general direction of Dr. Carl Wiman of the University of Uppsala.

The first extensive collection of Chinese invertebrate fossils made by a Chinese, was that made by Dr. V. K. Ting in 1914 in Yunnan province. This collection was sent to the United States for identification, but some years later it was returned only partially identified, and most of it has since been monographed here. Dr. Ting had conceived the plan of publishing a series of palaeontological monographs patterned in a general way on the *Palaeontologia Indica*, and to be issued under the general title of *Palaeontologia Sinica*. This was divided into four series as follows: series A, fossil plants; series B, fossil invertebrates; series C, fossil vertebrates; and series D, ancient Man.

When I was called to China in 1920 to develop the work in palaeontology and to train Chinese students in that science, I was given charge of series B, and in that series the first fascicle of the *Palaeontologia Sinica* appeared on April 28, 1922. This dealt with the Ordovician Fossils from North China, though previously two short papers on other Chinese fossils had appeared in Bulletin 2 of the Geological Survey. This was the beginning of palaeontological publications in China, and since that time 37 fascicles of the *Palaeontologia Sinica* have appeared, with a total of 3604 quarto pages, 796 text figures, 339 plates, 107 tables and 1 chart. A more detailed analysis of these monographs will be given later.

Other foreign palaeontologists have from time to time been called to China to carry on special work. First among these was Dr. Theo Halle[1] of Stockholm, Sweden, who made extensive collections of fossil plants though many of these were lost by the foundering of the ship on which they were being carried to Sweden. Dr. E. Norin, a Swedish geologist, made other extensive collections of plants, and still others were made by Dr. Andersson, and by a field party under my direction. Many of these have now been described by Dr. Halle in a completed volume of the *Paleontologia Sinica* Ser. A. and others are being studied, Dr. Norin also made extensive collections of invertebrates from the Carboniferous rocks of Shansi and these are now being studied in the Palaeontological Laboratory of the Survey, and several groups have already been monographed. For several years Dr. Otto Zdansky made collections of vertebrates in North and Central China, and since his return to Sweden he and other European palaeontologists have published a number of monographs on this material in the *Palaeontologia Sinica*.

In was during his exploration of the bone caverns of Choukoutien near Peking, first discovered by Dr. Andersson, that the teeth of a hominid, the now famous Peking man *Sinanthropus pekinensis* Black and Zdansky, were found. This led to an extensive exploration of these caverns, carried on jointly by the Geological Survey and the Rockefeller Foundation, and this has already brought forth many interesting remains of this remarkable early pre-human. To undertake a part of this work of exploration, another Swedish palaeontologist, Dr. Birger Bohlin, was called to Peking, and it was he who discovered the tooth that eventually was made the holotype of the new genus and species *Sinanthropus*

1　应为 Thore Halle（1884—1964）。——编者注

pekinensis. Two Chinese palaeontologists Dr. Young, and Mr. Pei, and several Chinese geologists have also been active in the exploration of these deposits. To carry on the work of preparing and studying these fossils a Ceno-Psychozoic Laboratory of the Survey was established and placed under the charge of Dr. Davidson Black who is describing the hominid material from these deposits, while the work on other fossil vertebrates will be directed by the eminent French palaeontologist Dr. Teilhard de Chardin. In this laboratory active work in the preparation of the vertebrate material is now going on, and this, with the large new palaeontological laboratory of the Survey on Ping Ma Sze, the center of work on Palaeozoic fossils, and the affiliated Peking Laboratory of Natural History on Kaka Hutung, where Tertiary and Recent mollusca are being studied, has concentrated the work of Palaeozoology in the Survey's hands in Peking. It is hoped that before long a part of the work in palaeobotany may be carried on here as well.

In this connection must be mentioned the recent establishment of a palaeontological department at Sun Yat Sen University in Canton, and the coming to China of a famous German palaeontologist, the late Professor Dr. Otto Jaeckel[1], of Greifswalf[2]. Dr. Jaeckel had labored successfully in his new field for nearly a year, when coming to Peking for the meeting of the Geological Society of China in February 1929, he contracted pneumonia, to which he succumbed, not however, before he had taken a most active part in the meetings of this Society. His death is a great loss to science and a greater one to Chinese palaeontology. The impetus, however, which his coming has given to palaeontology in the South, bids fair to have lasting results, and we may look to the time when another center of palaeontological research will be firmly established in South China. Finally in Central China, palaeontology is becoming an important part of the research work carried on under the auspices of the National Research Institute.

Nor may we forget the trail-breaking work of Peres Licent and Teilhard de Chardin in the palaeontology of Palaeolithic man, and above all the great work begun by Dr. J. G. Andersson in the collection and study of Aeneolithic man in

1 应为 Otto Jaekel（1863—1929），德国古生物学家、地质学家。——编者注
2 应为 Greifswald，Otto Jaekel 来华前为格赖夫斯瓦尔德大学（University of Greifswald）教授。——编者注

China, which has already led to the publication of a number of important monographs in series D, of the *Palaeontologia Sinica*, on the implements and artifacts by Dr. Andersson and others, and the human skeletal remains by Dr. Black.

It is thus apparent that palaeontological research, unknown in China ten years ago, has now become firmly established. Meanwhile active collecting is constantly going on in many parts of China, by the geologists and palaeontologists of the National Survey, as well as the various provincial surveys and a number of the institutions of higher learning. Nevertheless the field is vast, and its cultivation has only begun, and many decades must elapse before the palaeontological treasures of China are unearthed and will become adequately known.

The solidarity of Chinese palaeontology has further been enhanced by the founding of the Palaeontological Society of China, as a branch of the Geological Society of China, founded some seven or eight years ago; there are at present about 25 active Chinese-trained palaeontologists,[1] several of these having carried on graduate work in vertebrate palaeontology in Europe and the United States. These with the foreign palaeontologists, now resident in China, form the membership of this new society.

The training of independent research workers in palaeontology, requires, of course, much more than is given in the palaeontological courses included in the curriculum of the geological departments of our universities. This is not always understood by the educational authorities and sometimes not by the students themselves. All too frequently men are called upon to teach the subject of palaeontology which they had studied in courses during their undergraduate years. These courses should be considered merely as a preparation for palaeontological research, and no one should be required to teach the subject who has not spent a number of years in active research work in palaeontology.

It is an unfortunate fact that we have at present in China no real university, that is, an institution which gives not only lecture and laboratory courses in

---

1　The most brilliant of these, Mr. Y. T. Chao, has recently been murdered by bandits while engaged in field work in Yunnan. His untimely death is an irreparable loss to Chinese science and especially Chinese Palaeontology. He is deeply mourned by all his colleagues and all who knew him.

science, but also provides facilities for research for post graduate students. The Geological Survey, recognizing this lack of research facilities has organized what is essentially a department for graduate work in geology and palaeontology. The most promising undergraduate students are enrolled for two years of graduate study, and training in research under adequate supervision. If during those years, these students prove themselves competent and capable of developing the power of independent judgement, they are taken on as junior members of the staff and given the opportunity to carry on independent research work in co-operation with the chiefs of the departments.

This is properly the function of the universities and the next step in the higher education in China, should be the establishment of such graduate schools at institutions of higher learning. In geology and palaeontology, and in the sciences which are prerequisite to a proper preparation, China has probably a sufficient number of men to equip one university. It is hardly necessary to emphasize, that no one who is not himself an investigator and no one who has not already produced scientific work of merit, is fitted to be on a staff of a university where the training in research work is a primary object. The present tendency in China to scatter its able men among a number of institutions which call themselves universities, though not one has a real right to that name, is to be deplored. No real progress will be made in China in higher education and the training of research workers until China's first true university is established. Such a university must have an adequate staff qualified to direct and supervise the research work of graduate students. Nor must students be permitted to think that they are qualified to carry on independent research work after they have completed their four years of college training. Such students are all too frequently called to organize courses and give instruction in their subjects, at institutions ambitious to broaden the scope of their curriculum. Whatever may be said of other subjects, certainly in science, no individual is fitted to direct the activities of a department, who has only pursued the undergraduate work in his subjects.

During the nine years that I have been training Chinese students in palaeontology, I have produced two men whom I feel are adequately prepared to organize palaeontological departments elsewhere in China, and these two students have not only studied with me during their undergraduate years, but since their graduation have for many years carried on research work in the laboratories of the Geological Survey.[1] I have many other students who have completed their undergraduate work and who give promise of becoming leaders in Chinese palaeontology, provided they are given the opportunity to carry on research work for at least a number of years in connection with our unofficial graduate school at the Survey. But, if these men are called to other Chinese institutions to organize departments of geology and palaeontology, and become responsible heads of such departments before they have become thoroughly grounded as research workers, and have developed a research spirit which will remain with them through life, I for one refuse to be held responsible for the instruction and training given in these sciences at those institutions. I should deplore the fact that promising young men are ruined by being burdened with responsibilities which they are as yet unable to shoulder, and by being accorded scientific recognition which they have not earned. This is all the more to be regretted when these young men are well fitted to earn such recognition by the successful accomplishment of scientific labors, were they given the opportunity to undertake them. I am prepared to assert that China has many young men, capable of becoming eminent as scientists, but I am equally prepared to say that the great mass of these young men will never achieve eminence, if responsibility and recognition is thrust upon them prematurely. Nor do I think that the present practice of sending numbers of newly fledged graduates to America or Europe for graduate training in science, will bring results commensurate with the expenditure involved. Indeed I bold that too many promising young men are spoiled by premature contact with western educational institutions. No one should be sent abroad for scientific study, who has not already become well grounded in his particular subject, and has pursued it to the full extent possible here in China. Only such students should be sent abroad who are ready to profit by several years of contact with specialists in the subject which they have chosen. In other words, no student who has merely acquired knowledge and has not yet

---

1　One of these was Mr. Chao whose death has been noted above.

made a beginning in research work is fitted to profit by contact with foreign scientific men.

Again it is not the institution which the student should select when prepared to go abroad, but the men who can carry him forward in his own science. It is the men pre-eminent in the science chosen by the student, that should attract him, not the institution, no matter what its general reputation. The practice of the Geological Survey of sending abroad only proved men should be adopted by other institutions, and no institution of learning should be encouraged to send men abroad, unless it has a graduate department of its own in the subjects in which it trains its candidates for foreign study. And finally no student should be sent abroad unless he has a plan of study and is prepared to devote strenuous years to his particular subject.

Returning to the science of palaeontology, we may note that the intensive training which is judged necessary for the development of palaeontologists, has already produced very gratifying results. I have previously referred to the *Palaeontologia Sinica*, the first number of which appeared in 1922. Since then thirty-seven fascicles of this work have appeared, of which 13 fall under series B, "Fossil Invertebrates of China," with 5 others in course of publication. The total number of Chinese species and varieties of invertebrates described in these numbers is 526 of which 301 are new. If we add the number of species in the five fascicles now in course of publication which is 387 of which 295 are new, the total number of invertebrates described in these fascicles will be 913, of which 546 are new. The total number of invertebrate fossils previously described from China by foreign palaeontologists in other publications is 2,294 of which 651 are new.[1]

The results of less than ten years of palaeontological work in China, compares thus most favorably with that done by foreigners on China fossils during the last sixty years. Of the 13 fascicles of series B of the *Palaeontologia Sinica* 8 or about 62 per cent are the work of Chinese palaeontologists trained in China, and this number will be greatly increased when the monographs now in course of preparation are published. So far, most of the monographs in the other three

---

1 These data were kindly collected for me by Mr. Y. S. Chi, Secretary of the Palaeontological Society of China.

series have been written by foreigners, who were entrusted with this work by the responsible authorities. But a beginning has been made in the field of vertebrates where several important monographs have been published by Chinese palaeontologists, though only one of these has appeared in the *Palaeontologia Sinica*. Other are, however, in preparation, and with the establishment by the Survey of the vertebrate laboratory and the extensive collection of Cenozoic and Psychozoic vertebrate remains, including those of primitive man, there has come a new impetus for work in vertebrate palaeontology in China.

When it is realized that the field is of vast extent and the study of fossils in China has only been begun, it will be seen that we are just entering upon an era of fruitful palaeontologic work in China, and there is opportunity for many generations of adequately trained Chinese palaeontologists.

But the description of the Chinese fossils is only the beginning of the work in palaeobiology which must be undertaken in the future. As the fossil fauna of each geological system becomes fully known, its relation to the faunas of the same system in other parts of the world must be made the subject of intensive study. Foremost among the problems to be considered, is that of the center of evolution and migration of the faunas. We know now that throughout geological time most, if not all of the great oceans which still exist, were the centers of evolution of marine organisms. As the opportunity was given for the ocean waters to enter the geosynclines and epeiric seas of the continents, in the sediments of which the only remains of marine organisms known to us are preserved, each ocean sent its respective quota. Sometimes the fauna of one ocean basin predominated and at others that of another, and sometimes there was a comingling from two or more centers of distribution. The problem then becomes one of evaluation of the separate faunas. This can only be accomplished if the specific determination is a very critical one. No analysis of morphological characters that does not take into consideration the ontogeny of the individual, can serve this purpose. Often a complete restudy of the faunas of other regions becomes necessary, when the original characterization is based on superficial or adult characters. Nor must we leave out of consideration the evidence for palaeogeography of the period, furnished by stratigraphical studies, for this must give us a primary clue to the probable connection of the inland water-bodies that have furnished our fauna, to the great oceans of the period. The highest Palaeozoic or Permian faunas of China may serve as an illustration.

The original Permian system was differentiated in northern Europe, where it has long been known as the Dyas, because of its two-fold development. This is best known from Germany, where the lower part is an almost barren sandstone, the Rothliegendes, while the upper part is a limestone series, the so-called Zechstein which carries marine fossils in its lower part, while its upper portion includes the great salt and potash deposits which have made the formation economically as well as scientifically famous. In England red continental sandstone predominates, the limestone having a subordinate development. In the eastern end of the basin, along the western front of the Ural mountains in Russia, the series shows a more diversified development and it is from this region, the government of Perm, that the name Permian is derived. The series begins with a limestone which rests with a disconformity and hiatus upon Middle Carboniferous beds. The existence of this hiatus was not recognized by the early investigators and hence they concluded that these beds represented Upper Carboniferous. They, however, contain a fauna, which is almost wholly new to the region, although in the beginning, some form indigenous to the basin were mingled with the new element. Nothing was known of the origin of these new biological elements, until the corresponding deposits of China were studied, when it became apparent that this fauna was an invasion from the east, and represented late palaeozoic organisms of the Indo-Pacific realm, whereas those of the north European basin were at home in the Boreal realm, whence they sent migrants, not only into the Russo-German basin, but also into central North America. The pathway of this first great invasion of the Permian fauna from the east, has now been sufficiently traced for us to realize that it was one of wide extent. Not only did it submerge the Ural barrier, which up to that time had efficiently separated the Russian basin with its Boreal waters from the Asiatic region, but it also invaded the Mediterranean region of Europe, either along a pathway now occupied in part by the Himalayan mountains or through the Nan-Shan geosyncline. On the other side of the Pacific, these waters entered Texas and New Mexico, but did not join the interior seas which covered portions of the Rocky Mountain states and which were an extension of the Boreal realm of the time.

After this first wide-spread invasion by the Indo-Pacific seas, the Ural barrier again became influential in re-establishing the isolation of the Russian Basin. In the strata which succeed the Schwagerina limestone (the record of this first invasion) the Boreal fauna again becomes prominent. But the descendants of the migrants of the invaders from the east still lingered on, though in a diminished number of species, and their remains were buried with those of the indigenous fauna in the Artinskian sediments which follow upon the Schwagerina limestone since the latter had been erroneously thought to represent a closing Carboniferous formation, and since the Artinskian enclosed many of the same species, as well as those typical of the Boreal Permian, it was regarded as marking the transition from the Carboniferous to the Permian, and was commonly referred to as Permo-Carboniferous. Now that we know that the older formation is not Upper Carboniferous but represents the invasion of the early Permian fauna from the east, we recognize that the Artinskian is not Permo-Carboniferous, but marks the re-establishment of the reign of the Boreal fauna in the Middle Permian deposits of the Ural region. This Boreal fauna was thenceforth confined by the Ural barrier to the Russian basin and its extensions, a fact clearly indicated by the character of the sediments. The Indo-Pacific sea on the other hand, continued to cover various parts of China and hence we find in the higher Permian beds of China these Pacific faunas represented in their essential purity. It is the study of the Chinese Permian that has enabled us for the first time to differentiate and evaluate this distinctive fauna of one of the largest of the Permian oceanic basins.

The problem of the origin of faunal elements which make their appearance suddenly in the strata in many regions of the earth also calls for solution by the palaeontologist interested in marine faunas. A pertinent example is furnished by the class of graptolites, those ancient representatives of the modern hydrozoa, which make their appearance abruptly in the basal Ordovician strata, wherever these are known to occur. Since these organisms, at their appearance, already show a high degree of development, it is evident that they must have undergone evolution during a long period of time, preceding that in which they first appear, and that this abrupt appearance must be due to wide dispersion at the opening of the Ordovician period, from a region hitherto isolated, in which they had undergone their development. Moreover, the period in which such development took place must have been during Cambrian time, which is the period immediately preceding the Ordovician, unless we can show that there is a long time interval between the Cambrian and Ordovician deposits, which is nowhere represented by fossiliferous sediments. But if these organisms developed during the Cambrian period, their remains should be found in Cambrian strata, which is not the case. Hence the only conclusion that we can draw is that the Cambrian strata of the world as we know them, were deposited in basins from which the graptolites were excluded by efficient land or climatic barriers, for since these organisms led a floating life, nothing but such a barrier would prevent their entering the Cambrian seas, wherever these were open to invasion. Of the known basins in which the Cambrian strata of the world were deposited, none could be the home of the graptolites in Cambrian time, since their remains are absent from the Cambrian strata, though they commonly occur in the overlying Ordovician beds. Therefore, we must search for a basin from which Cambrian strata are still unknown and the only one that satisfies the requirements is that of the western Pacific. No Cambrian strata referable to the western Pacific are at present known, but it is known that the barrier which separated the Pacific from the Indian oceans throughout Cambrian time, disappeared at the opening of the Ordovician. We are therefore led to believe that it was in the Pacific basin that graptolites underwent their early development, and that it was the opening of the barrier between the Pacific and the Indian oceans, that permitted them to enter the latter basin and, from it, the inland seas, which then extended from China to Europe on the one hand, and to the Boreal region and the interior of North America on the other. Cambrian strata of the Pacific realm should contain not only a Cambrian fauna but also the remains of the ancestral graptolites, as well as those of many other groups of organisms, which make their abrupt appearance in Ordovician time, and whose genetic relations to the Cambrian types, have heretofore eluded investigation.

Still another problem of biological significance is that of the differential evolution of parts of a once wide-spread fauna, which have become separated by rising continental barriers. In the early Tertiary, a continuous water body extended from the Mediterranean Sea of Europe across Egypt, and the site of the present Red Sea to the Indian, and thence to the Pacific basins. Free migration was possible, and as a result there was more or less widespread uniformity of faunas, though of course local faunal groups existed everywhere.

But when the great world revolution which produced the Alps, the Himalayas and many other mountain ranges began, a barrier arose, separating the Mediterranean from the Indian basin. From that time onward, the basins have remained essentially distinct, and the dismembered faunas underwent independent development.

This is best illustrated by the molluscan elements of the fauna. If we except circumpolar types, which could venture both into the Atlantic and the Pacific, we find that the molluscan fauna, though related in the two oceans is nevertheless distinct, having not a single species in common, though most of the genera are represented in either basin. To understand this independent evolution, we must go back to the Miocene or Middle Tertiary era, where this differentiation of a former homogeneous fauna, first became manifest. It has been stated that in the Miocene and other Pacific regions, about 75 per cent of the Mollusca are still living, whereas the Miocene of Europe scarcely contains 25 per cent of living organisms. It is probable that the investigation into the ontogeny of the Pacific forms, both fossil and recent, will greatly reduce this number of species.

But such study to be of any value, and if it is to reveal the manner of differentiation of faunas, must be a comprehensive study of the entire Indo-Pacific Molluscan fauna from the Miocene to the present, and a comparison of the Miocene species with their Oligocene and Eocene ancestors as well. And since the personal equation invariably enters into the question of such a study, the entire problem must be investigated under the direction of a single individual, though the number of his aids need not be limited. But whoever undertakes such a comprehensive study, must be qualified for it by prolonged training in the ontogenic methods of investigations of the molluscan shells. Such work in order to be successful must be concentrated in a single scientific center, where vast collections of both recent and Tertiary Mollusca of the Indian and Pacific realms can be brought together. Here is an opportunity for China to undertake the investigation of a world problem, and one that will lead to the positive solution of questions concerning the method and perhaps the causes of evolution of organic forms.

Palaeontology has a great future in China. Not only because there is so much descriptive work of new species to be undertaken, but because the accomplishment of this work will lead to the solution of problems of migration, and dispersion and of evolution, not only of the older but of the modern types of many of the great biological realms.

——原载 *Lingnan Science Journal*, 1929, vol. 8

亚细亚和人类的进化

葛利普著 乔峰译[1]

　　人的摇篮一般皆公认在亚细亚洲，虽然究竟发源于那一处地方，至今尚是争辩的问题。如果有人能在他洲找到人类起源的中心地，最能当选的，当推亚非利加。但是如果我们不承认人种起于多源，而黑人式人种起于亚非利加，别的形式起于他洲，我们必须承认，亚细亚为人的出生地，实在最有理由的。不特因它的物理性质显著的发达，近年又发见了在那洲化石的古人了。

　　我们应当首先把亚洲后期发达的情形查考大要，看它和人族起源问题关系如何，后再考察原人居住在亚洲的进步。我们必须从"哺乳动物纪"讲起，此时即为第三纪或新生代。虽然人根本属于第四纪即心理时代，但他的根柢却深埋在第三纪岩石中，我们须从那里看他的发端。第三纪又分作许多时期，每期的时间计算起来约数十万年，最后一期到终了时，约有百万年，心理时代幕开处，人已出现于世界的舞台上，人类进化的剧已在开演了。这几期较近代的地史，照次序排列在下面，新者在前，下去最古：

第四纪即人的心理时代 { 全新世即近世，新石器人及近代人 / 最新世即冰河期，古石器人 / 多新世下第四纪，原始石器人

第三纪即哺乳动物新生代 { 鲜新世 } 中新世 } 新第三纪 / 喜马拉雅诞生期 / 渐新世 } 始新世 } 古第三纪 / 古新世

　　古第三纪和新第三纪间的界线是在渐新世和中新世之间[2]，是极分明的，

1　译自 A. W. Grabau, "Asia and the Evolution of Man," *The China Journal*, 1930, Vol. XII, No. 3, pp. 152-163. 乔峰即周建人。周建人（1888—1984）字乔峰，浙江绍兴人，生物学家、社会活动家，鲁迅三弟。——编者注
2　文中地质年代术语与现在已略有不同，请读者注意。——编者注

因为在古第三纪之末，世界上大起变化，结果，形成巨大的山脉，呈今日欧尔亚细亚洲的特性。我们可以举出亚尔泼司山[1]，喀尔巴阡山，高加索山，我们的研究上最重要的，是喜马拉雅山。

喜马拉雅山兴起之前，这地位本是一处沉积地。换一句话，那里本是大向斜层（geosyncline），所积的堆积物从南部古高地而来，其地为今日北印度的巨大印度恒河平原。大向斜层中间本有海股伸入，在早期长落不定，到始新世末，海水完全退去，其地遂成大平原，和今日的中国平原相似。摹想起来，这宗平原广阔平坦，有河流缓缓流行其间，向辽远的出路出去，以达于海。在南部，低地高起来，多少生起林树，有湿风从南部的海吹来，扫过平原。平原的性质如何，我们虽然没有直接证据，在现今的山脉未有之前怎样，但是我们可以推测，当很有些树木的，低地大概也如此，此即那时候的西藏，喜马拉雅大向斜层北部的边地。此等地方，那时没有障阻，故最大的兽，无角犀牛，即俾路芝兽（*Baluchitherium*），在渐新纪自由游行全部亚细亚，它的骨头曾在俾路芝坦（Baluchistan）发见，这地方现在已被喜马拉雅山脉把它和亚细亚别部隔开。俾路芝兽生活时，如果已有此等山脉存在，那兽便只能生于南亚细亚，因为中级的高的山，它便不能越过。但是俾路芝兽并不限于南亚细亚。它广远的游行，遍巴尔亚细亚（Pal-Asia）全部，巴尔亚细亚是第三纪及中生纪以前的亚细亚的名称。它的骨殖见于西部西伯利亚的托波尔河（Tobal R.），蒙古的沙石层，甚至西部中国的许多地方也有的看见。它是渐新世森林及平原中的巨兽，食树的枝叶，所游行的地方必有树木的。此种动物所以不入欧洲，由于有不能越过的海的障碍，古代吐加（Turgai）海峡，将巴尔亚细亚和巴尔欧罗巴（Pal-Europe）分开。

俾路芝兽的同时代的动物中，更有早第三纪的类人猿，那宗高等专化的亚人（sub-human），曾经被认为人的祖先。这等巨大类人猿，和他们的近代代表相似，也是旁出的进化系的个员，也是树林的居住者，大抵家住树中，不在平地。古代树林不特供给他们以藏匿之所，及居住地，又给他们食物及生活上的各种必需。这是用不着多大想象的，此等动物那时候居住山林中，搜罗食物，和别的动物极少妨害，和繁育家族，食物丰富是无疑的，又因为离开地面，住在枝上，危险也少。林树生长既密，他们可以在树枝上游行，他们能用强有力的手臂，从这枝荡到他枝，并用能把握的足的帮助，须要降

1　今作阿尔卑斯山。——编者注

地上来的时候极少，他们的进步当然极缓慢，他们步行起来是不大会走的。

但有不和谐的力，进入此广大的快乐的伊甸园，将地层折叠，叠起巨大喜马拉雅山脉，本来只是大平原，成喜马拉雅的大向斜面的表面。因不可抵抗的压缩力，山脉慢慢高起来，使南北两面渐渐接近，地层曲折生皱，并且破碎，彼此滑过，逐渐高耸。在进行中的时候，先后发生震动，地壳震颤，居住的生物大受惊恐，因为他们从不经验过这样大的扰动的。运动后来缓慢了，遂新生喜马拉雅山脉，于是将乐园，即巴尔亚细亚切断，并且分俾路芝兽及大类人猿的巡猎地为南北两部，北部包括西藏及中国本部，有蒙古及西伯利亚在其北，南部留存着印度半岛，没有大变动，只将它和亚细亚别部分开了。这除却，现在有古高地，本为沉淀物的贮积所，折叠入喜马拉雅山中，遂沉下复成为沉淀物的填积所，沉淀物由北部，新成的喜马拉雅山来的新河流带去。换一句话，它便变为后第三纪的希伐立克（Sivalik）的大向斜面及今日的印度恒河的大向斜面。

俾路芝兽经此变化不复生存，但留剩在印度的类人猿生活方法没有大更变的必要。真的，反给他们刺激，使他们继续树上生活，因为这地方，使山林长发的状况比从前更佳了。原因出于从南部温和的海吹来的挟带水蒸气的风扫过此地，触新成的喜马拉雅山的障碍，被迫的上升，遂将潮湿积下。所以喜马拉雅的南面变为湿润之地，有利于森林的滋长，足供生存者的后裔的居住，而且也少有理由用得着改变古代的生活方法。

但在大山之北，情形便不这样，林中的居住者一经和南方快乐的弟兄们隔离，命运即陷于悲惨。同是这携带湿气到南坡，使林树滋长的风，吹过喜马拉雅山，经过地面，湿气反被吸收。效果所至，使在风下之处，皆成荒漠。风吹来的一方面常常是湿润的。当风被迫的升入较寒冷的高处，一经密缩，从前所含的湿气不能含留。但经过障碍，重复下降，风遂扩散、变温，湿气的容量随即增加。他们很像从蒙古来的冬风，冷风吹过北部中国，将地面的湿气吸收，使尘土飞扬。

西藏从前是气候均一的地方，在中新世才干燥起来，它的地位就在新生的喜马拉雅的北部。它那时候还没有今日的高，最后的升高是在较迟的时代。但那时候它便受干风的苦，今日也是这样。森林毁坏的第一结果，是由于水层低落，直到多数的树，根支不能吸收水份。这等树于是死掉了，森林凋败，林居者家宅被毁，食物也减少了。到隐居的树木既已消灭，此等动物不得不走到旷地上来，于是面到许多新奇的状况。生活树林中的时候，从这树枝荡那树枝去是手足并用的，现在被迫的只好用足走，并且他们的本来适用于握物的足，渐渐发生变化。他们又不能如别的四足兽的走，因为他们的手臂长久已作别用，所以不能不直立起来，只用两足行动了。行走艰难，用断枝来拄。这是第一次用杖。

这是无疑的，所有孱弱的，缺乏克服新环境的困难，和缺少适应能力的个体，尽皆灭亡。他们的被扫除，因在唤起潜伏能力的时候，是较不起反应的，这是自然选择，孱弱既灭亡，生存得以前进，以适合于所面到的新环境。

这是习知的事实，气候干燥的地方温度也就大有变化，因为空气稀薄，水蒸气缺乏，昼间积热放散，夜间遂严寒。

我们可以想象，此等生物本惯生于相等的气候，受森林的保护的，现在改在冷暖剧变的温度下，于是弱者又受淘汰，较强的躲避在悬崖之下或穴洞中，直到大发见后，方有居住在旷野可能。

在那快乐的日子，此等先人（pre-human）尚居住树林中，他们最怕的是森林火烧，树被电击后，森林常常会起火，但到森林不作住所，火烧和他们没有直接关系。真的，炽燃的枯树成景子，此等生物得从远处安静的眺望，甚至使勇敢者能近前去研究了。一定的走近火烧的树林，火的喜悦的热即刻使这脆弱的生物受打击，并且这种发见给灵敏者从新的方面去观察火。但这只能使怯弱的不敢近前，群中最勇敢的会得拿了烧着的树干回来，使家族坐着喫嚇。他们即刻觉到喜悦的和暖，和看出孤单的火干的比较无害，他们会的急速加上树枝，使垂灭的火复燃。蓬帐中最初的生火是由这样差不多的情形来的。火神（Prometheus）并不被缚住。原始人民将从前的敌人为奴。现在妒忌的要守住这火了，于是绕火集合而成部落（clan）。局外人是不欢迎的，势力所及又有限制，为了求更能抵制外来者，部落遂组织起来，领袖选出来，这是原始集团的开端。

因他们每日需要粮食和食物供给的不绝减少，先人部落的雄者遂广远的游走，经过荒废的平原和草地，各各带着行杖，偶然当作战棒，当别部落的雄者来竞争食物的时候。因为食物供给减少，他们被迫的变成仇敌和竞争者，这地方在从前食品丰富时，他们的祖先本来亲和的住着的。自从那时候以来，竞争地上所产的食物，一直是部落间、集团间、国民间的战争的主要原因。

干燥起来给土地以破坏植物以外的别种效果。泥土失去植物的根的连络，变为风的游戏品，每季加甚。各处露出的地方，泥土吹去，下面的岩石露出来。温度昼夜的变迁，特别在沙漠地方，使岩石破碎，渐渐的，坚硬的岩层，破裂变成碎片。在这宗地面上，行走着劳苦的搜寻食品者，疲乏地将身靠在行杖上走，那杖因多用，已经劈开，岩石破片遂能嵌进去。现在如有敌出现，这小心的游行者便预备战争。他们用杖猛击，冲刺，及觑视，反复数次，这样的冲击出血，攻击者和被攻击者皆惊愕。敌既败走，检视染血迹的杖，看出有燧石嵌在裂开的地方。新的武器于是发明，原始石器，即原始的自然石片，于是开始应用了。并且原始的战士即刻学知此等破片可以改形，使愈适于装在杖上，于是获得了新的价值。

在这时候，气候加励，使原人向北游行，过不可居住的西藏草原，那里有广大的肥谷存在，当形成喜马拉雅山时起重压的时期造成的。这便是新疆的塔里木（Tarim）盆地，今日已成沙漠；但在晚第三纪及早第四纪是富于川流的，从有层累的黏土堆积物可以看出来，这层累今在沙漠之沙的下层。这地方便是晚第三纪人找到的新的快适的家，若将层累发掘，他的骨当保存在许多古黏土层中无疑。

原人从中国西部新发现的家出发，游行他处，他从一条路东行，此路从前有许多鲜新世动物走过的，后来到商业时期，为旅行队的通道。原人早在多新世和许多动物一道到中国东部，这事实，我们从离北平西南二十五英里，保存在周口店穴洞中的头骨、颚骨，及牙齿知道的。

他又向西移徙，过欧亚细亚洲，那时候北海尚没有，他便走到今日称为英格兰的地方，他的遗骨见于塞塞克斯（Sussex）的匹忒唐（Piltdown）地方。英格兰的匹忒唐人[1]（*Eoanthropus dawsoni*），和中国北部的北平人[2]（*Sinanthropus pekinensis*）同时代，而且亲族极近。两者都从更原始的祖先传来，他显然最初住于塔里木盆地。或者爪哇的猿人（*Pithecanthropus erectus*），也是这祖先的后裔，如果如此，他必早先便向东及南游行，及南部的契丹细亚（Cathaysia），物理现象平易，生活容易，没有努力的诱因，所以也没有使更发达的刺激。所以猿人虽和匹忒唐人及北平人同时，却比他们原始的，因为他们的环境有刺激的效能，结果加速的积极进化。

1　现一般译为皮尔当人或皮尔唐人，已被证实为一个骗局。——编者注

2　即出土于周口店的北京人，因当时首都南迁，故称北平人。1927 年步达生最先命名为"中国猿人北京种"（*Sinanthropus pekinensis*），后葛利普将之命名为通俗的北京人，被广为采纳。现在北京人的学名已调整为"北京直立人"（*Homoerectus pekinensis*）。——编者注

当然，爪哇猿人也可以认为从猩猩的祖先独立进化来的，那祖先留在大喜马拉雅障碍物的南部，进化向人形很缓慢，环境的刺激轻微。无论如何，爪哇猿人大概不能认为近代人的祖先的直接系统。

但是原人并不是在亚细亚洲上继续进化，没有阻碍的，因为地球的物理进化的进行上有着大变化，这起来在中第四世纪的开始，即洪积纪（Pleistocene）。这或因地轴变易，使北极变为南格林兰，或者因别的缘故，皆可以解释，事实上总之是以这地方为中心积起冰来，向四周漫延开去。在亚细亚旁边，这极方冰帽仅到北面交界处，但别一边，在欧罗巴及亚美利加，盖过大部分今日的陆地，有几处延扩到第 40 纬度，或者更低。斯干狄那维亚，不列颠大部分，北部日尔曼，及许多别的地方皆埋葬在冰下面，洪积纪人遂被驱逐到南部欧罗巴无冰之处。在较冷的时候，他找穴洞躲避，但在温和的间冰期，他到河边平地来，居住在旷野里。

洪积纪人即我们称为"古石器时代"的人。他学会弄碎自然石，造为更适用的器具，他们在学术上称为装饰，并且慢慢的发达起各种有用的用具。他能如此做，大概由于发现，当他向北移徙，塔里木洼地，由河道从昆仑山带来的石子都是浑圆的，如溪中的砾石，大抵这样，大概这样的石子掷在大漂石上，偶成破片，这遂使他知道可以拷出需用的器具的。经过许多试验，他大概学会，拣出可以斫成和弄薄，以做成最好的用具的石子。

古石器时代的文化时期的分类，以用具为基础，分为许多类，名目皆取自南部欧罗巴的地方名，因这时期首先发见于那里。至于人，我们也有如他们文化的知道的多，他们的骨骼在南欧许多地方发见了许多。有显然不同的二式，宁德退尔人（Neanderthal man），生活在前古石器时代，和克罗马农人（Cro-magnon man），生活在晚古石器时代，和别的形式（葛利马第 Gremaldi，勃隆 Brunn，桑斯拉特 Chancelade，及弗福葛莱纳 Furfoot-Grenelle 等），宁德退尔人和近代人不同，称为 *Homo neanderthalensis*，但克罗马农人是我们相像的，也是 *Home sapiens*。

宁德退尔人在南欧的穴洞中及古葬地遗物极多，东部亚细亚却极鲜少。在鄂尔多斯（Ordos）黄土堆积物基部只看到一个牙齿，克罗马农人不见过一点亚细亚的遗物。

见于欧罗巴的古石器人的文化时期，没有证据可以证明有继续发达起来的次序。反而他们指出的是每次从远处的中心地移来的，每一群移来，带有新的文化，这在祖先的家发达起来的。各方面指示出来，中央亚细亚大概是这祖先的家的地方，在多新世的时候，从那里出来，向东西移徙，达北平和匹兹唐。

在洪积纪的时候，人已进步到古石器时代，他何以不东来，却只是向西移徙的呢？这是毫无疑惑的，在洪积纪开始，几乎没有的移徙到中国，或者减到极微，经过详细的搜究，古石器人遗物及其文化的遗留物少有的看见。除上面所说的一个牙齿外，李桑（Pére Licent）[1]神父和推耶[2]神父（Teilhard de Chardin）在鄂尔多斯黄土的基部，及推耶和杨格[3]（Young）在沿黄河的有几个地方发现简单的文化时代的证据。纳尔逊在蒙古又看到可疑的物事。此等东西是代表摩司替利时期的（Mousterian stage），即宁德退尔人的文化时期之末，虽然这增加纯是人造品，属于欧罗巴的那时代的形式的。但没有证据，亚细亚的这文化时期，和欧罗巴的这一时期时候相同，因为在亚细亚比到欧尔巴去发达的早的多。

要企图解释这个问题，"古石器人在中国为什么这样少的，在欧罗巴却如此丰富，而散布的中心地却和前者近，离后者远？"据我想来，我们第一步必须认知黄土，黄土是特别的中国的堆积物。它是一种特别的黄色的土，由极细分子集成，这样堆积起来，没有成层的证据，并且压的很实，能成峭壁，在被溪流切断，或成路道的地方，这路经数世纪走行下来，低下成峡谷。

大多数的威权者皆一致赞成，黄土是由风带来的细分子积成的。此等质料从那里来的呢？它是岩石崩坏下来的最后的产物，它从岩石变来，但只是原来的岩石的极小破碎部分。我们曾经研究过此等岩石，它破碎下来的质像黄土，又能产下许多石英沙，并且产下黏土，这便是变成黄土的，沙粒混合在其中。把两者分开，须要自然的助力，其中最有力量的首当推风，风将细的质料吹去，沙粒剩在后面。

中国的黄土似乎由这样的微细质料积成的，岩石崩解后，这产物从沙分出，由风带到中国。如果确实如此，那么必有地方能看到纯粹的沙，这是风吹细质料时剩下来的，此种沙起原上当看得出是风吹沙的性质的。

这样沙的堆积我们已见于西部中国，新疆省[4]及别的地方，塔里木盆地，在南面昆仑山、北面天山山脉的中间，形成可厌的沙漠，是北沿塔里木河的。这条河，受从两个山脉来的间歇川的灌注，终于罗泼诺尔（Lop Nor）[1]盐湖内的盆地的东面。水分蒸散，盐遂沉积，这盐由溪水从山上带来，山上古代海成岩石中含化石及海水在石孔中，岩石分解，遂放出来，这沙漠称为塔克喇麻干（Takla-Maken）沙漠，全是沙丘的沙漠，沙丘高达一百密达以上。照海定[2]（Sven Hedin）说，此等沙丘安放在有层的黏土上，这黏土是多雨时期的堆积物，在现代的沙漠时代之先。在十分或然范围之内，现在成沙丘的沙，当初是河内的堆积物，被于大部分盆地，倘若如此的，这质料不是纯粹的沙，却是沙、细岩粉、及黏土的混合物，正则的河内堆积物是如此的，并且这又很明显，在此等堆积物形成的时候，其地并非沙漠，却属多雨气候的。

这样的地方变为沙漠，须经过干风的长入的作用。长住在北部中国的人知道的很明白，冬季从蒙古吹来，向中国较暖的沿海之地吹去，这是此等干风的模式的例。风从高压的地方吹出来，那里因为空气密集，湿气容量中等，一经吹到沿海，含湿气的容量即增加。因此所经的地方，土中的湿气皆被夺取。土地松散，容易被风吹扬，遂把它吹作尘土的云。这是一般的错误，以为北平的沙从戈壁沙漠来的。北平的沙无非平地上的细沙土，被风吹起，风是从戈壁来的，但不挟沙土，此等风先把湿气吸收，土地失却抵抗它的攻击的能力。所以北平的沙是一地方的松土被经过的风所吹起的。此等负带尘土的北平风虽然力量足以刺戟人，几致不能忍耐，然而比之于远在洪积纪时代吹黄土到中国来的风，是柔和的和风了。中国的黄土是在洪积纪堆积下来的，这时候欧洲和美洲正葬在冰下面，已成确定的事实，黄土成自风送的尘土，这也同样确定了的。

现在让我们把所发见的事实加以考验和使关连，堆积在中国的很多的黄土是风吹来的尘土的厚积层。这尘土是微细的质料，从较古的堆积物分离出来的，那古堆积物本由沙和尘土混合而成。在新疆省的塔克喇麻干及蒙古的戈壁的若干地方，大部分面积皆盖着沙，这便是从尘土分出来的。取各种事实来研究，原动力只有风。黄土显然是尘土，由风从塔里木及戈壁盆地搬来。但今日的风已太弱，不足搬运戈壁和新疆的尘土到内地。它只能在近地吹起些尘土。所以洪积纪的风，即在堆积黄土的时候，比今日最强的季候风必强

1　今译为罗布泊。原文为 Lap Nor。——编者注
2　即瑞典探险家斯文·赫定，曾于 1927—1935 年在中国西北地区组织了大规模的中瑞西北科学考查。——编者注

许多倍。近代的季候风是蒙古的大冷的产物，它的干燥力由于流入较和暖的地方所致。洪积纪的风它携带尘土从塔里木及戈壁到中国的，必定是更冷的产物，比之于今日亚细亚内地所流行。如果我们想起这个时期是欧罗巴被着冰的时候，我们就可以感觉到这种大冷的来源，如果结冰由于地极的地位变化，比现在倾南十五度，那么对面的赤道，便是在中国的沿海地方，必定比今日还要过北十五度。这样，北平所享受的气候，性质当像今日的菲律宾。

但是便是华北的纬度和今日一样的，欧罗巴冰地和中国沿海地之间的温度的不同也足以使气压发生极大的殊异，这就足以发生有力的风，运送黄土。使塔里木盆地变成沙漠的便是这等风，因为它的干燥效果首先觉到的必定是那里，搬运尘土直到中国平地来落下的也必定是此等风。今日有挟尘土的风吹起时，生活有些不方[1]便，洪积纪的尘风中，在中国生活必定近乎不可能。塔里木盆地是坏透的，当有冷风从冰田吹过时，不过这等风只拾起尘土，负载而去，所以还可以住人。然而，此等变化的状况，终于逼迫原人找寻气候较柔和，或者较有屏障可藉的地方。这样的保护地他便找到南欧的穴洞，并且他见气候更温和，在较热的间冰期，这时候冷风停止吹号。但在东部，冰风吹着，少有或没有保护可找，即使人能忍耐这寒冷，他也不能抵抗阵尘的破坏的效果。所以洪积纪人的不到中国是没有疑惑的，许多大动物和他移往南欧，也不住在中国。动物遗留物确实偶然有的看见的，但它的存在似乎和气候并不适应，气候想象起来是峻严的。此等动物如巨大的鸵鸟，即 *Struthiolithus*，它的卵见于中国的黄土。但大概此等动物是温和的间冰期游行到黄土地方来的，这时候挟尘土的风暂停止，并且它们从沿海带来，这在从前的时候在比今日更远在东方。因那时候，古契丹、亚细亚还不曾分割开，在黄土堆积的时期没有此等动物，也没有原人能长久住在华北。所以在欧罗巴，比较在亚细亚，我们得看到在中第四纪时代人的进化的记载。

到洪积纪终了时，冰河退缩了。大概北极回复了如今日的位置，因此早先覆冰的区域，冰又没有了。当冰帽消去的时候，强风也停止吹过亚细亚，黄土的堆积也告终止。于是新石器时代人能够东迁，占据了前代人所不能够容留的地方。华北、西伯利亚，及日本有新石器时代人存在的证据，和没有古石器时代人存留的证据，同样的惹人注意的，不特新石器人的制造品见于各处，且极丰富，便是人的遗骨也很多的见于古代新石器的葬地。这种可注意的事,是因安徒生[1](J. G. Andersson)的搜求阐明出来的,便是此种人的文化,经地中海到日本群岛根本相同。人住在亚细亚洲不满足,经过百令桥梁(Bering Bridge)到北美,住在那里,有些后裔,美洲印度人的末代,至今尚在异国人中间存留着,那人口是在历史时代从欧罗巴海岸移徙到亚美利加去的。

——原载《自然界》1930 年第 5 卷第 7 期

---

[1] 即瑞典地质学家、考古学家安特生。——编者注

The Outlook for Science in China

By Dr. A. W. Grabau

Of the Great Nations, China is the latest to come into the brotherhood of scientific workers of the world. The ancient learning, dominated by philosophy and literature, has at last admitted science to a full recognition in its curriculum. In this respect China is following the course of intellectual development of the West, for as is well known, science was the last of the great fields of knowledge to gain a foothold in the learning of the western world.

Little more than a century ago, the chief universities of Europe consisted of the faculties of Philosophy, Theology, Law, and Medicine, with concessions to Mathematics and perhaps Astronomy, which were tolerated as adjuncts of one or the other of the dominant faculties. Gradually natural science was admitted as the hand-maiden of Medicine, and only slowly as it gained maturity was it reluctantly given a status of its own. Even then, it was subordinated to, or at last joined with Mathematics, as is shown by the fact that many of the leading learned academies had a Mathemathish-Naturwissenschaftliche Klasse—a section of Mathematics and Natural History, which also included chemistry and physics, though later these formed a class by itself, a Chemisch-Physikalische Klasse. Institutional Learning in the 18th century, as before, was dominated by the respect for the classics if it was not in bondage to theology; and even in our own time these two tyrants of the mind have not wholly lost their strangle-hold upon the western world. The sciences, especially the natural sciences, were left to amateurs, that is lovers of science, and every student in these fields today, knows how much was contributed by these voluntary workers to the upbuilding of the temple of science in which we worship today.

It is true that the oldest Scientific Academy, the Kaiserlich Deutsche Akademie der Naturforscher, was founded in 1652, or nearly 300 years ago, but it was, until quite recently, a society without a permanent home of its own. Founded by medical men, it was as much a society of intellectuals as of active naturalists, though its avowed object has been, from the beginning, the investigation of nature. Among its membership it has counted such eminent naturalists as Celsius, Linnaeus, Alexander von Humboldt; Goethe, Berzelius, L. von Buch, de Condolle, Karl Ernst von Baer, Cuvier, Escher von der Lind, Wm Buckland, Alexander Brongniart, Johnnes Miller, Louis Agassiz; Rudolf Virchow, Charies Darwin, Hershel, Huxley, Charles Lyell; Richard Owen, John Tyndall, and a host of others; and to this day if has been active in its exalted object "the investigation of nature for the benefit of mankind", and remains the oldest existing Academy of Science in the world.

It is true an Academy for the study of Nature was founded in 1560 or nearly a hundred years before, in Italy, but it was soon suppressed by the theologians who then ruled all intellectual departments. The Royal Society of London was founded ten years later, that is in 1662, though the meetings of a private society "Occulta" which finally led to its founding, began in Oxford in 1645 and thus antedates that of the German Academy. The Academy of Sciences in Paris was founded four years after the Royal Society, that is in 1666, and all other scientific Academies thereafter began in the 18th or the 19th century.

In America, Science is scarcely a century old, though Benjamin Silliman, the leader of American Science in the early half of the 19th century, was already teaching chemistry at Yale, and making collections of minerals, at the opening of the century. But systematic instruction in the natural sciences did not begin until the fourth and the fifth decades of the 19th century. The oldest school of Natural History, devoted primarily to Geology and Botany, was founded by Amos Eaton at Troy in 1827. Here many of the older American geologists and botanists had their first training in the sciences. Though it soon after developed into an Engineering School, the Rensselaer Polytechnic Institute, geology and mineralogy at least, have always been taught there, and some seventy years after its foundation, these departments were in my charge. And it is worthy of note that this oldest American School of geology was only about fifty years younger than the oldest European school of geology at the Mining Academy of Freiberg in

Saxony, under the illustrious Werner, the father of geology, and the first teacher of the science.

But the real impetus to Natural Science in America was the coming of Louis Agassiz in 1846, followed during the next year by the founding of the Lawrence Scientific School at Harvard, and the Sheffield Scientific School at Yale. The small coterie of ardent naturalists at Boston, who in 1830 had founded the Boston Society of Natural History, hailed Agassiz as their leader, and young men from all parts of the country flocked to Cambridge to devote themselves to Natural Science under this great master. From his laboratory have gone forth most of the leaders in American Science.

It has been my good fortune to know a considerable number of these leaders of American science personally, though not the great Nestor of American naturalists himself. In point of fact, several of Agassiz' famous pupils, among them Alphaeus Hyatt the palaeontologist, Nathaniel Southgate Shaler and William Crosby, the geologists, and John Runckel the mathematician, were my own special teachers, while others, such as Edward S. Morse, Samuel Scudder, A. E. Verrill. and Alexander Agassiz, had a strong personal influence on my scientific career. When it is recalled that Louis Agassiz, my own "Grand Teacher"—as it were, was himself a disciple of Cuvier, it will be seen how near we are still to the founder of the sciences of comparative Anatomy and Palaeontology in Europe.

Even in my student days, Cambridge and Boston were still the centers of Science in America, and in Agassiz's day this was the training-field of American scientific men. Around him the eager students gathered and in mutual converse confirmed their enthusiasm, and ripened those powers which later enabled them to go out in all directions to establish centers of science throughout the land. But it must not be forgotten that these men were not young graduates from lecture courses, but experienced workers, drilled by laboratory methods which few students today would survive. They were firmly grounded in knowledge and enthusiasm before they quitted the companionship of their leader and of their fellow research workers to found scientific communities of their own. And for many years after, whenever possible, they would return to the fountainhead, there to draw renewed inspiration from the converse with their brethren in their own and allied sciences, and even in unrelated fields. It was just such need of mutual intercourse, felt by the geologists of Eaton's School, and the naturalists that came with, and were inspired by Agassiz, that led in 1847 to the formation of the American Association for the Advancement of Science, as many years before a similar need for mutual inspiration led to the founding of the British Association for the Advancement of Science, and, in recent years, this society, with similar aims in view.

But there is one difference that must not be overlooked· The disciples of Agassiz, who in America went out to carry the light of science into all quarters, were seasoned men of science—they had remained together long enough to have become firmly grounded in the habit of research. Here, where science is still young, the danger is that the young men who devote themselves to it, are scattered too early, almost as soon as their preliminary training is completed—and before they have had opportunity in mutual association and scientific fellowship, and under the guidance of the older men, to strengthen their powers of research, to allow their enthusiasm to gain deep roots, which will make possible resistance against the storms of adversity, and enable them to continue in unabated endeavor, even though the location of their field of work condemns them to intellectual solitude.

It would be of inestimable advantage to Chinese science, if there could be, for several decades at least, one recognized center of scientific research, or at most two such centers—say one at Peking and one at Nanking, —though at first, a single center in Peking, where all the ablest men should be gathered together, would be most effective in accelerating the development of scientific research workers. For, scientific men, more perhaps than those in other fields of human endeavor, need the stimulating contact with their fellow workers, not only in their own but in related, and even in unrelated fields. Isolation not only puts a damper on enthusiasm; it cramps the faculties, and leads the worker to look at his subject from too narrow an angle. Annual meetings like this one just closed, are indispensable, but they are not enough in a country where most of the younger scientific men have not yet been seasoned by prolonged contact with, and criticisms by, their fellows and the older, more experienced workers in their fields. When the young scientific worker has grown to intellectual maturity, and when his outlook is broadened, and his habits of scientific work firmly grounded, then, and not until then, is he equipped to go out into new field, there to create a scientific name of his own.

I have reminded you that science in America is still only in its early maturity. And so far as appreciation of science by the public is concerned it is younger still. When Sir Charles Lyell came to Boston in 1841 to lecture before the Lowell Institute on Geology, lectures which indirectly led to the calling of Agassiz to become the founder of American Natural Science, geology was so little understood by the public at large, that even in Boston, that oldest city of American culture, it was found profitable to ridicule the teachings of so eminent a founder of the science. One of the Boston theaters, devoted to popular vaudeville, swept clear its stage of all the actors, save only one, who was made up to represent Lyell, and hugely entertained the audience by a mock lecture on geology.

But if the people as a whole thus scoffed at Natural Science, individuals counted the coming of the great English Geologist an unprecedented boon. Not only were his lectures attended by crowds of interested and thoroughly sympathetic laymen, but young enthusiasts found it an opportunity not to be neglected, no matter at what cost or personal sacrifice. And even before his coming the scientific lectures at the Lowell Institute in Boston drew the eager seekers after knowledge.

The story is told, and it is authentic, that one young man, living at Hingham, a village near Boston, walked to that city, a distance of some twenty miles, to hear Silliman lecture on geology, and after the lecture walked home again through the night, because he was too poor to afford the cost of transportation. This same young man in 1830 walked to Troy, a distance 220 miles—mind you, all the way on foot, —to enter as a student in the school of Natural Science founded by Eaton. And this young enthusiast rose to become the leader of American science in the fields of palaeontology and stratigraphy, and today his name is a household word among the workers in that science the world over. This man, James Hall—you all know of him—for more than sixty years carried on the greatest State survey that America has known, the Geological Survey of the State of New

York. It was my good fortune in 1896 to be present at the joint meeting of the Geological Society of America and the American Association for the advancement of Science in Buffalo, when these scientific bodies celebrated the completion of Hall's 60 years of service to the State, and it was my good fortune to have this revered master of American Palaeontology in the audience before which I presented my first paper on palaeontology.

When I became interested in geology and in fossils, there were no institutions nearer than the Atlantic seaboard where instruction in those sciences were offered, and I was one of the first students to enter the newly founded course in Geology at the Massachusetts Institute of Technology in 1890. But I did not have to walk all the way. There were railroads by that time in America.

The Massachusetts Institute of Technology opened its doors in 1865, eighteen years after the founding of the scientific schools at Harvard and at Yale, but the oldest University in America, founded for advanced instruction and research, with the express understanding that science, pure and applied, should be on an equal footing throughout with literature and the other humanities, —that is Cornell University—is only one year older than I am.

And you know how recently the western world has become free from the shackles of theology. I remember the tempest that still raged, when I was a boy, around the controversy raised by Darwin's "Origin of Species" in 1859. And nineteen years ago I had the pleasure of discussing with the aged Ernest Haeckel, in his famous study at Jena, the progress of science since the day when he was the chief defender and protagonist of Darwin on the continent of Europe. And you all know, that the dragon of theological persecution of science has not yet been slain, and that it has dated, in recent years, to raise its poisonous head in certain sections of the United States, and that for a time it was successful in officially suppressing true science in some American institutions of learning.

Fortunately you in China are free from the hampering influence of theology, although you have your share of superstition to overcome. And with Andrew D. White, the founder of Cornell University, I wish to emphasize, that by theology I do not mean religion. This society of which I am proud to be a member, is doing its part, and a big one it is, to spread scientific knowledge throughout this nation.

It is indeed the Chinese Association for the Advancement of Science, the C.A.A.S., which with the American association, the A.A.A.S., and the British association, the B.A.A.S., forms the great A.B.C. of that scientific movement, which, with others of like aim, is endeavoring to free the minds of men from intellectual bondage.

Let me attempt to outline what I conceive to be the modus operandi of such a society as this. We may perhaps classify the main work of this society, as of the others of its kind, under the following three categories, taking them in the order of increasing importance as well as of natural sequence.

I. The creation of an environment favorable to science, by the development of an interest in, and appreciation of, science among the people at large.

II. The unification, spreading and widening of scope, and the improvement of scientific education.

III. The fostering of scientific research, and of cooperation among scientific men the world over.

Let us consider these categories somewhat more at length.

Science can not flourish, any more than any other human endeavor, under an alien or adverse environment. Fortunately we have nearly outgrown the baneful influence of theological teaching that has so hindered the development of science in the past, an influence of which you have been largely free. But you share with us of the west the depressing influence of popular apathy, and the retarding influence of the superiority complex of business and politics. So long as a nation is only tolerant of science, as a harmless occupation for what it regards as minds of inferior ability, incapable of aspiring to the superior altitudes of trade and politics, so long science will languish in the shade of the ranker growth of those more primitive human endeavors. Only when men of science are accorded their right to come out into the sunlight of universal recognition, to take their place as the natural leaders in any advanced community, only then can science and the nation flourish. So long as teachers and scientists must worry about their rent or grocerybills, —or whether or not their monthly salary, small as it may be, is forthcoming, —so long the progress of science, and with it the progress of the world, will be retarded.

The undisputed leadership which Germany has so long exercised in the fields of science is in no small measure due to the esteem in which such work has always been held in that country, and though even there emoluments were all too often inadequate, yet the higher reward of intellectual leadership could safely be counted upon.

In America, though teachers and research workers are better paid (not always however) while those who apply the discoveries of the worker in pure science, are sometimes munificently rewarded, still the scientist is all too often regarded as the employee of the business men or politicians who control the intellectual output to a large degree, by virtue of the superiority which their acquisitional ability gives them in the popular estimate, as well as in their own.

When a Van't Hoff, undisputed leader in an important field of chemical research, must read in a reputable journal, which described the banquet given to inaugurate his lecture course at an American University, that he had the honor to sit at the right hand of John D. Rockefeller, he must realize that in the land of munificent endowments, the millionaire still ranks above the scientist in popular estimation. Let us, however, be generous, and assume that the great philanthropist himself was duly conscious of the honor which was his, by being permitted to entertain, as his guest, one of the world's master intellects.

To make the business and political leaders of the nation appreciate the man of science and his work, as they should be appreciated, you must first of all get the populace to acclaim such work and those responsible for it.

Vox populi est vox dei may be paraphrased in these days *vox populi est vox rectoris*, the voice of the People is the voice of its ruler, or will be, if the rulers know what is good for them. And as the business and financial leaders are after all human, supermen though some are accustomed to be acclaimed, they will in the end heed the voice of the people on whom in the last analysis they must rely.

A great novelist or a great musician gets his full recognition from the people, because they can at least value, if they do not wholly understand his art. Why should a great scientist be without honor in his own country or abroad? He would not be if the people appreciated his work, though they might not understand it. Einstein is honored the world over, as Newton never was in his day, yet there are probably not more than a dozen or two individuals living today, who can fully understand and appreciate Einstein's work. Of course he never got the popular acclaim that Lindberg got for flying across the Atlantic, but then people are stirred to enthusiasm by the display of daring and courage, while intellectual achievements leave them cold, however much they may admire them.

But the man of science does not aspire to be a popular hero. What he does want, and wants legitimately, is appreciation of, and respect for his work, that the people will accord him once they understand its significance.

It then becomes a paramount duty for a society like this, to spread a knowledge of the importance of scientific work among the populace, and to create an interest in it, so far as that is possible. I know of no better way than is furnished by the introduction of nature-study in the schools, provided it brings the pupil in contact with living nature, not with preserved relics, or with nature-lore in books. To rouse an intelligent love for nature in the young will prepare the soil for a future intellectual harvest that will nourish the world. But the older generations must also be reached, and especially so, because they will help to cultivate the field that is to bring forth the intellectual harvest, provided they understand the value of such a crop.

Hence, I suggest, the method to be employed should be the organization of local Natural History Societies in which all, are encouraged to take part. These societies must be kept on a non-technical plane, and those of you who would guide their meetings, should keep in mind that they must be popular enough to interest the people, while at the same time they do not depart from a strictly accurate, that is, a scientific basis. The success which the Peking Society of Natural History, an organization of this type, has achieved in a few years is an earnest of the success that awaits such societies elsewhere.

Coming now to the second of our categories, that of improving scientific education, we must realize that it is not sufficient to found an institution and equip a laboratory. We must first and foremost have the men. And to produce these men, intensive and prolonged training is necessary. Rarely if ever, is a young man ready at the end of his four years college training, (or university training, if you prefer that somewhat ambiguous term) to become an able and inspiring teacher of other young men or women. Graduation at the end of 4 years training is not the end of education but its beginning. That is why it is called "Commencement" in American institutions. No one can successfully teach a subject until he has mastered it to such a degree that he feels at home in it. When a successful teacher of English is asked to teach mathematics, history, and physics, and perhaps chemistry as well, he is bound to make a failure of most, if not all these subjects, for students soon tire of remasticated book-learning, especially when they find that the teacher keeps only one lesson in the book ahead of them. Again when a graduate from the geological course is set to teaching languages in the middle school we can not well blame him if he loses enthusiasm and soon falls into an uninspiring rut, from which his pupils suffer as much as he does.

It has been well said by one of your intellectual leaders, that education in China—and for that matter everywhere—must begin at the top, you must first of all produce your good teachers who know their particular subject thoroughly, and who have made that subject their own, not because they hope to earn a living by it, but because they would rather work in that field than do anything else. Love of your subject is the first pre-requisite to success in it, and when you are an enthusiast in your own subject and have the opportunity to teach it to others, you can not help being a successful teacher, at least for those in whom you succeed in rousing a similar enthusiasm.

But, and here is the rub, you must have the opportunity to teach your own subject, and here is one of the adjustments to which this society should address itself. I know of no sadder waste of intellectual powers than that which results from the compulsion which forces a man to teach a subject in which he has no interest, while at the same time he is debarred from the field in which his enthusiasm and ability would make him a leader.

Not only then, must we not permit our young men to become responsible teachers in a subject which they have not yet entirely mastered, and in which they have not yet acquired the right, (mind I do not say the opportunity to exercise, but the intellectual right, by virtue of attainment), to form independent judgment; but we must also see to it that those who have thus achieved, will be given the opportunity to transmit as well as to augment and apply the power they have gained, and are not rendered intellectually sterile by being compelled to labor in an alien, and to them a stony, field.

But in order to train men successfully in a given department of science, we must avoid one-sidedness and limitation to a few special fields. It is a great mistake to scatter the available forces over the land. China has, I will not say too many Universities, but too many institutions claiming university rank. America has passed through a similar course, and China should profit by its experience. It would be far better to have one or two real Universities, Universities where all departments are adequately staffed by the ablest men available, where men could be thoroughly trained, trained intensively as well as broadly. And not until such well-trained men are available, should departments in their subjects be opened in other institutions of learning. If you have a good mineralogist, a good chemist and a good physicist at one institution, you may train students in those sciences, but you will not make good geologists or biologists. Again if you have a good specialist in one branch of zoology at one institution and a good structural geologist or petrographer at another, but only inadequately trained, no matter how enthusiastic physicists, chemists, botanists, mineralogists, or palaeontologists at either, you can not train a good all-round geologist or biologist. You need first of all an institution where you have at least one good man in each of the important branches of science, and if there are not good men enough to equip more than one institution, be content to start with one, and in a few years you will have the men to equip a second, and thereafter it will go forward in geometrical progression. And when you have such a galaxy of good men at one institution, see to it that they remain, that the conditions surrounding them are favorable, that they are free from financial worry and interference, and their equipment is adequate.

Concentrate your force, and in that way, and that way only, will you progress in the building of the intellectual edifice whose influence will spread over the land and reproduce itself a hundred-fold to the glory of the nation.

And now I come to the third of my categories enunciated above, namely Research.

Research is the glorious fruition of intellectual endeavor. To push forward the boundary of knowledge, to add to the sum of intellectual achievements of the world, what more glorious destiny can one look forward to. What a pitiful figure is an Alexander, or a Napoleon, a Genghis Khan, or a Tamerlane, conquerors of people and of territory, when compared with him who achieves a conquest over nature. Their names may linger for a while in legend and in history-books, but their deeds are nullified, their achievements have been superseded. But a Galileo, a Newton, or a Darwin, and the rest of that great host who have labored for truth and the advancement of knowledge, are immortal, and their intellectual conquests have become milestones on the road of human progress.

Great naturalists are born not made, but they appear oftener than we are wont to imagine. Given the right environment, given the opportunity, they will emerge. It is true, the towering intellect will rear its head even above its most squalid surroundings. But when you see such a giant struggling to arise, it is your duty to lend him a hand, to cut away the bondages with which a Lilliputian army endeavors to hold him down, and above all, to give him an opportunity to take his rightful place. If envy and jealousy shear the locks of Samson, to deprive him of his strength, and then chain him to the treadmill, remember that regained energy may bring the roof crashing over the heads of the Philistines.

But let us not wait for supermen to arise in our midst. There are unsuspected powers in all of us. What is needed is plan and method, as well as desire to achieve. The man who wastes his energies in small and scattered endeavors will not produce results of lasting value. It may be proper for a qualified teacher to select a subject for research for the beginner, but the man of experience should know the problem which he prefers to attack above all others. And if you have found your line, persist in it. Many years ago the great Charles D. Walcott, the American Palaeontologist, director of the United States Geological Survey, and subsequently Secretary of the Smithsonian Institution, said to me: "Select your special field, and stick to it, as I have stuck to Cambrian palaeontology ever since I collected my first trilobites from Cambrian rocks", I was myself studying Devonian fossils then, having collected my first fossils in Devonian strata. Well I have tried to heed Dr. Walcott's advice, and I am working on Devonian fossils even now.

That does not mean that you should limit yourself absolutely to one field. An intellectual diet thus restricted would be as fatal as a diet limited to rice alone. You can't afford mental Beri-Beri or whatever the name of the ailment produced by such a restricted diet may be. But just as rice or bread is the staff of life, so a single field should be your intellectual staff—be your additions to it ever so varied. Only remember that too rich a diet persistently indulged in, is equally harmful.

To drop all metaphors, it is incumbent on the man of research to lay out definitely the field and to plan a definite method of procedure. And don't be too modest in your planning. If you are an Ichthyologist, plan the systematic study of the fishes of China. You may never complete it, most probably you will not cover more than a fraction of the field, but if you have planned the problem, you can pursue it systematically and leave it for others that follow you to carry to completion, and thus build a monument to your farsightedness and your breadth of vision. And by systematic study of fishes I do not mean mere identification and naming of species, but everything concerning them, their morphology, embryology, classification, bionomy, distribution, habits, and uses to man, and whatever else you can discover. And publish the results of your researches in an adequate manner and preferably in a series where the same work can be continued in the future. In other words, avoid scattering your results among the one thousand two hundred and one periodicals of a zoological nature with which the scientific world is now afflicted, as my friend Dr. Ping tells me.

I may encourage a promising young palaeontologist not to limit his interest to brachiopods, but to try his hand at corals, pelecypods, or even trilobites. But I also urge him to make himself master of one field, and if he comes to know more about Productus than any man living or dead, he will be a better asset and a greater honor to the Palaeontological fraternity.

I would then say to all researchers: Specialize, make yourself absolute master of one subject, no matter how limited. Then, if you ever keep that goal in view, you may make as many side trips, nibble at as many tidbits that come your way, as long as your philandering does not divert you from the pursuit of your true love.

And what about cooperation? No man of science and no group of scientists can be self-sufficient. Science is international, and scientists, above all men, should be internationalists. The brotherhood of scientific men knows no barrier of country, of nation, or of race. The advancement of science, not the glorification of a group, or a nation, is the primary aim of the scientist, and if in the achievement of his high aims, he brings glory to his city or his nation, then his city and his nation should rejoice and do him honor. But they must not forget that he has built on the foundations which a great host of workers before him has laid, and that that host is drawn from the nations of the world, and all are united in the democracy of intellect.

And now I want to say a parting word to the young men, those who are beginning their career, and have just left or are preparing to leave the intellectual nursery—their alma mater. Don't try to be a Bacon and take all knowledge for your province. That may have been possible in Bacon's time—though I doubt it—but certainly is not now. Don't plan, so one young man has planned, to take a year to reexamine all the laws of physics and of chemistry, and reshape them to your own advanced conception. I doubt if even a Bacon would have found a life-time sufficient, in spite of the more limited number of laws then known, and you, even if you possess superior ability, should at least allow a few years for this Herculean task, considering the many discoveries that have been made since Bacon's time. Seriously speaking, don't try to encompass an impossible field. Hitch your wagon to a star by all means, but first pay attention to the spokes and axles of your carriage, and don't forget the axle grease.

Begin with a small and circumscribed portion of a problem and devote yourself strenuously to it. It is better that you should describe one Palaeozoic coral or one living starfish thoroughly, and find out all that you can about it, than that you write a treatise on palaeontology or zoology, with philosophy and logic as your chief guides, or a book on evolution with data gained by browsing in a library. And don't attempt to write a text-book. Wait till you have taught a dozen or a score of classes in your subject and then ask yourself if you know enough to write a text book. Even if a misguided publisher of text-books urges you to do so, have regard for your reputation, and wait until you feel you can do the work with success. I signed the contract for my text-book of geology many years before I felt ready to write it, and that only because the editor tempted me. I wish I had waited until a few years from now; think of the tidbits of Chinese geology that I might have included in it. Fortunately there may be a chance for a second edition if I can ever find the time for it.

And a final bit of advice, which fortunately is unnecessary for most of my younger hearers, but you may pass it along. Avoid hasty and superficial work, and don't regard yourself as a master before you have completed your apprenticeship. That may sound harsh, but it is intended in the kindliest of spirit. It is because I know that temptation besets you on all sides, that I warn you. There are too many responsible places waiting for you. You may be called to a professorship in one of your institutions of learning, because you are a promising young man, and the best one available. You may accept, and I should hesitate to dissuade you, if circumstances compel you to accept, but don't forget that a title is not always an index of achievement. You may be destined for great things, but if your election should lead you to the belief that you have already achieved superiority, it were better that you declined.

If you can make the position an opportunity to develop your powers, accept it by all means, but don't regard the position as an achievement, which puts you on a par with those who have long labored in the field. And if possible, postpone acceptance until you have had an opportunity, by more intensive work, to qualify yourself for leadership.

To all of you, students, fellow workers in science, and my special colleagues, you who have come together to tell of the achievements of the year, and to encourage one another to renewed endeavor, to all of you I give as a parting salutation the motto of the oldest Academy of Science in the world, the Kaiserlich-Deutsche Akademie der Naturforscher, to which I have the honor to belong.

"*Nuuquam Otiosus!*"—Be Ceaselessly Active

Let us cease not in our labors to advance the frontiers of knowledge each in his respective field, for so alone we shall be worthy followers of the great ones who have labored before us. But let us, while valuing the achievements of the past, not be shackled by tradition, but be fearless to defend new truths, if it be our fortune to discover them, nor hesitate to accept new interpretations in the place of those long sanctioned by tradition, if our researches show us their validity.

New occasions teach new duties,

Time makes ancient good uncouth.

They must ever on, and upward,

Who would keep abreast of truth.

——原载 *The Chinese Social and Political Science Review*, vol. XIV, 1930.

中国科学的前途

葛利普先生
在民国十八年八月二十五日中国科学社年会公宴的演说词

任鸿隽译

在现今各大国中，中国最近始加入世界科学团体。中国古时的学问，偏重于文哲两科，最后始承认科学在学课中的位置。在这方面，中国亦遵着西方智识发展的路径，因为在西方的学术界，科学也是最后方得插足的。百年以前，欧洲主要大学的课程，不外哲学、神学、法律、医药四科，或者容忍算学与天文，做一个其他某重要科目的附庸。

渐渐地，自然科学成了医学的伴侣，及后自然科学到了成熟时期，才慢慢的得到一个自己的地位。就是这个时候，自然科学还是算学的附属品，或至少可说是与算学有关的，因为在那时任何的重要学术院中，都有一个算学与自然历史组（Mathematisch-Natur Wissenschaftliche Klasse）。化学与物理学也包括在这一组之中，后来才分离起来，成了一个物理化学组（Chemische-Physikalische Klasse）。十八世纪的学术研究，也和以前的时代一样，即使不受神学的束缚，也免不了古典学的影响；就在现代这两个人心的专制魔王，尚未完全失掉西方的领域。科学，特别的自然科学，只让了玩意家即科学的爱好家去照管，现今凡研究过这些科学的，晓得我们目前所崇拜的科学庙堂，有好多地方是由这些自告奋勇的研究家出力造成的。诚然，世界最古的科学会即独逸皇家科学会（Kaiserlich Deutsche Akademie der Naturforscher）[1]，在一六五二年即约三百年前已经成立，但直至最近，他才得到一个永久的会所。这个学会系学医的所创设，虽其目的自始即为自然界的研究，实则说是自然

1　德国皇家科学院，其前身是成立于 1652 年的自然好奇心学院（Academia Naturae Curiosorum），是德国最早的科学学会，比伦敦皇家学会（1660）和巴黎皇家科学院（1666）还早。1677 年以神圣罗马帝国皇帝利奥波德一世（Leopold I，1640—1705）命名，后规范为德国利奥波第那科学院（Deutsche Akademie der Naturforscher Leopoldina）。2008 年改组为德国国家科学院（Nationalen Akademie der Wissenschaften Deutschlands，German National Academy of Sciences）。——编者注

科学家的学会可，说是一般智识界的学会亦可。在他的会员中有不少的自然
科学家，如摄尔西（Celsius）[1]、李理亚（Linnaeus）[2]、方洪博慈（Alexander
von Humboldt）[3]、贵特（Geothe）[4]、倍随理斯（Berzelius）[5]、方布徐（L.von
Buch）[6]、德康多尔（de Condolle）[7]、方柏耶尔（Karl Ernst von Baer）[8]、居维耶
（Cuvier）[9]、方德邻（Escher von der Lind）[10]、布克兰（Wm. Buckland）[11]、布朗
几亚（Alexander Brongiart）[12]、米纳（Johnnes Miller）[13]、亚嘉祉（Louis Agassiz）[14]、
费尔周（Rudolf Virchow）[15]、达尔文（Charles Darwin）、赫尔薛（Herschel）[16]、
赫胥黎（Huxley）[17]、乃耶耳（Charles Lyell）[18]、阿文（Ricard Owen）[19]、丁达

1　安德斯·摄尔修斯（Anders Celsius，1701—1744），瑞典天文学家、物理学家，以发明摄氏温度计而著
　　称。——编者注

2　林奈（Carolus Linnaeus，1707—1778），瑞典生物学家，创立动植物双名命名法，影响深远。——编者注

3　亚历山大·冯·洪堡（Alexander von Humboldt，1769—1859），百科全书式的科学家，德国地理学奠基者
　　之一。——编者注

4　歌德（Johann Wolfgang von Goethe，1749—1832），德国著名作家、思想家、艺术家、科学家。——
　　编者注

5　贝采乌乌斯（Jöns Jacob Berzelius，1779—1848），瑞典科学家，现代化学的奠基者之一，尤其对有机化学
　　有重要贡献。——编者注

6　利奥波德·冯·布赫（Leopold von Buch，1774—1853），德国著名地质学家。——编者注

7　德康多尔（Augustin Pyrame de Candolle，1778—1841），瑞士植物学家，为确定植物属间的自然关系建立
　　了科学的结构标准。原文作 Condolle 有误。——编者注

8　卡尔·恩斯特·冯·贝尔（Karl Ernst von Baer，1792—1876），德裔俄国胚胎学家，比较胚胎学的创立者，
　　提出胚胎法语的"贝尔法则"。——编者注

9　现一般译为居维叶（Georges Cuvier，1769—1832），法国动物学家，比较解剖学和古生物学的创立
　　者。——编者注

10　应为 Arnold Escher von der Linth（1807—1872），瑞士地质学家。——编者注

11　威廉·巴克兰（William Buckland，1784—1856），英国地质学家、古生物学家，以研究动物粪便化石而
　　著名。——编者注

12　布隆尼亚（Alexandre Brongniart，1770—1847），法国博物学家、地质学家，首次在地质年代表中描述了
　　第三纪（Tertiary Period）。原文作 Alexander Brongiart 有误。——编者注

13　应为约翰内斯·穆勒（Johannes Müller，1801—1858），德国生理学家、比较解剖学家。原文作 Miller 有
　　误。——编者注

14　路易·阿加西，详见上文。——编者注

15　鲁道夫·魏尔肖（Rudolf Virchow，1821—1902），德国病理学家、人类学家、政治学家。——编者注

16　此处不知指的是威廉·赫歇尔（William Herschel，1738—1822）还是约翰·赫歇尔（John Herschel，
　　1792—1871），他们父子同为著名的天文学家，且均为德国皇家科学院的院士。——编者注

17　赫胥黎（Thomas Henry Huxley，1825—1895），英国生物学家、解剖学家，以宣扬达尔文进化论而闻
　　名。——编者注

18　赖尔（Charles Lyell，1797—1875），英国地质学家，现代地质学奠基者。——编者注

19　理查德·欧文（Richard Owen，1804—1892），英国解剖学家、古生物学家，最早将恐龙另立为一科，还
　　以反对达尔文进化论而闻名。——编者注

耳（John Tyndall）[1]，并其他等等。即至现在，其学会对于其"研究自然以利人生"之目的仍甚活动，并为现存世界最古科学会之一。

诚然，约一百年前即一五六○年，在意大利曾有研究自然科学会的设立，但不久即为箝制一切智识的宗教家所解散。英国的皇家学会，成立于一六六二年[2]，即在十年以后，虽然这皇家学会前身的私人团体在一六四五年，已在奥斯福聚会，比德国科学会的聚会还要早些。巴黎的科学会[3]成立于一六六六年，即英国皇家学会成立四年之后，其余的各科学会，则在十八世纪或十九世纪方才逐渐成立。

美国科学的年纪，仅仅不过一百年，虽然做美国科学领袖的西里曼（Benjamin Silliman）[4]在十九世纪的前半，已经在耶律大学[5]教授化学，并且在十九世纪的初年，已经采集矿石标本了。但自然科学的有系统的教授，直到十九世纪的四五十年中方才开始。最初的一个自然科学学校，以教授地质学及植物学为主要的，系伊东（Amos Eaton）[6]在杜雷（Troy）所设立，时为一八二七年。美国的许多老辈地质学家植物学家，皆曾在此受过训练，不久这个学校就发展成了一个工程学校，名列斯勒高等工业学校（Rensselaer Polytechnic Institute）。在这个学校中尝有地质矿物两科，在成立之后约七十年，我曾做这两科的主任。可注意的，这个最老的美国地质学校，不过比地质学始祖，科学教授第一人有名的维尔纳（Werner）[7]曾经教授的欧洲最老的地质学校，即沙克孙里（Saxony）的弗来堡（Freiberg）矿业学院，年轻约五十岁。

但美国自然科学得到的真正激进，是一八四六年亚嘉祉的来美，及次年哈佛大学洛能斯科学院（Lawerence Scientific School），耶律大学习非而

1　约翰·丁达尔（John Tyndall，1820—1893），爱尔兰物理学家，长期居住在英国。——编者注

2　英国皇家学会（或称伦敦皇家学会）实际成立于 1660 年，1662 年获英国国王查理二世颁发特许状，成为受国王认定的学术团体。——编者注

3　即巴黎科学院成立于 1666 年，1699 年法国国王路易十四对其进行改组，正式制定章程，被更名为巴黎皇家科学院。——编者注

4　本杰明·西利曼（Benjamin Silliman，1779—1864），美国地质学家、化学家、教育家，1796 年毕业于耶鲁大学，后长期在耶鲁大学任教，1818 年创办美国科学与艺术杂志（*American Journal of Science and Arts*），在科学与教育方面做出重要贡献。其子西利曼（Benjamin Silliman Jr.，1816—1885）亦为著名化学家。——编者注

5　耶鲁大学。——编者注

6　阿莫司·伊顿（Amos Eaton，1776—1842），美国植物学家、地质学家，曾跟随西利曼学习，伦斯勒理工学院的创立者之一。——编者注

7　维纳（Abraham Gottlo Werner，1750—1817），德国地质学家，水成论创立者，长期在弗莱堡矿业学院任教。——编者注

（Sheffield）科学院的成立。波斯顿的自然学者，在一八三〇年，已经成立了波斯顿自然历史学会，即推奉亚嘉祉做他们的领袖，全美国各处的青年，也群涌到康桥来在这些大学者领导之下，研究自然科学。大半美国的科学领袖，都是他的试验室出来的。

我最幸运，曾与多数的美国科学领袖认识，虽然不曾认识这个最大的主教。实际说来，许多亚嘉祉的有名弟子，如古生物学家的赫梯（Alphaeus Haytt），地学家的席列尔（Nathaniel Southgate Shaler）、克乐斯贝（William Crosby）、算学家的伦克耳（John Runckel）都是我的业师；其余如摩尔斯（Edward S. Morse）、石卡德（Saumel Scudder）、费利耳（A. E. Verril）及亚力山大·亚嘉祉（Alexander Agassiz）都于我的科学生涯大有影响。当我想着我的"大先生亚嘉祉就是居维耶的门人，觉得我们离开欧洲比较解剖学及古生物学的始创者，还是不远呵。"

即在我做学生的时候，康桥与波斯顿尚是美国科学的中心，而在亚嘉祉的时代，则为美国科学家的训练所，在他的四周，围满了热烈的学生；他们的言谈，使他们的热心愈加坚固，并且使他们的能力成熟，可以四出到全国去设立科学中心。但是我们要记着，这些人不是由讲堂毕业的青年，他们乃是有经验的作者，是由现今学生所不易通过的实验室方法养成的。他们在离开他们的领袖和同道的朋友以建立自己的科学社会以前，已经有坚固的智识与热念。即在多年以后，只要可能的话，他们总得回到他们的老源头，去和他们同道或不同道的朋友谈话，以期得到一些重新的神助。正是为了彼此交际的需要，才引得伊东学校的地质学家和与亚嘉祉同来并热心的自然科学家们，于一八四七年发起美国的科学促进协会（American Association for the Advancement of Science），正如在几多年前，因彼此友助的需要而引到发起英国的科学促进协会，及近年来中国科学社以同样的目的而发起一样。

但是有一个不同的点不可轻轻看过。亚嘉祉的学生在美国的各处去散布科学之光时，他们已经是成熟的科学家，他们曾经长久住在一处，使他们研究的习惯坚固确立。在中国，科学的年纪尚属幼稚，青年的危险，乃是分散太早。他们最初的训练，一经完了，便分头去干自己的工作，不曾有一个机会使他们得到相互的交际，科学的友谊，和在老辈指导之下，强健他们研究的能力，使他们的热念生下一个深根，即使他们将来遇见了不幸的事情或孤立的智识生活，也不能使他们灰心懒意，而继续不断的抵抗奋斗。

我曾经告诉过你们美国的科学尚在青年时代，说到平常人对于科学的尊敬，他更是幼稚。当一八四一年乃耶耳爵士到波士顿在洛维尔学社（Lowell Institute）讲地质学，间接的引起亚嘉祉到美国成了美国的自然科学始祖的时候，地质学在美国还不大为一般人所认识，就在以文化自诩的波斯顿城内，还有人嘲笑如乃氏的大科学家以敛钱的。波斯顿城里有一家专做笑剧的戏园，把他的演员除了一个之外，全体解职了，这一个就是曾经假扮乃耶耳，学他的讲演地质学以引逗听众笑乐的。

但一般人虽对于自然科学不知尊敬，有些人却看这个英国大地质学家的来游，是一个空前难遇的机会。他的讲演不但为注意的同情的众人所乐听，青年热心的人，并觉得无论出什么代价与牺牲，这个机会是不能失掉的。即在乃耶耳未到美国以前，洛维尔学社的科学讲演，已能吸引热心求智的人众了。

有一个故事，并且是可靠的，说有一个青年，住在离波斯顿不远，名叫亨罕（Hingham）的乡村，步行二十英里，到波斯顿去听西里曼的地质学讲演，听完了，在晚间步行回家，因为他太穷了，没钱雇代步。就是此人，在一八三〇年，步行二百二十英里到杜雷去进伊东的自然科学校。此人终久成了美国古物学及地质学的领袖，今日世界上研究这两种学术的人们，没有一个不熟悉他的名字的。他就是郝尔（James Hall）[1]——你们都认识他——做了美国最大的省的地质调查——纽约省的地质调查——六十年。我很幸运，在一八九六年，曾在水牛城到过美国地质学会及美国促进科学协会的联合会，那时这两个团体正庆祝郝尔服务纽约六十年满期，我在这个会里宣读了一篇论文，并很幸运，得这个美国古生物学名师，为我听众之一。

当我对于地质学及化石发生趣味的时候，在大西洋海岸以外，没有教授这些科学的学校，一八九〇年，麻省工科学校[2]设立地质学系时，我也是最初入学学生之一。但此时已有铁道，我不须步行全途了。

麻省工科学校于一八六五年开门，比哈佛、耶律的科学馆迟十八年。但美国最老的学校，为促进高深的教育及研究而设立，并承认所有的纯粹及应用科学，都该与文学及其他人文学立在同一地位的，要算康乃尔（Cornell）大学，——他比我恰长一年。

你们晓得西方脱离神学的羁绊，不过是最近的事体。我还记得一八五九年间达尔文的物种由来所激起的争论，那时我才是一个小孩。十九年后我很

---

1 詹姆斯·霍尔（James Hall，1811—1898），美国地质学家、古生物学家，长期主持纽约州立调查局工作，编撰《纽约自然志》，其中包括动物学、植物学、矿物学、地质学、农业、古生物学各卷。——编者注
2 即麻省理工学院，下同。——编者注

愉快的和耄年的赫克耳（Ernst Haeckel）[1]在他的有名的耶拿书斋内讨论自他在欧洲宣传达尔文以来科学的进步。你们晓得压迫科学的神学毒龙还不曾完全制服，他们最近还在美国的某处，抬起头来，藉着官力，把有些学术机关的真科学压抑下去。

幸而你们在中国没有神学的拘制，虽然你们有你们的迷信应当铲除。我所谓神学并非宗教，这一点我和康乃尔大学的创立人惠特博士是一致的。中国科学社对于传播科学智识于中国尽了不少的力，我自己是一个社员，觉得是很光荣的。这个社也可以称为 Chinese Association for the Advancement of Science（中国科学促进协会），简写为 C.A.A.S.，恰合美国的科学促进协会简写 A.A.A.S.，英国的科学促进协会简写 B.A.A.S.，成了一个科学运动的大 A.B.C.；这些和其他有同样目的的组织，都是要解放人心的束缚的。

让我现在把我心中所认为这样学社的重要工作大概举出来，我们或者可以把本社及同样学社的工作，分为三种，依着他们的重要和天然的次序，分列如下：

（一）科学环境的造成，其道在唤起一般人众对于科学的兴趣及领会。

（二）科学教育的统一、普及，及其范围的推广，并科学教育的改良。

（三）科学研究的奖励以及全世界科学家的合作。

我们且就此三项略加讨论。

科学在逆境之下不能发达，正如其他人类的努力一样。往昔神学教训的恶影响，阻碍科学的发达的，幸而已成过去了，这种影响在你们国内自来即不存在。但是你们国内普通人对于科学的冷淡和政治实业对于科学的藐视，于科学的发达也颇不利，是和西方一样的。一个国家，如其认科学为无足轻重，只有能力低弱，够不上希望实业政治高位的人才去投身其中，那么，科学在这些比较的野蛮势力繁盛之下，只有日即于委顿的一途。只有科学家得到一般人的注意，承认他在任何进步的社会中，占天然领袖的地位，那科学和国家才能发达。如其科学家和教员对于他们的生活费都有朝不保暮的顾虑，那科学的进步以及世界的进步是无希望的。

德国在科学上的领袖地步是无可置疑的，他所以能如此，正是因为德国人的崇拜科学；虽然金钱的报酬常常不见得正确，但较高尚的智识领袖的荣誉总是可靠的。

1　恩斯特·海克尔（Ernst Haeckel，1834—1919），德国动物学家、进化论者，以提出胚胎重演律而闻名。原文 Ernest 有误。——编者注

在美国教员和研究家虽然薪入较丰（也不是常常这样），至于应用纯粹科学发明的人，其报酬尤为丰厚，但是科学家常被认为实业家或政治家的雇佣；实业家和政客们，靠了他们占有的力量，使他们自己及在一般人心目中，都处于优越的地位，因之对于智识的出产，也能有大部分的宰制。

当化学大家范霍夫（Van't Hoff）[1]到美国某大学来讲演，在欢迎席上他坐在洛克斐尔（John D. Rockefeller）的右手，某报记载此事，说范霍夫不胜荣幸之至了。这可以见得在捐资兴学盛行的国度中，普通人对于科学家的尊敬还不如大富豪。但如我们大量一点，也可以说这大富豪得招待世界最大智慧者做他的客人，才真正是他的荣幸呢。

要使实业和政治领袖尊崇科学家及科学事业，必须先使一般民众称赞此等工作及对于此等工作负责的人。

欧洲有一句古语说，"民众的声，就是神的声"（Vox populi set vox dei），也可以说，"民众的声就是君主的声。"若是君主晓得什么于他们有利的话，实业家及政治家虽然有时自以为超人，实则他们还是人，他们对于民众的声是不能不注意的，因为民众是他们最后的倚赖者。

一个大小说家或大音乐家，常能得到民众充分的认识，因为他们即使不能了解他们的艺术，至少还能评价。何以一个大科学家在自己国内或国外不能得到相当的荣誉。若使民众能崇拜他的工作，即使不能了解，他就不会那样了。安斯坦[2]全世界都知道尊崇了，牛顿的时候就不如此，但现在活着的人类中，不过一两打能够了解安斯坦的工作罢了。自然，安斯坦从来不曾得到如飞过大西洋的飞行家林德堡那样的欢迎，但此时的人众是为他的冒险勇气所激动，至于智识的成功，他们无论如何的赞服，总是冷的。

但是科学家绝不希望做一个民众的英雄。他所要的，也是他应得的，只是对于他的工作的领会与尊敬，这在民众知道他的工作的重要以后，是必定与他的。

唯其如此，所以如中国科学社这样组织的最大义务，即在传播科学智识的重要于民众，并且造成他们的兴趣，要达到这个目的，莫妙于在中小学校中设立自然研究科，但学生不可仅靠标本读本，必须和活的自然相接近。此时能引起学生于对于自然的爱好，将来自然有智识的收获，这个收获是培养

1　雅各布斯·范特霍夫（Jacobus H. van't Hoff，1852—1911），荷兰物理化学家，因化学动力学理论于1901年成为第一位诺贝尔化学奖获得者。——编者注
2　即爱因斯坦。——编者注

世界所必需的。但老辈的人也不可放过，因为只要他们能了解这种收获的价值，他们也能帮着耕耰这智识收获的田地。

在这里，我提议的方法，是在各处设立博物学会，鼓励他们都来入会。这些博物学会，必定要非专门的，而且主持会务的人，必须记着他们必须与普通人发生兴趣，同时又不至于离开正确的科学的基础。北平博物学会这几年来的成功，是他处同样的学会所值得效法的。

现在说到我们的第二种工作，即科学教育的改良。关于这一层，我们必须晓得设一个机关添一个试验室是不够的。我们必须先有人。要造就人才，必须有深刻的长期的训练。一个青年，在四年大学教育之后，便能做其他青年的有能耐有精神的教师，是绝无仅有的事体。大学生四年毕业，并非教育的终了，乃是教育的开始。所以美国的大学，毕业叫做"开始"（Commencement）。无论何人，设非对于他的专门完全了解，他教授此门功课必不能成功。设如叫一个好英文教员，去教算学、历史、物理学，甚至于他学，他于这些功课的大多数——即使不能全数——非失败不可。因为学生不久，就觉的反嚼式的书本学习的可厌，特别的他们晓得教员所预备的功课，比他们的仅多一课。又如学地质学的叫他去教语言，设如他不久便对于功课失去热心而成一种无精神的故辙，使他与他的学生一样的吃苦，我们能怪他吗？

你们智识界的领袖某君说得好，中国的教育——旁的地方也是一样——必须从顶上做起。你们必须先制造好教员，他对于他的专门学科有彻底的了解；他选择他的学科，不是因为他要靠以谋生，乃是因为他喜欢这门工作，比甚么都多。一个人在他的专门要成功，必须先爱他的专门学科，你既对于你的专门学科有甚大热心，又有机会去教人，你虽欲不做成功的教员而不能，至少在你能收发同样热心的几个人以内。

但此处有一个紧要的关键，你必须有机会去教你自己的专门学科，这一层中国科学社是应该尽力的。

我所知道的智识的耗费无过于强迫一个人去教他无趣味的功课，而同时又使他对他有热心、能力，能有领袖资格的学科无暇从事研究。

所以我们不但应该阻止我们的科学家，对于他还不曾纯熟，并且还不曾得到独立裁断的权利（注意！我说的是智识上由修养得到的权利，不是机会）的学科，去做负责的教师；但我们也应该注意，既得到这种权利的人，必须有机会去传授并且增加应用他们得到的能力，而不至强迫去在陌生的石田上工作，使他们在智识上不能有所生长。

但要训练一个人成为一个有用的科学家，我们必须避免偏狭与限制于几个特别科目。把有用的能力分散于全国，是一件大错。中国的大学，我不能说是太多，但名不副实的大学，实在是太多。美国也曾经过这样的阶级，中国可从他的经验得到一点好处。最好的办法，是先有一两个真正的大学，校内各科都有最好的教授，学生来学的，也可以得到十分深广的训练。除非这种人才已经训练出来，旁的学校对于这些科目，可以不必开办。如你在某一机关，有一个好矿物学家，一个好化学家，一个好物理学家，你可以在这几种科学里训练出好学生，但你不能训练出好地质家或生物学家来。再说，你如有一个好的动物学专门家在一个学校，一个好的地质构造学家或岩[1]石学家在另一个学校，而任何学校没有受过相当训练的物理学家、化学家、植物学家、矿物学家，或古生物学家，你不能造成全副本领的地质学家或生物学家。最要的，是最少每一重要科目有一个很好的教授，如所有的好教授不够分配，那么，你先以一个学校为满足，几年之后，你将有人去充实第二个，以后就可以照几何级数进步了。如你在一个学校有了那样一群的好教授，你再留心，他们不会他适吗？他们的环境好吗？他们的经济充裕吗？他们的设备足用吗？

集中你的力量，靠了这个方法，只有靠了这个方法，你才造成智识的杰构，他的影响，行且普及于全国，而且尊生百倍，以为国家的光荣。

现在说到我们的第三种工作研究。

研究是智识努力的光荣结果。世界上的事业，还有什么比推进智识的疆界，增加智识的总和再要光荣的吗？征服人民和土地的英雄，亚力山大、拿破仑、成吉思汗、塔吗仑，若和征服自然的科学家相比较，真是可怜得很，他们的名字容许在历史小说上暂时存留，但他们的事业，则早已彼此相消了。但一个盖理略[2]，一个牛顿，一个达尔文，同其余无数的为真理及智识的进步而工作的科学家是不朽的，他们智识的势利，是人类进步的界碑。

大科学家是天生的，非人力所能造成，但他们的出现，也没有我们想象的稀少。只要有了适当的环境，有了机会，他们就会出来。诚然，出群的天才，即最浊的环境，也不能把他埋没。但如你看见一个天才在挣扎出头的时候，你的义务就在助以一臂，把他的横身的束缚解放了，给他一个机会，使他得到他相当的地位，如让羡慕与妒嫉，夺去他的能力，使他不能不屈就苦

1　"岩"原文无，现据句意补。——编者注
2　伽利略。——编者注

役。你要记着，他一旦回复了他的能力，定要弄得栋折榱崩，把你们这些俗物的头一齐打坏。

我们不要死等超人由我们中间出现。我们人人具有相当的能力。我们所需要的不过是计划、方法和成功的愿望，一个人把他的能力分散在许多小事体上，决不产生有永久价值的结果。也许一个教员，替初入门的学生选出一个问题，是应该的；但是有经验的科学家，应该晓得他要研究的问题。如你已经选得一个专门，就得锲而不舍。许多年前美国地质调查所所长，斯密生学会秘书古生物学家瓦尔可提君（Charles D. Walcott）告诉我说："选定你的专门，并且粘着不放，如我粘着寒武纪（Cambrian）的化石，自从起始由寒武纪岩石采集第一个三叶虫一样。"在那时候，我正从地阿尼[1]（Devonian）地层中采集了第一个化石，研究这一类的化石。到现在，我听了瓦尔可提的话，还在研究地阿尼的化石。

这不是说你须绝对的把你自己限制在一个学科以内。一个单调限制的智识菜单，不能令人生活，正如一个人只食米饭一样。若是这样，你的心上容许要生脚气病，那是不得了的。但正如米饭或面包是生命的正粮一样，一个专门也是你智识生命的正粮，不管后来你加入什么作料，只要你记着，过于丰美的肴馔，太吃久了，是一样有害的。

直捷言之，一个研究家应该把他研究的地方和研究的方法，正正确确的计划出来。但你的计划中不要太谦逊了。要是你是一个鱼类学家，作一个研究中国所有鱼类的计划，你容许不会完成你的计划，大分你仅能做到计划中的一部分；但你有了计划，你就可有统系的进行，即使你自己不能完成，他人可以继续你的工作，而你的远见阔视，也可以永垂不朽了。我所谓有统系的鱼类研究，并非仅指鉴定和命名而言，我的意思是包括一切有关系的问题，如形态学、胎生学、分类学、生理学、分配情形、习惯、对于人类的效用，及其他你可能发见的问题在内。你研究的结果，最好以正当的形式在一个丛刊里面发表，使将来有继续刊布的可能。换言之，不要把你的结果，在那一千二百零一种的动物学杂志上分散了，这个数目，是秉农山[2]博士告诉我的。

我可以奖励一个有望的青年古生物学家，不要把他的兴趣限制于两手蚌类[3]，但对于珊瑚类、斧足类或三叶虫类，也不妨试一试。但我也劝他对于一

1　泥盆纪。——编者注

2　秉志，字农山。——编者注

3　疑为腕足类。——编者注

个科目必须完全了解，如他对于长身贝比任何生存或死去的人知道的多，他在古生物学会中就是一个较大的光荣和较好的资产了。

我现在要对于所有的研究家说几句话：专精一科，对于一科无论如何狭隘，须有绝对的了解。自后，如你能不失这个目的，你可以旁行斜出，在路上遇见美味你可任意取食，只要你的眉来眼去，不把你的真爱抛弃就得了。

关于合作的问题又怎么样？没有科学家或一群的科学家可以说能绝世独立的。科学是国际的，科学家比什么人都应该为国际的人。科学家的结合，没有国家人种的隔阂。科学家应该以科学的进步而非以一国或一群人的光荣为目的，如因达到他的目的，同时也使他的国家或城市得到光荣，那么，他的国或城应该高兴，并且去尊敬他。但是他们不要忘记，他的成功是建筑在一群工人所筑的基础上，这一群工人是由世界各国智识共和的团体中找出来的。

现在我要对于将要离开智识的襁褓，即他们的母校的青年们说几句话。不要想做培根，把所有的智识都放在你的范围以内。这样的事在培根的时代容许可能——我仍不能无疑——在现在是绝对不可能了。不要如某青年的计划，划出一年来从新审定所有的物理学及化学上的定律，并且从新把他们改定使合于你的深奥的思想。我不信培根觉得他一生光阴能做这样的事，虽然培根的时代科学律并不多；而你虽然有过人的能力，至少须要好几年才能把这样巨大的事体做了，试想自培根以来科学上的发明又有几多了。老实说罢，不要圈定不可能的范围，认定你车上的南针，但先须注意你车辐与车轴，并且不要忘记了车轴的油。

从一个问题的一小部分着手，用你的全力去从事，你若能详尽的叙述一个古珊瑚虫或一个生的星鱼，并且尽你的力量找出他的底细，比你用哲学及逻辑作引导而写一本古生物学或动物学，或在图书馆寻些材料写成一本进化论好得多。不要写教科书。等你教过一二十班学生之后，再问问你自己是不是知道的够写教科书了。即使书馆的人要求你写，你自己也应该顾全名誉，等到你有成功的把握，再写不迟。我也曾经受过书馆编辑人的引诱，订约写一部地质教科书，但我答应之后，几年不敢动笔。我希望我能等到现在的几年以后，更可以把中国的地质放些进去。现在只好等到再版的机会，加以订正，要是我有时间的话。

最后的一个劝告，今天听众的大多数，是用不着的，但你可以传递给旁人去，不要匆遽及肤浅，在你的徒弟还未做完以前，不要就自命是师父。这个话看似鲁莽，实则是由好意流露出来的。因为我晓得你的各方都有引诱，所以不能不警告你。有许多重要的地位正等着你。你或者被请去做某学府的教授，因为你是一个有望的青年，并且除你之外没有第二人。你或者就接受这个请求。若是情境逼迫你非去不可，我也不能劝你不接受；但不要忘记，一个头衔并非就是成绩的索引。你容许有成大人物的可能，但若是你的选任，使你觉得已经成了大人物了，那我想你还是辞去的好。

如你能使地位成为发展你的能力的机会，当然不迟疑的接受，但不要以为地位就是你的成功，使你同在此道中长久工作的前辈们同等了。如是可能的话，暂时不忙，等到你有了机会用功深造，真正成功了领袖之后，再行接受不迟。

对于此次到会的各位同事，同道及学生们，你们此次到北平来报告一年来科学的进步，和鼓励大家前进，我谨举自己有关系的一个世界最古的学术团体——独逸皇家科学会的一句格言，作为临别赠言："不断的努力。"

我们须不断的努力，去推进我们每一种专门智识的前线，因为如此，才不愧为科学先进的继承者。我们尊敬过去的成绩，同时不要为前人的传述所束缚；我们若幸而发见新真理，须要猛勇无畏的去拥护他；我们也要不迟疑的去承受反乎传述的新理论，若是我们的研究证明他的无误。

新境教我们新义务，

过时的美物都成了古怪；

你若要和真理并驾，

必须不断的前迈。

——原载《科学》1930 年 14 卷 6 期；

亦载《东方杂志》1930 年 27 卷 13 期

Contributions to Geologic Science by Graduates of the National University: An Address Delivered at the Celebration of the Thirty-first Anniversary of the Founding of the University

By Professor A. W. Grabau

The intellectual development of China today is along lines formerly undreamt of by her leaders in thought. In old China philosophy, literature, and history occupied the attention of the intellectuals, while science was represented by mathematics and astronomy, both of which were more or less subordinated to philosophy. Today the Natural Sciences have come into their own in China as elsewhere, and they are beginning to take their place in the forefront of intellectual endeavor.

This new development was in no small measure directed by the National University of Peking, which has always been China's foremost institution of higher learning. I believe it is an unquestionable fact, that this university has been the center of training in science in China, and that in the majority of cases, where the natural sciences are cultivated in other institutions in this country, the work is in charge of men who have graduated from, or were connected in a teaching capacity with the National University of Peking. This is true at any rate so far as the geological sciences are concerned, and these are the only ones I am in a position to speak of. And it is the graduates of this university, together with those of the geological school of the Survey, which for a time took over the work of the University in that field, which have developed geological sciences in China and brought it to its present stage of advancement. And much of this has been accomplished during the last decade of intensive work, stimulated by the activities of the National Geological Survey, under its present and its former directors.

Today, a large proportion of the men active on that Survey and on the provincial surveys, which have been created in various parts of China, graduated from this university during the last ten years. One of our present students in the geologic department, Mr. C. S. Kao, of the class of 1930, has compiled for me a list of the graduates in the Geological department from 1920 to 1929 inclusive. These total 144, a number which must not be compared with the graduates in this science in the American or European universities of today, but rather with those of one hundred or one hundred and fifty years ago, when this science was first taught in the university of the west. The oldest school of geology was connected with the school of mines in Freiberg in Saxony, which is in the south of Germany. It was in 1775 that Johann Gottlob Werner[1] was appointed Professor of Mineralogy in that school, and he not only taught that subject, but virtually founded the science of geology, of which he was its first teacher. When in 1910, I visited this institution, I found it still very active in this field, and since coming to China I have realized that Freiberg has a special interest for us, aside from the fact, that it was the ancestor of all geologic institutes, for it was in Freiberg, that our own Professor Wang Liegh[2], the dean of our geologic department had his European training.

I do not know how many students Werner had, but I doubt if the number during the first decade of the existence of his school was equal to that of our own institution.

We know that many of the famous geologists of the early days began as students of Werner at Freiberg, and moreover "men already distinguished in science studied the German language and came from the most distant countries to hear the great oracle of geology" (Lyell). But we must not forget that geology in those days was an embryonic science. Indeed, it was little more than mineralogy transferred from the laboratory to the world at large, and to the history of human civilization, and much of this was speculative, having the old philosophic flavour. It was its novelty and the attractive way in which it was presented by Werner, that made it at once popular and instrumental in raising a small school of mines,

---

1　应为 Abraham Gottlob Werner（1749—1817）。——编者注

2　此处指的是王烈（1887—1957），长期担任北京大学地质学教授、系主任。——编者注

formerly unheard of in Europe, to the rank of a great university.

But today geology is an exact science and the century of investigation and thought given to it has developed it to such an extent, that no single mind can encompass it all, and that to acquire the mastery of even a small branch, requires years of intensive study, such as none of the older geologists were confronted with.

Indeed, as I look back on my own career in geology, I realize that in my student days, geology was not nearly so difficult a subject as it is today, and that text books were much simpler and geological literature much less voluminous than is the case at the present time. I have seen the first text book of geology that was published in America by Professor Amos Eaton, the first American teacher of geology and it contained a little more than forty pages.

It appeared in 1818 but was preceded by Maclure's *Observations on the geology of the United States*, published in 1809, and Silliman's *Mineralogy of New Haven* in 1806. But neither of these two were of the nature of a generalized text book as was that of Amos Eaton.

It was this man Amos Eaton who founded the first school of geology and Natural History in America. This was in 1827, little more than one hundred years ago and the school still continues as one of the most famous scientific schools in America, the Rensselaer Polytechnic Institute, at Troy, New York. Thirty years ago, when the National University of Peking was founded, I was professor of Mineralogy and geology at this oldest American school of science.

It may interest you to learn that the early classes in that school of sciences were no larger than those of the department of geology in this university today. Thus, the class of 1832, which graduated five years after the foundation of the school had only four members, and only one of these, James Hall, became a distinguished geologist, but he justified the existence of the school by becoming a leader in that science. A second graduate of that class, also trained as a Naturalist, was S. Welles Williams, who later developed into the distinguished diplomat and oriental scholar, who spent a large part of his active life in China. He was the most eminent student of the Chinese language and dialect that America has produced, and in his later years he became professor of Chinese and Oriental literature in

Yale University. The other two members of that class were of no importance in the development of science. One became a physician and the other a railroad president.

The first class to graduate however, that of 1829 was larger, for it contained eleven men. Of these, two died early, three became physicians, two lawyers, one a teacher, one a minister, one a civil engineer and only one a geologist, but he also became eminent in American science.

Another great scientific American school, that of Harvard University of which I myself am a graduate, and which was founded twenty years after the Rensselaer school, had only three graduates in the first year, only one of which became an eminent geologist, while the graduates in the department of geology of the Massachusetts Institute of Technology, of which I was one in 1896, were only three in number, but all continued in geology. And one other, besides myself, came to China, at least temporarily, namely M. L. Fuller, the oil geologist, who carried on oil explorations in Shensi some years ago.

If then we look at the number of graduates in geology during the past ten years, 144 in number or an average of over 14 for each class, we realize that the beginnings in China are in no way behind those in America or Europe. And one third of these graduates, or an average of nearly five for each class, are still active in geology in different parts of China, and some of these at least have made important contributions to this Science. If I can mention only a few of these, it in no way reflects on the others, but these few have had exceptional opportunities, and they have been active long enough to make accomplishment possible. For one of the great lessons which have to be learned, is that science is a difficult mistress, who has to be wooed long and earnestly before she admits the wooer to her confidences. Four years of instruction in geology and kindred science is not sufficient to make a geologist. It only prepares the student for the task of making a geologist of himself, for after all no teacher can ever make a scientist, the scientist must be made by the forces that lie within himself. The teacher can only give opportunity for expression and development, and guide the students to the sources of knowledge, and aid him in making the most of his individual powers. What China needs above everything is an institution of higher learning, which has not only undergraduate courses, but has a graduate department in science, where those who have completed their preliminary work can develop to the full the spirit of research, and acquire the proper knowledge of their own powers for independent work and develop these powers to the state of effective productivity. Only such an institution can truly claim the name of a university and it is into such an institution that Peita must develop, if it would retain the leadership in science as in other subjects in China. When it has so developed, those of its graduates who propose to continue in a scientific career, will no longer be compelled to go abroad to get, not so much the opportunity for advanced work, for that lies right here, but rather the recognition of such work as embodied in the degree of Doctor of Philosophy or Doctor of Science, for then this university will itself confer such a degree upon its worthy graduates. And there is no doubt that this Institution has produced such worthy graduates in the past and will produce them in the future. Some of those of former years have taken their degree abroad, but others have found their work so absorbing here that they have been unwilling to interrupt it long enough to obtain a foreign degree, and still others are steadily climbing upwards to that eminence, the attainment of which is signalized by the conferring of the highest academic degree. And here too we might mention the fact that this Institution has become known sufficiently elsewhere to bring two foreign students to its halls, both of which carried on advanced work. One of these, a Japanese, worked in geology and is now editor of the Geological Survey of the Manchurian Railway. The other from Czechoslovakia received a doctor's degree in Prague on the strength of the work in palaeontology done here. Other foreign students had planned to come to Peking to study at this University, but were prevented by the disturbed political conditions.

Let me then mention by name a few of those who have contributed to the geological literature on China even though in their modesty they would prefer to remain unnamed! But it is by their work that we will know them, and such work deserves popular recognition on an occasion such as this. I shall of necessity refer only to those who have published their contributions, or otherwise have done work with which I am directly familiar, but you must understand that there are many others who are doing excellent work, some as teachers, others as members of surveys, while some are still working abroad to prepare themselves for more specialized work.

Of the older graduates I shall only mention two, but the work of these has reflected credit on their Alma Mater. Foremost among them in geology stands our own Professor L. Wang of the first graduating class. Prof. Wang went to Freiberg, where he devoted himself primarily to mineralogy and petrography, though not neglecting other branches. Ever since his return from Germany he has been a member of the staff of the geological department which he helped to organize, and for a number of years he has been the dean of the department and guided it safely through all the difficulties of the recent years. As teacher and administrator he has been a prominent influence in the development, not only of the department of geology, but of the University as a whole. As a public lecturer and writer (chiefly in Chinese) he has made his influence felt in wider circles. His varied activities have been fittingly recognized by his recent election to the Vice Presidency of the Geological Society of China.

Dr. C. Ping was also an early graduate of the University, afterwards entering on graduate study at Cornell University, in the U. S. A. Though he devotes his energies primarily to recent Biology, publishing many important contributions to zoology, and successfully directing the activities of the Biological Laboratory of the Science Society of China, and those of the Fang Memorial Laboratory[1] in Peking, he has recently entered the field of palaeontology and has already produced two important monographs for the *Palaeontologia Sinica*, one on the Fossil Cretaceous Insects of China, and the other (in press) on terrestrial Mollusca from North China.

Coming now to the class of 1920 which graduated before my connection with the National University, we find that several of that class continued in geological work, and with two of these I have been personally associated. Foremost in this class I would mention Dr. Y. C. Sun whose excellent work as teacher and investigator is well known to you. Dr. Sun has already several monographs to his credit and has published many short papers. His monograph on the Cambrian faunas of China is a contribution of first rank, and was moreover

---

1　应为 Fan Memorial Institute of Biology（静生生物调查所）。——编者注

the first monograph on Palaeontology published by a Chinese, a performance which was duly celebrated at the time. Dr. Sun was Chinese delegate to the International Geological Congress held in Madrid, and acted as Vice President and chairman of the meeting on several occasions. Dr. Sun graduated before I came to China, but for a number of years he was my assistant, and later my colleague at this university, besides holding professorships at several other institutions in China, and being a leading member of the National Survey, and for a number of years the secretary of the Geological Society of China of which he is an original fellow. Dr. Sun, while abroad, not only obtained the degree of Doctor of Science at Halle, but also made extensive acquaintance among the geologists and palaeontologists of Europe, with whom he made many field excursions, acquainting himself with the classical grounds of European stratigraphy. The results of his studies are now benefitting our students in the advanced courses.

Another member of this class Mr. J. Y. Wang, now holds a professorship in Anwhei University, but it has never been my fortune to know him well, or to be able to know his work, which I am sure reflects credit on his Alma Mater.

Mr. S. T. Chien, another member of this class however, has keen from the beginning one of my associates on the Survey, where he efficiently fills the roll of librarian. Although a graduate when I came, he attended my first course in palaeontology, and he and Dr. Sun accompanied me on my first geological expedition to the Kaiping Basin, where we obtained much interesting material, as well as clearing up some obscure stratigraphic points.

Perhaps the best way of summarizing the contributions to geology and palaeontology made by the graduates in this university, will be by calling attention to the various papers that are published in the journals and Survey memoirs, and similar scientific publications, and the elections to membership in scientific societies. The Geological Society of China was founded in 1921 and the first volume of the bulletin appeared in 1922. It is of interest to note that among the charter members of the society elected, this university was represented in addition to the officers of the geological department, by Dr. Y. C. Sun, while the following students of the National University (Class of 1923) were elected as associate members at the first meeting.

1.	Dr. S. C. Chang	
2.	Y. T. Chao	
3.	T. F. Hou	
4.	C. C. Tien	
5.	Dr. P. Tsai	
6.	Dr. K. M. Wang	
7.	Dr. C. C. Young	

Of these the first and the last three received their doctor's degree abroad all except Dr. Tsai, who received it in America, taking their degree at a German University.

At the second annual meeting in 1923 the following 13 students of the Geological Department of the National University were elected associate members.

Elected in 1923

8.	C. T. Chang	Class 1924
9.	Y. S. Chang	1924
10.	Z. Y. Fong	1925
11.	H. T. Han	1924
12.	K. H. Hsu	1925
13.	Y. T. Hsu	1925
14.	Y. T. Liu	1924
15.	W. P. Sun	1924
16.	C. C. Wang	1924
17.	S. S. Yoh	1924
18.	C. C. Yu	1924
19.	H. T. Yu	1925
20.	H. S. Yuan	1924

Making a total of 20 students in the membership of the Society, subsequent elections were as follows:

Elected in 1924

21.	Y. J. Liao	1925
22.	N. Y. Mo	1925
23.	C. Wang	1925

24.	H. J. Chen	1926
25.	Y. T. Liu	1924
26.	W. P. Shu	1924

Elected in 1925

27.	K. H. Chang	1925
28.	T. C. Wu	1925
29.	C. T. Chu	1926
30.	M. C. Liao	1925
31.	H. K. Sun	1925
32.	H. C. Sze	1926
33.	T. H. Ting	1926
34.	T. Ting	1925
35.	C. Wang	1925
36.	S. W. Lo	1926
37.	S. Y. Yih	1926

Elected in 1926

| 38. | T. L. Lee | 1927 |

Elected in 1927

| 39. | T. W. Young | 1928 |
| 40. | T. K. Huang | 1928 |

Elected in 1928

41.	Y. S. Chi	1930
42.	Y. C. Wang	1927
43.	W. C. Pei	1927

In the following list, only those papers are given which have appeared since the beginning of 1920, but it must be noted that there are a number of papers now in preparation, both by former and by present students of the University. An analysis of this list shows that the graduates of the year 1923 have contributed the largest number of papers, five of these graduates having contributed 34 papers. Among these, the work of Mr. Y. T. Chao stands out strikingly with 18 titles, four of these representing monographs, in the *Palaeontologia Sinica*. That Mr. Chao should be taken from us in such a tragic manner, is an irreparable loss, not only to the University of which he stood forth as one of its most brilliant sons, but to science in China and to geological science as a whole. Mr. Chao died in the performance of his duties and his last act was an attempt to save invaluable data from destruction. His energy, his devotion to his science, and his tireless activity in the advancement of geologic science in China, must forever be an inspiration to those of us who have known him. It is such men as Mr. Chao that China needs for the development of its science, and no higher tribute can be paid to his memory than by the determination of his fellow graduates, and of all the young men, who take up this science in China to follow in his footsteps, to devote themselves heart and soul to the science of which he was so able a champion, to resolve to add as much as possible to the sum of knowledge in the accumulation of which he was one of the pioneers, and to aim at that accuracy of detail and that soundness of deduction which so eminently characterized his work. Let his life and his death be an inspiration to the sons of Peita, and help them to give the best that is in them to their Alma Mater, their country and their science.

Although his name does not appear in the list of publications given here, the work of Mr. W. C. Pei of the class of 1927 deserves especial mention. It was largely through, his indefatigable industry and devotion that the epoch-making discoveries at Chou-Kou-Tien have been made; for though he was not the first to discover the remains of the Peking man *Sinanthropus pekinensis* his were the discoveries of some of the most important of these remains, including the now famous skull, which represents Mr. Pei's latest find.

Partial List of Published Papers of Graduates of the National University of Peking from 1920 to 1929

Dr. Y. C. Sun. 1920

Contribution to the Cambrian Fauna of North China. *Palaeontologia Sinica.* Ser. B. Vol. I, Fasc. 4.

Upper Cambrian of Kaiping Basin. *Bull. Geol. Soc. China.* Vol. II, No. 1-2.

Cambrian, Ordovician, and Silurian of China. Extrait du Compte-Rendu.

XIVe Congrèe Géologique International 1926. Madrid.

Mundsaum und Wohnkammer der Ceratiten des Oberen deutschen Muschelkalks. Leipzig, 1928.

History of the Palaeontological Research work of China. *Scien. Quart. Nat. Univ. Peking,* Vol. I, No. 1.

Dr. C. C. Young. 1923.

Fossile Nagetiere aus Nord-China. *Palaeontologia Sinica.* Ser. C. Vol. IV, Fasc. 3.

Topographic Features of Nankou Range in the Vicinity of Nan Kou Pass. *Bull. Geol. Soc. China.* Vol. II, No. 1–2.

Dr. K. M. Wang. 1923.

Die obermiozänen Rhinocerotiden von Bayern. *Palaeontologische Zeitschrift.* Bd. 10., Heft 2. Berlin.

Die Fossilien Rhinocerotiden des Wiener Beckens. *Mem. Inst. Geol.* Shanghai. No. VII.

Ein Versuch zur Neugruppierung der Europaeischen Dinotherium-Arten nach den Zaehnen. *Ibid.* No. VII.

Y. T. Chao. 1923.

Stratigraphy of Lin-cheng Coal Field, Chihli Province. (with C. C. Wang and C C. Tien) *Bull. Geol. Surv.* No. 6.

On the Stratigraphy of the Tze-chow and Liu-Ho-Kou Coal Fields of S. Chihli and N. Honan (with C. C. Tien) *Ibid.* No. 6.

Geology of I-chang, Hsing-shan, Tzekuei and Pa-Tung Districts, W. Hupei. (with C. Y. Hsieh) *Ibid.* No. 7.

Succession of the marine beds in the Chang-chiu Coal Field of Shantung. *Ibid.* No. 8.

Carboniferous Stratigraphy of South Manchuria. *Ibid.* No. 8.

Geology of Western Chekiang. *Ibid.* No. 9.

Geology of Kaiping Basin and its Environs. (with C. Y. Lee and T. F. Fou) *Ibid.* No. 12.

Productidae of China. Part I. *Palaeontologia Sinica*. Ser. B. Vol. V, Fasc. 2.

Productidae of China. Part II. *Ibid*. Ser. B. Vol. V, Fasc. 3.

Fauna of the Taiyuan Formation of North China. Pelecypoda. *Ibid*. Ser. B. Vol. IX, Fasc. 3.

Carboniferous and Permian Spiriferids of China. *Palaeontologia Sinica*. Ser. B. Vol. XI, Fasc. 1.

The Structure of the Nankou district. *Bull. Geol. Soc. China*. Vol. II, No. 1-2.

A Study of the Silurian Section at Lo Jo Ping. W. Hupeh. (with C. Y. Hsieh). *Ibid*. Vol. IV, No. 1.

The Mesozoic Stratigraphy of the Yangtze Georges. (with C. Y. Hsieh). *Ibid*. Vol. IV, No. 1.

Geology of the Gorge district of the Yangtze from Ichang to Tzekuei with Special Reference to the development of the Gorges. (with J. S. Lee) *Ibid*. Vol. III, No. 3-4.

On the age of the Taiyuan series of N. China. *Ibid*. Vol. IV, No. 3-4.

Classification and correlation of Palaeozoic coal-bearing formations in N. China. (with J. S. Lee) *Ibid*. Vol V, No. 2.

Brachiopod Fauna of the Chihsia limestones. *Ibid*. Vol. VI, No. 2.

C. C. Tien. 1923.

Stratigraphy of Lin Cheng Coal Field, Chihli Province. (with C. C. Wang and Y. T. Chao). *Bull. of Geol. Surv*. No. 6.

On the Stratigraphy of the Tzechow and Liu Ho Kou Coal Fields of S. Chihli and N. Honan. (with Y. T. Chao). *Ibid*. No. 6.

Carboniferous Crinoids of China. *Palaeontologia Sinica*. Ser. B. Vol. V, Fasc. 1.

Stratigraphy and Palaeontology of the Sinian Rocks of Nankou. *Bull. Geol. Soc. China*. Vol. II, No. 1-2.

Study on the Stratigraphy of the upper Palaeozoics in Central Hunan. *Mem. Inst. Geol*. Shanghai. No. VII.

Report on the Shuikou Shan Lead and Zinc Mine, Hunan. (with C. C. Liu and C. Y. Ou Yang) *Geol. Surv. Hunan. Bull*. No. 1. Econ. Geol. No. 1.

A Study of the Devonian Sections in Changsha and Siangtan districts,

Central Hunan. *Ibid*. Bull. No. 1. Geol. No. 1.

Report on the Shang Wu Tu Manganese Deposit, Central Hunan. (with H. C. Wang). *Ibid*. Bull. 4. Geol. No. 2.

Report on the Siangtan Gypsum and Salt Deposit Central Hunan. (with S. Y. Kuo and H. C. Wang). *Ibid*. Bull. No. 4. Geol. No. 2.

Report on the Pan-Hsi Antimony Mine, Yiyang. (with S. Y. Kuo and H. C. Wang). *Ibid*. Bull. No. 5. Geol. No. 3.

Hsichih Chang. 1923.

Die Funktion des Kauapparates bei den Proboscidiern. *Palaeobiologica*— Jahrgang II. Bd. II. 1–3. 1929.

T. F. Hou. 1923.

Geology of Kaiping Basin and its Environs. (with Y. T. Chao and C. Y. Lee) *Bull. Geol. Surv. China*. No. 12.

K. C. Yu. 1924.

Coal Fields of Puchi, Kiayu, Hsienning, Chunyang and Wuchang districts, Hupeh Province. (with C. Li and V. P. Shu) *Mem. Inst. Geol.* Shanghai. No. IV.

Geology of Siangyang, Nanchang, Ichang, Chingnen, Chung-hsiang and Chingshan districts, North Hupeh. (with V. P. Shu) *Ibid*. No. VIII.

S. S. Yoh. 1924.

A Geological Reconnaissance from Chung Ching, Szechuan to Kueiyang, Kueichou Province. *Bull. Geol. Surv. China*. No. 11.

Geological Reconnaissance of W. Kueichou. *Ibid*. No. 12.

Geological Reconnaissance of S. Kueichou. *Ibid*. No. 12.

Notes on the Geology of the Pou Mu Chung Old Field near Kueiyang, Kueichow Province. *Ibid*. No. 12.

On a new genus of Syringoporoid coral from the Carboniferous of Chihli and Fen tien provinces. *Bull. Geol. Soc. China*. Vol. VI, No. 3–4.

On the Occurrence of Lyttonia fauna in the vicinity of Kwei-yang, Kwei-chow province. *Ibid*. Vol. VI, No. 1.

H. S. Wang. 1925.

Igneous Rocks of Miao Fengshan and Tiao-chi shan in the Western Hills of Peking. *Bull. Geol. Surv. China*. No. 11.

Geology and Mineral Resources of Moling and Mishan district, Kirin Province. *Ibid*. No. 13.

Geology along the valley of the Neng River, Heilungkiang Province. (with H. C. Tan) *Ibid*. No. 13.

The Tayeh iron deposit. *Bull. Geol. Soc. China*. Vol. V, No. 2.

Rectanguler graphs as applied to the Proximate analyses of the Chinese Coals. *Ibid*. Vol. VII, No. 2.

C. L. Ho. 1926.

The Igneous Rocks of Yangsin, Tayeh, and O-cheng districts, Hupeh Province. *Mem. Inst. Geol.* Shanghai. No. II.

T. K. Huang. 1928.

Geology of the coal field of Fou Hsin Hsien, Jehol Province. (with C. C. Wang). *Bull. Geol. Surv. China*. No. 13.

A Preliminary Report on the Geology of the Hsiao-Shih Coal Fields, Penchi Hsien, Fengtien Province. *Ibid*. No. 13.

On the Cambrian and the Ordovician Formations of Hsishan or Western Hills of Peking. *Bull. Geol. Soc. China*. Vol. VI, No. 2.

Some notes on the contact between the Yangyang Granite and the overlying Tiao Chi Shan Beds. (with S. Chu). *Ibid*. Vol. VII, No. 3-4.

C. Y. Lee. 1928.

Geology of Kaiping Basin and its Environs. (with Y. T. Chao and T. F. Hou) *Bull. Geol. Surv. China*. No. 12.

S. Chu. 1928.

Upper Palaeozoic Formations and Faunas of Yaoling, Chenhsien, S. Hunan. *Bull. Geol. Soc. China*. Vol. VII, No. 1.

Description of two species of Chaetetes from the Moscovian of North China. *Ibid*. Vol. VII, No. 3-4.

Some notes on the contact between the Yangfang Granite and the overlying Tiao Chi Shan Beds. (with T. K. Huang) *Ibid.* Vol. VII, No. 3-4.

V. P. Shu. 1929.

The Result of a Geological Survey of the Hongshan Intrusion, North Honan. *Bull. Geol. Soc. China.* Vol. III, No. 2.

Coal Fields of Puchi, Kiayu, Hsienning, Chunyang and Wuchang Districts, Hupeh Province. (with C. Ki and K. C. Yu) *Mem. Inst. Geol.* Shanghai. No. IV.

Geology of Siangyang, Nanchang, Ichang, Chingnen, Chunghsiang and Chingsha district, North Hupeh. *Ibid.* No. VIII.

Y. Akasegawa. Post Graduate.

Notes on the microstructure of some coals from N. China. *Bull. Geol. Soc. China.* Vol. V, No. 2.

F. N. Kolarova. Post Graduate.

The Generic Status of "*Tripleria*" poloi. *Bull. Geol. Soc. China.* Vol. IV, No. 3-4.

——原载 *The Science Quarterly of the National University of Peking,* 1930

北大毕业生对于地质学之贡献

葛利普博士在北大三十一周年纪念大会上之演说词

胡伯素译

韶光易逝，转瞬又是一年。曾忆三十一周年纪念日，葛博士着其三十年前受哈佛大学博士学位时之博士服，登台演说，口若悬河。声似鸿钟，命意尤恳挚警惕，以见博士爱护本校之忱，无微不至。当时全场听众莫不欢呼感动。嗣彼复为文节录其演词，发表于《北大自然科学季刊》，译者读竟，觉其语重心长，激人至深，感人亦至切。顷阅报载，美人顾林[1]在中华教育文化基金委员会席上，提议年拨二十万元赠与本校聘请研究讲座及专任教授，以救济大学教育，是其所期望于本校者亦复甚大。因译此文，私意盖欲使爱我北大者，能闻风兴起，知所当务，庶三十三周年时博士之一片热望，不致再如三十二周年之落空，并期有以副一般外人关心我国教育者之美意耳！又篇末原附之毕业生著作表关于地层学及古生物学方面者居多。现代科学研究愈精，分门亦愈细，如古生物学即有离地质学而别成一独立科学之势，至地质学三字则另有所指；葛先生原文为 Geologic Science，而不言 Geology 者，要以地质学三字含义过狭未足以尽括一切，故本题以译"北大毕业生对于地质学的科学之贡献"为宜，但为简略易读起见，暂从现译。再该一九二〇年至一九二九年间北大毕业生著作一览表其中文名称可查北平地质调查所印行之地质图书目录，内有中外名称对照表，甚为了然。至诸短篇作品在地质会志上刊出者，皆不见于该目录，则可检阅该会志，亦附有中文译名。本篇之末，恕不附译。——译者识

中国今日智力发达正趋向于昔日思想界领袖所梦想不及之途径；曩者中国几惟哲学文学与历史为智识界注意之的，而科学以数学与天文代表之，此

---

1 即顾临（Roger S. Greene，1881—1947），长期担任美国中华医学基金会（China Medical Board）和北京协和医学院负责人，并长期在中华教育文化基金会中任职。——编者注

二者亦仅哲学之附庸而已。至今日则自然科学之进展，一如他国，已蓬蓬勃勃，形成独立之地位，且有居学术界优越境地之趋势焉。

北京大学为中国最高学府，此种新的发展受其指导之力居多。余尝深信无疑，北京大学乃中国科学之发祥地，诸如其他大学，虽多致力于自然科学，但一考其工作人员，则或为北大毕业生，或曾在北大执鞭主教，类皆具有深切之关系。余他所非知，即就地质学方面而言，可谓地质学由萌芽而进步以达今日之兴盛时期，皆北大毕业生与北京地质调查所地质研究班毕业生之劳绩（在先北大尚无地质系，故所中之学生曾称盛一时，今则该研究班已停办，而北大毕业学生日多，已成并驾齐驱之势矣。）虽然，于此得力于国立地质调查所前后诸所长之热心领导，亦匪浅鲜也。

时至今日，该所及其他诸省立地质调查所内之中坚分子，多为北大十年来之毕业生。吾生一九三一级高振西君为余作得一自一九二〇年至一九二九年间地质系毕业人数表，计达一百一十四人。此数目只可拟之于百年或一百五十年前地质学方在西方诸大学设教时毕业生之人数，自不足与今日欧美诸大学毕业人数相提并论也。至言研究地质最早所在地，要推德国萨克索尼（Saxony）之佛莱贝吉（Freiberg）矿务学校。一七七五年维纳（Johonn Gottlo Werner）[1] 被任为该校矿物学教授，顾彼不特讲授是科，并于地质学亦树其基，而为教授地质学者最早之一人。一九一〇年余参观该校知于地质方面仍努力弗懈，待余来中国始感佛莱贝吉大学与吾人尤富兴趣，此则非独因其为地质学校之始祖，抑以我地质系主任王烈先生，即曾在其中受过洗礼，而为先生之母校也。

维纳究有若干学生，余莫得其详，但彼校最先十年之人数，或与我校十年来人数相等，亦未可知。

维纳之名，当时诚赫赫不可一世。吾人知许多著名老地质学家，慕维纳之学，皆负笈归之，重新开始为其门生。如来耶路（Lyell）所云："凡有科学智识者，皆研究德文，来自四方，敬聆此地质大贤之创见雄言，如水赴壑，盛极当时。"惟吾人所应毋忘者，即彼时地质学尚在胚胎时期，未见发达，实言之所谓地质者，至多不过将矿物学自实验室中移而公之于世，于人类文化史中略备一格而已。其大部份仍不外理想虚构，有旧哲学之意味，而无真正科学上之价值，所以令人惊异，哄动一时者，或在因维纳提升一素不知名之

1　当为 Abraham Gottlob Werner（1749—1817）。——编者注

矿务学校为大学，正式研究地质，使成有用之科学，独创一帜，先开风气，为可钦耳。

顾今日地质学已成为真正之科学，而集百年来群策群力，研究弗息之结果，已包罗广大，推敲入微，故非一人之心力所能穷其究竟；即欲精通一分门，亦非若干年潜心玩索不为功。此种困难，老地质学家，绝无遭逢者。

盖余回顾过去个人从事地质之经历，知余在学生时期之地质学，并不若今日之为一种繁难科学，教科书既单简，而地质书籍亦非如今日之卷帙浩繁，长篇累牍，余曾见美国伊顿（Amos Eaton）教授所著之第一本地质教科书（美国第一个地质教师），要仅五十余页耳。

此书在一八一八年始出世，但在其前出版者：一八〇九年已有墨克莱（Maclure）[1] 著之《美国地质之观察》（*Observations on the Geology of the United States*）与一八〇六西里曼（Silliman）之《矿物学》（*Mineralogy of New Haven*），此二者则皆不若伊顿所著者之有系统有精彩也。

伊顿即美国地质学与自然历史之首创者。此为一八二七年事，距今已百有余年矣。而此校今日仍为美国有名之一科学学校，即纽约之乌雷生拉多科学校[2]（或高等工业专门学校）（Rensselaer Polytechnic Institute）。三十年前北京大学肇始时，余正为此美国最早科学学校之矿物及地质教授。

该校最早诸班，并不较今日北大地质系人数为多，如其一八三二级在该校成立后五年毕业，仅有四人。而惟赫鲁（James Hall）已然崭露头角，成为有数之地质学家，该校之名亦因以大显。其次当推维里蒙（S. Wells Williams）[3] 氏，维氏初为博物学家，后渐成为外交家及东方文学家，在中国居留甚久，颇负盛名，允称美国人通晓中国语及中国文之第一级人物；晚年为雅礼大学之汉文及东方文学教授。其他二人，一为医士，一则为铁路经理，于科学发展上皆无关重要者也。

惟该校第一班毕业者计十一人，尚不为少，其中二人早亡，三人业医，又其二为律师，一为教师，一为公使，一为土木工程师，而为地质家者仅一人，今在美国科学界上亦有声誉。

再哈佛大学为美国之另一大科学学校，成立后于乌雷生拉大学二十年。余即为该校之一毕业生。忆初年毕业者三人得成为有名之地质学家者一人，

1　William Maclure（1763—1840），苏格兰裔美国地质学家。——编者注
2　今一般译为伦斯勒理工学院，下同。——编者注
3　即著名外交官、汉学家卫三畏，返美后，1877 年被聘为耶鲁大学汉学教授。——编者注

而在麻沙朱萨次工业专门学校（Massachusetts Institute of Technology）地质学毕业者（余为该校一八九六年毕业生）仅三人，现皆继攻地质，除余外尚有飞纳（M. L. Fuller）[1]，乃石油地质学家，数年前曾到中国陕西探查油矿，盘桓颇有时日。

　　故吾人苟一追忆过去十年间地质系一百十四毕业生之人数，知每班平均超过十四人，是中国起始情形，不较欧美任何学校为差，而此中三分之一，即每班殆平均有五人，仍在中国各地从事地质工作，已有足纪之贡献。现余试就少数人言之，固不足以表扬其他，不过此少数人赋有特殊机会，并工作甚久，其成就亦自不同，盖无何足怪也。诚然科学一物，如百求莫得之情妇然，苟非潜诚求之弥坚，彼将永不使求婚者有获其信托心之日，此实为求学者之一莫大教训，故在学校研习四年地质功课，决不能成为地质学家，仅使学生预备能作成一地质学者所应作之工作，因一教师，并无造就一科学家之能力，科学家惟利用自己特有之能力以自为之耳。为师长者，仅能领导学生，使有发展自我能力之机会，并示以求学捷径，使学生循之以寻其欲得之果，而不致虚掷其能力。在中国今日所最需要者：即为一寻求高等学问之学校，其中不仅须有大学部，且必具有科学研究院。彼初步求学工作已完毕者，可以继续充分发展其研究精神而由各人性之所近，自由从事专门研究，因此种特殊能力之充分发展，自有意外之美果可言。如此学校方名副其实而不愧称一大学。北大将欲仍在中国学术界保持领袖地位，则必建立于此种情况之下而后可，设已臻此境地矣，彼毕业学生愿继续作高深之研究者，不致迫而出洋，以度彼留学生活，于研究之机会，并不甚多；而探求学位之心理，则反甚迫切，因彼等常认研究工作完全公于科学博士或哲学家博士等学位之中，故每不能得实学。同时北大亦可择成绩优异者，自行授予学位，北大过去诚已造就此种成绩卓殊之毕业生，而将来人才更将济济辈出，亦可预言。北大毕业生固已有在外国取得学位者，或在国内研究孟晋，不愿中辍，以求外国学位者，更有正在国外力图上进，以冀荣膺大学最高头衔者，皆奋电所学，殊堪嘉慰也。附此更有一事，足值一述者，即有二外国学生，震于北大之名，特远道来学，此二人皆曾进行研究工作，一为日人，现充南满铁路地质调查所编辑，一为捷克斯拉夫人，因在中国研究古生物，著为论文，在巴拉加[2]

1　Myron L. Fuller（1873—1960），美国地质学家、采矿工程师，曾于 1913—1915 年与 Frederick G. Clapp（1879—1944）在陕西、甘肃地区进行地质考察。——编者注
2　今译布拉格。——编者注

（Prague）得博士学位，更有外国学生原拟来北京就学者，但为政局不定所阻，多作罢论矣。

现余请就研究中国地质，刊有论著者题名介绍之，虽以彼等之谦虚有加，其名或不甚显，惟吾人将由彼等所成就之工作，而必能予以认识，此等工作，亦殊值趁此机会，使诸君知之。但予于此拟仅择刊有著作，或与余共事甚久，相知最稔者，数人言之，此外尚多贤者，或执教鞭，或为地质调查所人员，正着手其伟大事业；亦有方在外国，一心探讨专门学问者，不能于此一一予以介绍，私心引为至憾！

关于老毕业生，余将仅述二人，其工作皆足使其母校，引以为荣，其最著者：首推第一班地质系毕业生现任教授王烈先生，先生初去德国佛莱贝吉大学习矿物岩石学，同时对于他科亦恒加注意，自德国归后，即服务地质系，襄助改革诸大务，而为地质系主任亦历有年所，本系饱受艰困，先生皆苦心维持，不辞劳瘁，先生近来复掌学校行政方面事务，故不特于地质系之发展诸多擘划影响，就整个北大言，亦多蒙其嘉赐，彼为一讲演家与著作家（大半为中文），故其声名远播，近年被选为中国地质学会副会长，足证先生之努力，已引起一般人之敬仰也。

秉志博士亦北大最早之毕业生，后入美国康乃耳（Cornell University）研究学院肄业。彼初专攻近世生物学，发表动物学方面论文颇多。历任中国科学社生物研究所及北平静生生物研究所所长，近则致力于古生物学，已为《古生物志》刊有重要论著二种：一名为《中国白垩纪昆虫化石》，他一册则在印刷中，名为《中国北方之淡水田螺》（译者按，此书现已出版）。

一九二〇级适于余抵中国前毕业，其中数人仍继习地质，与余私人关系较密者二人，一为孙云铸博士，彼之学术教材，已为诸君所习知，孙博士已有专著数种，短篇文品尤多，其《中国寒武纪化石》一书允属出版物中上选，亦关中国人著古生物学者之先导，彼曾被派为中国代表，出席马得里[1]世界地质学会会议，并数为大会副主席及分组主任。孙博士虽在余抵中国前毕业，但在北大充余助教有年，近与余同掌教席，并在他校兼任教授，亦为北平地质调查所之主要人物，自始即为中国地质学会会员，居秘书职数年。在外国时得哈鲁（Halle）大学科学博士学位，与欧洲著名地质学家及古生物学家会见殆遍，多相往还，并常相偕作野外旅行，凡欧洲大小各地有地层学上研究

1 即西班牙首都马德里。——编者注

之价值者，莫不亲身跋涉以赴之，故其考察所得，殊与后学以无穷之裨益也。

与孙博士同班者，尚有王若怡先生，现为安徽大学教授，惜余从未与彼谋面，或睹其佳作，但余信其亦深足为母校增光也。

同班钱声骏先生，自始即在所中为余之一助手，而司图书管理之责。余初次教古生物课，彼即来听讲，后彼与孙博士随余赴开平之第一次地质旅行，所得重要材料甚多，关于地层学上疑难之点，亦多所析明与更正也。

总之，欲明北大毕业学生对于地质学及古生物学之贡献若何，不若就各种杂志《地质汇报》及他种科学刊物上发表之文章，及选为科学会社会员者二事予以注意，不难推想其余。缘中国地质学会成立于一九二一年[1]，而第一卷《会志》[2]即于次年出版，地质系教授及孙云铸博士皆特许为会员，而下列诸学生（一九二三级）则在第一次会议通过允为会友：

1. 张席禔博士	5. 蔡堡博士
2. 赵亚曾	6. 王恭睦博士
3. 侯德封	7. 杨钟健博士
4. 田奇瑰	

其中除蔡堡君系在美国取得博士学位外，余则皆取自德国各大学。[3]

一九二三年举行第二次年会，地质系学生被选为会友者十三人其名如下：

8. 张竞择（1924 级）	15. 乐森璕（1924 级）
9. 张永寿（1924 级）	16. 袁熙绶（1924 级）
10. 韩修德（1924 级）	17. 方仲仪（1925 级）
11. 王庆昌（1924 级）	18. 徐光熙（1925 级）
12. 刘元斗（1924 级）	19. 许源道（1925 级）
13. 舒文博（1924 级）	20. 余新都（1925 级）
14. 俞建章（1924 级）	

至此前后入会者共二十人，以后年有增加，一九二四年有下列诸人入会：

21. 廖友仁（1925 级）	23. 王震（1925 级）
22. 莫迺炎（1925 级）	

1　应为 1922 年。——编者注

2　《中国地质学会志》。——编者注

3　此处翻译疑有误，葛利普原文谓：上述第一和最后三人（即张席禔、蔡堡、王恭睦、杨钟健），除蔡堡在美国取得学位，其他均从德国取得学位。参见本书第 264 页。——编者注

一九二五年入会诸生：

<table>
<tr><td>24. 张国祥（1925 级）</td><td>29. 斯行健（1926 级）</td></tr>
<tr><td>25. 伍廷琛（1925 级）</td><td>30. 丁道衡（1926 级）</td></tr>
<tr><td>26. 朱鉴堂（1926 级）</td><td>31. 丁同（1925 级）</td></tr>
<tr><td>27. 廖鸣基（1926 级）</td><td>32. 罗绳武（1926 级）</td></tr>
<tr><td>28. 孙锡琨（1925 级）</td><td>33. 叶向荣（1926 级）</td></tr>
</table>

一九二六年 H. J. Chen。

一九二七年加入者，有杨曾威、黄汲清二君，皆一九二八级生。

一九二八年加入者：计荣森（1930 级）

　　　　　　　　　王有中（1929 级）

　　　　　　　　　裴文中（1927 级）

（译者按：上列诸人毕业及加入学会年期，皆略有错误，且人数亦有遗漏，如观会志上所载自一九二二年该会成立起至一九二五年止之会员一览表，尚有王恒升、谭德勋、梁尚志、陈旭等多未列入，一九二九年又有常隆庆、李陶、郁士元、高振西、潘钟祥及胡伯素六人正式入会，葛先生亦未提及，而H. J. Chen 则又不知所指为谁，此皆须增入或更正者也。）

再在篇末所附著作为自一九二〇年以来已见刊行者（从略）。此外尚多毕业及在校学生正在从事著述，未及发表者，概未列入。吾人试分析此表，知一九二三级毕业生贡献独多。其中仅五人已有著作三十四种，而赵亚曾君一人又占十八种，四种为《古生物志》中专刊，诚有足为吾人所钦佩无已者，如赵君者，乃遽尔惨离吾辈而逝，又非特北大之损失，抑亦中国科学界及整个地质学所受之一大打击也，赵君之死，盖为其职责而死，其最后之挣扎，亦无非为恐丧失其零碎记录之残稿而已，赵君此种从事科学之毅力与决心，及其对于地质科学不断之努力，吾人诚应永矢弗谖，引以自励。

中国欲发扬科学，诚非赵君一类人才莫属。凡此后之毕业学生及中国青年从事地质学者若欲纪念赵君，则最大之纪念品莫若抱定决心，会集精神，一志所学，使地质学多所发明，亦须注意观察之精细，推断之明确，以步赵君之后尘，使赵君之生前死后皆足为北大诸子之楷模，果能因此刺激，倍加努力，为母校、国家及科学竭尽其能，斯亦毋负赵君矣。

再一九二七级裴文中君之名，虽不见于著作表中，其工作亦不能不特别一述。周口店第三纪人类化石之发见，实裴君不辞劳苦，备尝艰辛所得之珍贵代价，裴君虽非发见北京人遗骨之第一人，但其所发见者，实为此遗迹之最重要部分，即现在著名之北京人头颅骨，此不得谓非裴君最后惊人之成绩也。

——原载《北大学生》1931 年第 1 卷第 3 期

Palaeontology[1]

By Amadeus W. Grabau

Palaeontology is a new science in China, but fossils have been known to the Chinese since ancient times. In fact, the true nature of fossils was recognized by Chinese philosophers some 300 years before Leonardo da Vinci pronounced fossils to be the remains of once living organisms, and by his insistence upon this interpretation, put an end to the fanciful theorizing about the nature of these bodies and the wild speculations of the old philosophers of the Western world.

Leonardo can no longer be given the credit of being the first to understand the true nature of fossils, for Chu-Hsi wrote in A.D. 1200 as follows: "In high mountains there are shells. They probably occur in the rocks, which are the soils of older days, and the shells once lived in the water. The low places became high, and the soft mud turned into hard rock."[2]

It is noteworthy that Chu-Hsi recognized the fact that the mountains had been elevated since the day that the shells enclosed the living animals and were buried in the soft mud of the water bottom. Leonardo, on the other hand, supposed that the shells found in the Apennines indicated that the sea once stood at that level. He evidently had no notion of the origin of mountains as due to deformation and elevation after the formation of their rocks.

But while Chinese philosophers had reasoned correctly regarding the true nature of fossils at such an early date, the scientific study of palaeontology has lagged considerably more than a hundred years behind that of Europe. Hardly a decade has elapsed since the first scientific study of fossils was undertaken in China, though for many years previously, fossils collected in this country by

---

1 原载陈衡哲主编 *Symposium on Chinese Culture*（Shanghai, 1931）一书的第九章；此书 2009 年由王宪民、高继美翻译，以《中国文化论集》为名由福建教育出版社出版。——编者注

2 《朱子语录》：尝见高山有螺蚌壳，或生石中，此石即旧日之土，螺蚌即水中之物，下者却变而为高，柔者却变而为刚。（原注无标点。——编者注）

foreign expeditions, were scientifically described in foreign publications, and occasional specimens, brought to Europe by travellers and missionaries, were identified by European palaeontologists.

When however the Geological Survey of China was organized in 1913, it was felt by those in charge of the undertaking that the time was ripe for the development of palaeontological science in China, and the collection and description of Chinese fossils by native geologists and palaeontologists.

The first collection of fossils made by a Chinese scientist was the result of a field expedition into Yunnan by Dr. V. K. Ting in 1914. At that time there was no one in China ready to undertake their study and they were sent to America for identification. For several years, however, they received only cursory attention, and identification was merely attempted by students, the older men being too much occupied with their own problems.

It was not until active work in palaeontology began in China, toward the end of 1920, that these fossils were recalled, having meanwhile suffered preliminary identification and misidentification to a moderate degree; however, most of the specimens were returned in their original state, when they had not suffered the loss of their locality label. Since then, all but a very small part of this, material has been studied, and the results published, while the remainder will soon be identified.

Active fossil collecting in the field of invertebrate palaeontology was begun by the Survey palaeontologists in the spring of 1921, and has been carried forward on an ever increasing scale since then. The collecting of vertebrates and plants, however, had begun earlier.

From a collection which could be housed in a few drawers, the material has grown, until at present it requires several thousand drawers to store the unstudied portion, while that which has been described and determined, occupies scores of museum cases. Every season new material is brought to the Survey laboratory, and the palaeontological staff finds it ever more difficult to keep abreast of it. The staff at present numbers nine palaeontologists, one artist, three section cutters and one preparator, besides mechanical assistants. Though some of these palaeontologists are also active in other departments of the Survey, more than half their number devote all their time to the study of invertebrate fossils, and it is hoped that when some of the men now active in the field return, those, whose training has been along palaeontological lines, will be able to undertake the study of some of the material they have collected.

When it is remembered that China is a hundred years or more behind other countries in the study of its fossils, and when it is further realized that neither accurate work among the Paleozoic rocks, nor the mapping of the territory and the determination of the geological structure is possible until the fossils have been studied, it is evident that there is at present no more pressing problem in Chinese geology than the description of its fossils and the determination of the index species characteristic of the successive horizons and subdivisions. For a long time collections of fossils have come from individual and scattered formations, and while it was possible to determine the larger horizons to which these subdivisions belonged, and to designate those species which appeared to be most characteristic of those formations, the proper determination of the inter-relation and succession of the various members of the larger series has not been possible. Heretofore, comparisons had to be made with European and North American standards, there being so far no Asiatic standard for the succession of the geological formations. But a successful beginning in the development of a local standard time-scale for this continent has been made during the recent extended expedition into South China under the leadership of Dr. V. K. Ting, and the independent explorations by the late lamented Y. T. Chao and his associate Mr. T. K. Huang. The collections made in these expeditions are now undergoing critical study, and in the course of time we shall have a standard scale which shows the succession of the fossil associations or faunas from the earliest to the latest division of the great Palaeozoic era. And we shall likewise learn more about the range in time of many genera and species, when they made their appearance, and when they left the stage to their more specialized successors.

Already we have many a hint of great surprises in store for us. Already we begin to realize that the firmly fixed European and American standards, which guided the work of western geologists, are inapplicable to Eastern Asia. Here we find that many types, which previously we have looked upon as confined to very definite levels in the geological succession, because they showed such definite time limitations in the older standards, appeared in this country long before they reached Europe, and continued here long after they had joined the limbo of dead and forgotten species in the Western world.

There is reason for this remarkable phenomenon; there is an actual and perfectly scientific explanation, and we are beginning to see what that explanation is likely to be. It is too early to dogmatize, assuming that dogmatizing ever is desirable. Still we may say that the evidence so far accumulated, points to a great centre of developing marine life through the Palaeozoic eras in the basin of the ancient Pacific, and that it was from this source that many a faunal assemblage was derived, that temporarily occupied the shallow seas which covered various parts of Europe, at each successive period in the earth's development. If this surmise proves correct, if the Pacific Ocean of Palaeozoic time was the center of continuous development of marine organisms, we need not be surprised to find that such organisms entered with the waters that transgressed over the shallow depressions in what now is China, and lived bere long before their undifferentiated or but slightly modified descendants could reach Europe, which was only possible when the inland waterways had become extended until they stretched from one end of the Eurasiatic continent to the other.

Since Western America also bordered the Pacific, the problem of migration to that continent was less formidable. But throughout Palaeozoic time, the inter-relations between America and China were chiefly confined to the Pacific borders of the former continent, or to such extensions of the Pacific Ocean as were able to penetrate some of the west American lowlands or geosynclines. For the most part, America's relationship seems to have been with an ocean that lay where now the polar ice-cap rests, but in Palaeozoic time this Boreal Ocean had temperatures high enough to permit the growth of coral reefs. And it was through this Boreal sea that Palaeozoic America communicated with Palaeozoic Europe, for into both countries the sea extended its arms, and sent its migrants at successive periods.

But this is a mere outline of a vast problem, the problem of the origin and distribution of the ancient marine faunas, in the solution of which it will be the privilege of the Chinese palaeontologists to take the leading part.

That Chinese fossils may be available in the study of this world problem, as well as for the more immediate purposes of the correlation of the formations in different parts of the country, and for the determination of the geological structure, it is not sufficient to collect and label these fossils, and to store them in the museum. It is imperative that they be described in the greatest detail, and illustrated by accurate figures, and this is a work that requires the attention of well-trained men. Fortunately for Chinese science, the founders of this branch in China were far-sighted men, and realized at once the importance of this work and set about to provide an adequate medium of publication. This is the *Palaeontologia Sinica*, which from the outset was planned to be issued in four different series, though of uniform size and make-up. Series A is devoted to fossil plants, Series B to the fossil invertebrates, Series C to fossil vertebrates other than man, and Series D to ancient man. The first fascicle appeared in the spring of 1922, and since then more than 50 fascicles have appeared, ranging in bulk from 20 to 500 pages or more and in number of plates from 1 to 50 or more. At the present writing there are some half-dozen fascicles passing through the press and every few months sees the completion of a manuscript for additional fascicles in one series or another. This productivity is unprecedented in the history of palaeontological science.

The great *Paleontology* of New York, in 12 quarto volumes, required more than 60 years for its completion. The Palaeontographical Society's *Memoirs* have appeared annually since 1847, and the great work of Barrande[1] on the Palaeozoic fossils of Bohemia, the first volume of which appeared in 1852, is still being issued at intervals. Barrande was a man of independent means. The British publication was sponsored by a wealthy society of Lovers of Nature, but Hall's struggles with a refractory legislature, to secure the means to carry on his great work, are historic. And history too tells how for years he kept the work alive at his own expense and at great personal sacrifice.

China is fortunate that this is an age of enlightenment in science, an age of

---

1 Joachim Barrande（1799—1883），法国地质学家、古生物学家，1831 年起定居布拉格，对波西米亚的古生物进行了广泛而深入的研究。——编者注

appreciation of the tremendous importance of Palaeontological work, and so China has made, and will continue to make, adequate provision for its continuance, until the fossils of China are as well known as those of other countries.

Though some of the fascicles of the *Palaeontologia Sinica*, especially many of those dealing with vertebrates, plants, and man, are the work of foreign men of science, Chinese palaeontologists nevertheless have taken a very active part in the production of these monographs, especially those of the invertebrate series, of which by far the larger number is to their credit.

It was an event of no small importance in Chinese Science when the first palaeontological memoir ever produced by a Chinese scientist was published. This was in December 1924, when Dr. Y. C. Sun[1] published his monograph on the Cambrian faunas of North China, comprising 110 pages in English, 24 in Chinese and 5 large plates. In this work Dr. Sun described 55 species of Cambrian fossils from China, of which 45 species and varieties were new to science, despite the fact that Cambrian fossils from China had been so extensively described by the late Dr. Charles D. Walcott and other foreign palaeontologists. Dr. Sun also described and named 8 new genera of trilobites. This publication was an event duly celebrated at a dinner at which many Chinese as well as foreign men of learning were present and offered their congratulations. Since then Dr. Sun has produced two other memoirs for the *Palaeontologia Sinica*, one on Upper Ordovician and Silurian graptolites of China, and one on the Ordovician trilobites of Central and South China. Both of these are ready for the press and will be published soon. In addition to these Dr. Sun has published many shorter palaeontological and stratigraphical papers in Chinese as well as in foreign journals.

Among the graduates of the class of 1923 of the University, that is those that began their palaeontological studies at the opening of the past decade, three men have produced memoirs of outstanding merit which have been published in the *Palaeontologia Sinica*. These are Mr. C. C. Tien[2], Dr. C. C. Young[3], and the late Mr.

1　孙云铸（Y. C. Sun，1895—1979），古生物学家，曾留学德国哈勒大学。——编者注
2　田奇㻹（C. C. Tien，1899—1975），古生物学家。——编者注
3　杨钟健（C. C. Young，1897—1979），古生物学家，曾留学德国慕尼黑大学。——编者注

Y. T. Chao[1]. Mr. Chao was unquestionably the most brilliant palaeontologist that China has produced, and his untimely death at the hands of bandits, while engaged in geological research in Yunnan, was a blow from which Chinese science will be long in recovering. Fortunately Mr. Chao was a man of great mental force and personal magnetism, and during his short career as student, investigator, and teacher, he succeeded in inspiring a number of young men who hold his memory in reverence, and their ambition to reach the eminence where Chao stood, is the force which carries them through days of arduous labour in the science in which their beloved friend, comrade and teacher had laboured so unremittingly and reached such outstanding success.

Mr. Chao's contributions to Chinese Palaeontology include four great monographs in the *Palaeontologia Sinica*. The first of these was a monograph on the Productidae of China, Part I, published as Fascicle II, of Vol. V, on Sept. 1927. It comprises 244 pages of English text, 23 of Chinese text and 16 large plates. This was the first time that these difficult brachiopods of the Chinese rocks had been described in great detail, only a few species having heretofore been known from China. In this monograph Mr. Chao described 61 species distributed through 9 genera and sub-genera. Twenty-two of these 61 species were new to science and 3 of the sub-genera were erected by Mr. Chao. This work had gained him recognition among scientific men in all countries, and had he lived, this and his subsequent work would surely have brought him official recognition from foreign scientific bodies.

His second contribution appeared only three months later, on December 10th, 1927, as Fascicle III of Vol. IX. It is a study of the pelecypods of the Taiyuan formation of North China, and comprises 64 pages of English text, 10 of Chinese, and 4 plates. In it Mr. Chao described 30 species, distributed through 20 genera. Twenty-three of these 30 are new to science, and most of the others were previously unknown from China. He also gives a table showing the distribution of these fossils in China, and a bibliography of 33 titles. The excellent plates are reproductions of photographs of these fossils, taken by Mr. Chao himself.

1　赵亚曾（Y. T. Chao，1898—1929），古生物学家。——编者注

In his third contribution, Mr. Chao returns to the Productidae. This was published as Part II of the Productidae of China in October 1928, in Fascicle III of Vol. V. It comprises 81 pages of English text, 5 of Chinese text, and 5 plates reproduced from photographs taken by Mr. Chao himself. In this monograph Mr. Chao described 14 species of *Chonetes*, 9 of them new. He further described a new species of *Aulacorhynchus*, two of *Strophalosia* one of them new, one of *Aulosteges* and a new species of *Tschernyschewia*. He also described additional species of Productidae and amplified the classification which he previously devised.

After completing the manuscript and plates of his fourth memoir, a critical study of the Carboniferous and Permian spiriferids of China, and reading part of the proof, Mr. Chao left for two years of field work in South China, an expedition from which he never returned alive.

This work was published as Fascicle I of Vol. XI of series B of the *Palaeontologia Sinica*. It comprises 133 pages of English text, 6 of Chinese, and 11 plates, reproduced from photographs taken by Mr. Chao. He never had the satisfaction of seeing this work in print, but, like his other memoirs, it constitutes a monument to his zeal, accuracy of work, scientific acumen, and phenomenal productivity.

The number of species described in this work is 42, of which 30 are new to science, and very few of the others had previously been known from Chinese rocks. In conformity with recent methods of refined work, he subdivided the old genus *Spirifer* into eight divisions, which are given the rank of genera, and he discussed in detail the basis on which these subdivisions were made, and the characters of these genera, one of which is his own. He also gave a table showing the distribution and range of the *Spiriferids* in the Upper Palaeozoic of China. Among many other outstanding merits, this work takes account not only of the internal character of these fossils, but also of the variation in the different stages of development, and the changes in proportions of the various dimensions. Not only then, does this work make known many of the *Spiriferids* of the later rocks of China, but it also sets a standard for future work of this kind.

Mr. Chao had planned a number of other monographs on Chinese fossils, for which the material had been brought together, but it was his plan to add extensively to the collection before he undertook further study of them.

In addition to the four monographs mentioned, however, he has published fifteen other stratigraphical and palaeontological papers in the *Bulletin of the Geological Society of China* and in the Bulletins and Memoirs of the Geological Survey. Many of these embodied the results of his own field work, but among them there are also some important palaeontological contributions, notably his description of the brachiopods of the Chihsia limestone, a formation of the greatest significance in Chinese stratigraphy. In recognition of his many contributions to Chinese Palaeontology, the China Foundation awarded him its $2,000.00 prize in 1928.

In April 1926, was published the monograph on Crinoids from the Taiyuan Series of North China, by Mr. C. C. Tien. This appeared as Fascicle I in Vol. V of Series B. and constitutes the second monograph published by a Chinese palaeontologist, preceding by some months Mr. Chao's first monograph. This is a notable contribution with 58 pages of English text, 5 of Chinese and 3 plates. Although the number of species described is only 12, they were all new except one, which was not specifically determined. One of the 4 genera is also new, including 6 species. But what makes the paper especially noteworthy is the success which Mr. Tien had in reconstructing the crinoid heads from isolated plates, which was all that was obtained in the field. This required a most careful analysis and detailed study of the plates, and much skill and patience in fitting them in place. These are the first crinoids ever obtained from China. Mr. Tien has also published a number of shorter papers on stratigraphy.

The third member of the class of 1923, Dr. C. C. Young, published in August 1927 his memoir on Fossil Rodents from North China. This was issued as Fascicle III of Vol. V in Series C, and comprised 82 pages of German text, 2 of Chinese, and 3 plates. These studies were made in Munich, Germany, under Professors Schlosser and Broili, and represented his dissertation for his doctor's degree conferred on him by that university. In it he described 31 species, 13 of which are new. The work is of interest, as representing not only a study of these abundant small vertebrate fossils from the later formations of North China, but also because it is the first monograph on vertebrate fossils produced by a Chinese palaeontologist.

Dr. Young has a second monograph on the press which will appear before long, and he has also published several shorter papers.

Two other members of the class of 1923 have published palaeontological papers on vertebrate fossils, but these have appeared in foreign publications and not in the *Palaeontologia Sinica*. Of these Dr. K. M. Wang[1] has published two papers on European fossil Rhinocerotids, and one on European Dinotherium, while Dr. H. C. Chang[2] has published one on Proboscidians, also in a foreign publication. All these papers are in German.

Both Drs. Young and Chang were members of the General Asiatic expedition in Mongolia[3], where they collected vertebrate fossils, while Dr. Wang is engaged in field work in central China, as is also Mr. Tien.

One of the most important monographs on Chinese fossils was published by Professor J. S. Lee, in September 1927, as Fascicle I of Vol. IV, Series B. This is on the Fusilinidae of North China, and comprises 172 pages of English text, 9 of Chinese text, and 24 plates, mostly of thin sections. Although similar fossils have been monographed by the palaeontologists of Indo-China, many of them from Yunnan, this is the first time that Chinese species of one geological province have been brought together, and the first work of this kind done by a Chinese palaeontologist. In this work are described 61 species, a considerable number of them new, and a great number of them heretofore unknown from North China. But the work is more than that. It includes the revision of the structural elements, and develops improved methods of analysis of these organisms.

Professor Lee, who is now the director of the Geological division of the National Research Institute[4], as well as head of the department of geology of the National University of Peking, is continuing his work on these organisms, and has indeed already published many preliminary papers. It is hoped that another monograph from his pen will soon be ready for the press. Among the

---

1　王恭睦（K. M. Wang, 1899—1960），古生物学家，曾留学德国慕尼黑大学。——编者注
2　张席禔（H. C. Chang, 1898—1966），古生物学家，曾留学德国慕尼黑大学。——编者注
3　指 20 世纪 20 年代由美国自然史博物馆组织的中亚考察团（The American Central Asiatic Expedition）。——编者注
4　即成立于 1928 年的 Institute of Geology, Academia Sinica。——编者注

multitudinous duties of his offices, and the many geological problems on which he is engaged, Professor Lee has found time to publish a number of papers of the greatest significance in Chinese stratigraphy.

But not only geologists have entered the ranks of palaeontologists; zoology and botany have also supplied recruits for the study of fossil animals and plants. Dr. C. Ping, the director of the Biological Laboratory of the Science Society of China at Nanking, and of the Fan Memorial Institute of Biology in Peking, has already published several monographs in the *Palaeontologia Sinica*. One of these on cretaceous fossil insects of China, appeared at the end of September 1928, as Fascicle I Vol. XIII Series B, of the *Palaeontologia Sinica*. This comprised 56 pages of English, and 7 of Chinese text, 3 plates, and 27 text figures. 18 species were described, of which 13 were new to science. Cretaceous insects are hardly known from other parts of the world, and hence this contribution is of value not only from the point of view of Chinese palaeontology and stratigraphy, but because it has begun to make known the insect fauna of a period from which hitherto little was available.

Dr. Ping's versatility as a palaeontologist and zoologist is shown by the fact that from insects he passed to fossil molluscs. His second monograph deals with fossil terrestrial gastropods from North China, and was published as Fascicle V, Vol. VI Series B, of the *Palaeontologia Sinica*. This comprises 30 English and 6 Chinese pages of text, and 2 plates, and in it 14 species are described, of which 12 are new to science. A second monograph on terrestrial molluscs by Dr. Ping is now on the press. He has also published descriptions of fossil turtles.

Dr. Hu[1], assistant director of the Fan Memorial Institute of Biology, and himself a leading botanist in this country, has also undertaken the study of some plant fossils, and some of his students are likewise engaged in the study of fossil plants both at home and abroad. It should be added that the first palaeobotanical paper ever published by a Chinese palaeobotanist, and indeed the first palaeontological paper ever published by a Chinese, was on fossil cretaceous plants by Mr. T. C. Chow of the Geological Survey. This appeared in 1923 in

---

1　指胡先骕（Hsen Hsu Hu，1894—1968）。——编者注

Bulletin V part II of the Geological Survey, and comprised 7 pages of text and 2 plates. In it Mr. Chow described 11 species, 5 of them new to science.

Shorter papers in which new species of Chinese fossils were described were also published by Mr. S. S. Yoh and Mr. S. Chu, and a monograph on the Ordovician cephalopods of the Yangtze Valley by Mr. C. C. Yü, appeared in the *Palaeontologia Sinica*, as Fascicle 2, of Vol. I, Series B, with 71 pages English text, 20 Chinese, and 9 plates.

Other monographs completed, or in course of preparation for this series by Chinese palaeontologists are:

1. On the Lower Carboniferous Tetraseptate Corals of Kweichow by Mr. C. C. Yu.

2. On the Permian Coral Fauna of South China by Mr. T. K. Huang.

3. On the Species of Chinese Syringoporidae by Mr. Chi.[1]

4. On Ordovician Graptolites by K. C. Hsu.

5. On Carboniferous Foraminifera of Kweichow by K. C. Hsu.

6. On Devonian Corals from South China by Dr. Y. C. Sun.

7. On Brachiopods of the Orthis bed of the Neichia Formation by M. S. Chang.

8. On Lower Triassic cephalopods of S. W. China by C. C. Tien.

9. On the Coral Fauna of the Chihsia Limestone by S. S. Yoh and T. K. Huang.

Modern requirements of palaeontological work are vastly more exacting than those of such work in the older days, where descriptions were at best a brief enumeration of external characteristics, and where a few genera were made to do duty as the receptacles of hoards of species. Moreover, variation was thought to be haphazard, and it was largely a matter of predilection where specific boundaries were to be drawn. Today, with our new conception of the meaning of species, and our recognition of the significance of the individual as a member of a definite genetic series, it devolves upon the palaeontologist, not only to draw his specific limits with much greater precision, but to base his species on a single

---

1 后发表为：Yungshen S. Chi（计荣森），*Lower Carboniferous Syringoporas of China, Palaeontologia Sinica*, Ser. B, Vol. 12, Fascicle 4, Peiping, 1933. ——编者注

individual, the holotype, and diagnose his other individuals in terms of genetic relationship to the holotype, and the degree of acceleration or retardation in the appearance of its various morphological characteristics, in relation to the species unit. This requires a much more detailed examination and evaluation of morphological characteristics, and an understanding of their significance in terms of organic growth, and this cannot be acquired by mere attendance at lectures and elementary laboratory exercises such as at best can be given to the undergraduate student. It requires intensive post-graduate work for a prolonged period of time, under proper direction and with opportunity for discussion of the problems with fellow workers in the same or related fields. Such work can only be carried on in a centre of research, such as at present exists only in Peking, and cannot, for many years to come, be established in other parts of China, because at present the number of those engaged in such work is too small, and most of the younger men are not in a position to carry on this work in isolation. When Chinese palaeontology has reached the point where its present young workers have acquired the power, knowledge, and self-reliance, that so eminently characterized the late Mr. Y. T. Chao, then, and not until then, will it be possible to found other centres of palaeontological research in China. And any attempt to predict when this may come to pass, would be premature, as it would leave out of consideration the personal factor, and the rate of mental evolution of the individual. With some it may be rapid; with others it may be slow, but neither the one nor the other can serve as a measure of the intellectual capacity of the individual.

There is another very important factor that must not be overlooked, and that is opportunity for access to palaeontological literature, and to collections of types and properly identified material, not only from China, but from other parts of the world as well. To attempt working without such aids means courting disaster. And when it is realized that the vast palaeontological literature of the world is scattered through such an immense number of publications, that even a well-equipped library finds it difficult to include it all, it must be apparent that the founding of separate palaeontological centres of research is not to be considered lightly, and without due understanding of the equipment that is imperatively required.

Fortunately, the library of the Geological Survey has made such progress in acquiring palaeontological literature, that it may be favourably compared with many a Western institute where research is being carried on. But that it is by no means fully equipped for its task, is only too well-known to the workers in this field. To make this one library as complete as possible should be one of the first aims of those who have the development of scientific work in China at heart, and available resources should first of all be turned in this direction, rather than used in an attempt to found partial and inadequate collections of such books and journals elsewhere.

That complete collections of literature should be acquired in other centres, where in due time palaeontological research is to be undertaken, goes without saying, but for such creation of other centres, and the adequate development for them of proper facilities, the time is not yet ripe. First and foremost must be the training of Chinese palaeontologists, and for that, time must be allowed and equipments perfected.

It need hardly be said that before fossils can be adequately studied, their collection and proper preparation must be undertaken. Fossil collecting is not an undertaking that every one is fitted for, nor can its technique be acquired from books or lectures. Intensive field training is necessary, and one of the prerequisites is a clear understanding on the part of the collector that the picking of a specimen here and there, especially without regard to its stratigraphic position, is worse than useless. Collecting must be systematic, and proper attention must be given to keep the collections from successive formations distinct. Moreover, collecting must be extensive as well as intensive, for the greater the number of individuals of each species, the more complete will be the result of their study.

Many of our Chinese students have already received good field training, but in the majority of institutions such training is not given. Moreover, according to the object to be collected, the training must vary. A good collector of fossil invertebrates is not qualified to collect vertebrate remains, which require a wholly different technique. Fortunately, several of our students have had an opportunity for training in this field, by men who have made it their life work, and one outstanding result of this is seen in the work of Mr. W. C. Pei at Chouk'outien, with the culminating achievement of the extracting and bringing to the laboratory the now famous skull of *Sinanthropus pekinensis.*

To master any field of science is an arduous process, and it requires endless years of labour and patient devotion, with but little hope of adequate reward other than intellectual achievement. China has many young men, willing and eager to devote themselves to pure science, and it becomes the foremost duty of the intellectual leaders of the nation to see to it that these young men are given the opportunity for full development, and are not forced to enter the arena inadequately trained and handicapped by the lack of proper equipment, to be vanquished in the struggle for achievement, with the inevitable stunting of their powers, and the resulting inestimable loss to Chinese science.

——原载 Sophia H. Chen Zen, ed., *Symposium on Chinese Culture*
(Shanghai: China Institute of Pacific Relations, 1931)

中国之古生物学[1]

葛利普著　张鸣韶译

古生物学在中国是一种新科学，但中国人知道化石为时确是很早，在赖拿豆达汶斯[2]解释化石为生物遗迹之三百年前，中国哲学家已论及化石之真正性质，朱子《语录》云："尝见高山有螺蚌壳，或生石中，此石即旧日之土，螺蚌即水中之物，下者却变而为高，柔者却变而为刚。"当时朱子不仅承认化石为生物遗迹，并且暗示山脉为海底綑绉而成，较之汶斯仅假设爱皮泥山[3]之蚌壳为海面高时所遗留，似又更进一步矣。

化石在中国虽有很早之认识，而古生物学之发达，确较欧洲迟一百余年，除以前有少数化石为外国调查团发表外，化石研究在中国不过仅有十年历史而已。

民国三年北平地质调查所成立后，中国地质家始有采集及研究之机会，然因当时古生物人才缺乏，所得标本，又须运至美国请人鉴定。自十一年后，地质调查所每年采集化石极多，即以未研究之材料而论，已由数十增至数千抽屉之多，现古生物部虽有多数专家及助手从事研究，而结果仍不免物多人少。

古生界地层构造及界限之确定，处处以化石为标准，中国化石，因研究稍晚，以致古生界地层迄无详细及精确之分类，故有时不得不用欧美地层为标准也。最近经丁文江博士及已故赵亚曾先生同黄汲清先生在中国南部之调查，古生界地层不久即可瞭然而有一种基本之分类矣。但由此次所采集之化石，经研究所得之结果，知欧美地层标准，似不能用于东亚，因在亚洲各层化石生存时间较在欧洲为早而且久，其原因概不外乎太平洋为古生代各种生物发育之地，此种生物到达中国因停留而变迁后，始能由内海徙至欧洲。美

1　节译自 Amadeus W. Grabau, "Palaeonotology," in Sophia H. Chen Zen ed., *Symposium on Chinese Culture*, Shanghai: China Institute of Pacific Relations, 1931, pp. 152-165，见前文。——编者注

2　即列奥纳多·达·芬奇。——编者注

3　即亚平宁山。——编者注

洲西部亦地滨太平洋，故当太平洋所到地方，皆与中国有相当关系，然在古生代美洲大部似被一温暖北极洋所浸绕也。

北平地质调查所鉴于中国化石之重要，及与世界各国关系之密切，遂刊行《中国古生物志》，内容共分四种：甲种为植物化石，乙种为无脊椎动物化石，丙种为脊椎动物化石，丁种为人类遗迹。自十二年第一册出版后，至今已印行三十余册，现尚有数册在印刷中。出品之多，较著名之美国纽约省及英国古生物志，有过之无不及焉。

《古生物志》中虽有少数外国科学家之著作，然多数仍为中国人之撰述，兹列举重要者如下：孙云铸博士之《中国北部寒武纪动物化石》，于民国十三年十二月出版，内述五十五种寒武纪化石，其中有四十五种为新发见者，孙先生近著有《中国奥陶纪上部及志留纪笔石化石》《中国中部及南部奥陶纪三叶虫》，不久想可付印。

次为北大地质系十二年毕业生赵亚曾先生，赵先生为人聪敏勤学，为中国近年来古生物界最优秀之人才，去岁在云南调查，为匪所害，诚为中国地质界莫大之损失也。《古生物志》中赵先生著作有四：第一为《中国长身贝科化石》卷上，十六年九月出版，册厚二百余页，讨论长身贝科化石至有九属六十一种之多，内有三新亚属二十二新种。第二为《中国北部太原系之瓣腮类化石》，十六年十二月出版，鉴定化石则有二十属三十种，其中有新属二新种二十三。第三为《中国长身贝科化石》卷下，十七年十月出版，所述化石有戟贝亚科十四种，内九种属新种或族；长身贝亚科七种，内二种属新种；小介贝系化石五种，内三种属新种；李希霍芬贝亚科二种，内一种属新种。第四为《中国石炭纪及二叠纪石燕化石》，十八年六月出版，在此著内，赵先生取石燕化石内部构造及各部进化程序，为分类基础，不仅详言石燕化石在中国之分布情形，且其研究方法亦可作后人之借镜也。

《中国北部太原系海百合化石》，为田奇瑰先生所著，十五年四月出版，内述海百合化石新属一，旧属三，新种八，新变种凡四。种类虽不甚多，而整理散板及再造工作，诚为不易。杨钟健博士之《中国北部啮齿动物化石》，于十八年八月出版，是著为杨先生在德国明星大学[1]博士论文，内容对于新种之讨论甚详。关于脊椎动物化石，在中国人著作中，当以此为嚆矢。

1　即慕尼黑大学。——编者注

　　李四光教授所著之《中国北部䗴科化石》为《中国古生物志》中重要著作之一，于十六年九月出版，详述䗴科化石六十一种，其中多为从前所未见者。李先生对于䗴科构造及研究方法，均有特别讨论。先生现任中央研究院地质研究所所长，虽职务纷繁，而于中国地层仍不时有重要之贡献也。

　　秉志博士为动物学家而兼治古生物学者，博士著有《中国白垩纪之昆虫化石》，十七年十二月出版，又有《中国北方之田螺化石》，十八年十一月出版。此外如俞建章先生之《中国中部奥陶纪头足类化石》《贵州石炭纪珊瑚化石》，黄汲清先生之《中国南部二叠纪珊瑚化石》，计荣森先生之《中国管状珊瑚化石》，徐光熙先生之《奥陶纪笔石化石》及《贵州石炭纪之有孔虫化石》，或在研究中，或脱稿在印，不久当可刊印行世也。

　　窃谓近来古生物研究之趋势，已与从前迥不相同，以前分类仅照化石外表性质，而今则须明白化石进化程序；以前分类甚简，而今则甚详。是以此种经验学问，绝非大学四年读书所能得到，非毕业后在一研究机关受高明学者多年之指导不可。目下除北平地质调查所外，别无此种机关，又因领导人才缺乏，他处何时始能设立，亦甚难言。

　　又余谓研究此种学问，有一极重要之事，即为书籍。书籍设备，为研究各种学问之先决问题，若研究古生物而无参考书籍，则更无法进行，然欲集世界各国出版品于一处，则亦殊为不易。地质调查所所设图书馆，虽可与欧美同样机关相比，然内容仍不能谓之十分完全，故募集巨款，扩充设备，使该馆成为一最完美之图书机关，诚为当今之急务也。

　　科学似海，毫无际涯，吾人治任何学问，必须有坚忍之毅力，及经久之研究。迩来中国青年之欲以科学为终身事业者，固不乏人，甚望智识先觉，予以相当之指导与鼓励，使勿入歧途，则中国科学幸甚。

——原载《科学》1931 年第 15 卷第 8 期

中国对于西方之贡献

葛利普博士在北平欧美同学会演说摘译

会程诸委员派我演讲"中国对于西方之贡献",可惜我不是史学家、汉学家、语言家、统计家或经济家,我只是一考古生物学者,但是诸君未必注意化石。我可告诉诸君最古的北京人(Sinanthropus),他的生活简朴,无器具,无火,也无遮盖,只不过藏于周口店洞中。他很快乐,如同爱丁(Eden)花园里的亚当(Adam),还有易徹[1](Eve)伴住。他不耕织,也不穿衣,或者即是因为这样,他就快乐满足。

从原人到现在,中国对西方的贡献很多,因为我不是上述的各种专家,我也不敢多说,又因为时间很短,我就中国的许多贡献中,只提起教育问题一端,希望中国发展成一较新较好的大学制。中国的大学还在胚胎时期,究竟模仿欧洲或美国大学制,尚不能决定。我以为中国不应采取两者,中国须创一新意思的大学贡献世界。美国人做事常喜欢照一标准办法,适于甲者以为亦适于乙,故人人均受同样教育。美国专门学校大致相仿,学生亦如出一辙。实则专门学校系为普遍人设立,大学则限于优秀分子。譬如专门学校注重运动及社交,大学则功课应当艰深,大学学生尤须经严格之天然选择,考试尚不济事。吾人可在大学工作中,发挥学徒的意思,只靠听讲或书本,没有什么特效,不论师傅如何负盛名,也不论他的口才如何流利,学生如跟随他工作,总比听讲好得多。阿格薛慈[2](Agassiz)叫他的门徒随他工作,虽是吩咐他们的功课很困难,但他们很有兴趣,他们有许多成为生物学家,享国际盛名。大凡学习必须分等级,起初不妨较简单,渐次高深,因此有天然淘汰,最适者才能生存,不及格者中途被摒。好尔[3](James Hall)教导初学古生物学的人洗刷化石,分类标明号数,然后派他们赴田野实习,采集标本,最后方叫他们辨识化石及笔记等等,从他的实验室毕业出来的,现在均为美国

1　现译为夏娃。——编者注
2　指瑞士籍美国博物学家路易·阿加西(Louis Agassiz,1807—1873)。——编者注
3　现译为霍尔。——编者注

科学界的领袖。我个人从我的老师甘心习徒多年，我的门徒分布三洲，他们也同样的从我习徒多年。

大学不仅靠房屋设备好，须要有人才能鼓励及指导学生，这班人才极少，不论何时何地有的，就要使他们环境适宜，利于工作，同时选择一群学生，准备学习，追随他们，组成一大学单位。每一地方须有一专门学校，使人人均有机会受教育，但大学数目须有限制，——实则何不限于一大学？譬言中国大学，其大学单位则随地兴起，总部则设于非政治非商业的中心。

全国任何地之人民选习工作，孜孜研究，吸引有志青年，将来成为领袖，且日与其老师接触，更足发扬领袖才能。然后一群学者，不论老少，同爱一种学问，不可分离，且互相激励，开拓人类知识之境界，——此为将来之大学——庶乎为中国对世界之贡献。

上述须勇敢卓识之士，废弃陈旧意思，破除保守性，或人以为此乃破坏举动，殊不知有时破坏极好，只要有较美大厦，不妨拆毁旧屋，新时机有新职务，古时以为善良者今日或非，须不绝前进，与真理并驱。（珣[1]）

——原载《科学》1931 年第 15 卷第 4 期

---

[1]　应为长期担任《科学》编辑的姚国珣（1902—1981）。——编者注

Why We Study Geology

By Amadeus W. Grabau

Why do you study geology? What can you possibly find of interest in the rocks? That is the sort of question that is often put to the student of the earth science. Some of those who ask such questions, think nothing is worthwhile that dose not bring in money. Others think that the highest type of intellectual life is that devoted to literature or philosophy.

We are all more or less egocentric. The things that concern man seem the most significant, and the more remote they seem to be from his immediate needs, the less worth-while do they appear. In the old days of ignorance, when much of that philosophy which Hu Shih calls "bad science" was compounded, man regarded himself as the center of the universe and thought that everything that was created was made for his benefit. The world around him was merely a back-ground for the portrait of himself, and the universe was the frame of the picture, in which he formed the central figure.

That old belief is shattered. We now more or less believe what we are told by the intellectual leaders of today, that man is only a part, and that a very small one, of a single unit, in a universe of inconceivable dimensions.

Still, though we may accept the classification, it is extremely difficult to think ourselves into the true relationship with the rest of the world, which that classification implies, and that is largely due to the fact that we do not know the world in which we live, nor the universe of which it forms a part. We know there are rocks and stones and soil; mountains, hills and valleys; rivers, lakes and oceans; trees, shrubs, and herbaceous plants; beasts, birds and insects on the land, and fish and other creatures in the water; and that is a fair summary of our knowledge of the earth. The vast variation that characterises the integers of our crude classification, their interrelations, history and mode of origin, are not only beyond our ken, but we hardly realize that they can be subjects for inquiry. And when such realization is obtruded on our attention, we brush it aside with that phenomenal lack of interest that is like nothing else than a manifestation of an undeveloped if not arrested infantile mentality.

There lies the mainspring for much of our behaviorism. An undeveloped mentality. A persistence of the nepionic state of the human mind. Often this is due to inadequate intellectual endowment, in other words, to the primitive mind, the mind of our primitive ancestors, in essentials, the neolithic mind. But far more often it is due to our faulty education. We lay the emphasis where it does not belong. We concentrate on a minute fraction of a part, before we have comprehended even the existence of the whole. We do not develop, we merely expand. Our childish interests and activities expand to become the vocations and avocations of our adult lives, and slowly as our mental arteries harden, we find it more and more difficult to get out of the groove which our life-habits have worn, and end by regarding the borders of our trench as representing the limits of the universe.

There are few who recognize that our modern educational system is an inheritance from the age of the egocentric obsession. We call this an age of science, but our western education still follows, to a large extent, the paths laid out in the pre-scientific eras. Nor has the almost indiscriminate and wholesale adoption of the western methodology by the east, proved a happy experiment, for here as there, it implies woeful neglect of fundamentals.

Can a man be a vital member of a common-wealth, when he knows neither its laws and its practices, nor its history and tradition? And can a man be a citizen of this earth in the fullest sense, when he knows not his own place in nature and has no understanding of the world he lives in, nor of the laws which have governed its development, and to which he, in common with all living beings, is subject? Why is it that superstition has held sway from the dawn of human consciousness and why does it still rule more than ninety-nine percent of the human, race, and dominates their mental, moral and social outlook? Why is religion, or at least the dogmas into which it has crystallized, such an all but universal force in the world of men, and why are fear, and its baseborn children, greed and cruelty, such dominant forces in human life. Why? Because man is ignorant; however educated he may deem himself. For education still means to most of us the development of the mind in a limited humanistic field. It does not imply understanding of, or putting ourselves in right relation with, the universe of which we form a part.

But you will say: can the study of rocks fit us any better in this respect? No, for the study of rocks is no more the whole of geology than the study of the alphabet is the whole of literature. Geology is the study of our earth, the "Science of the Earth" as the word implies in its orginal Greek derivation. And looked at in this comprehensive manner, it is one of the two great fields of human knowledge, the other being astronomy or astrology, the science of the rest of the heavenly bodies, and of the laws which govern them.

Let us try and analyze the earth, and separate it into its component units.

Most familiar to us is the earth's crust the *lithosphere* on which we live. But if we could see this sphere as an observer on the moon might see it, we would note that more that seventy percent of it is covered by water. Much of this constitutes the sea of which the oceans are component parts; but there are also the lakes, ponds, and rivers, and the ground water which permeates the upper layers of the crust, sometimes near to, at others distant from the surface. Taken all in all, the water visible and hidden, forms an essentially continuous layer over the entire earth, and this is the *hydrosphere*. Outside of it and enveloping the entire earth is the *atmosphere* or sphere of gas, and these three, atmosphere, hydrosphere and lithosphere, form the inorganic spheres open to examination. But after all, these are only superficial, and there are other parts, inner spheres, in the earth, the nature of which is revealed to us only indirectly, but of which we know enough even now, to purge it of the libel that it shelters evil spirits, or a Danteesque Inferno[1], for the punishment of errant mortals.

Then in addition to the non-living spheres we have the sphere of life or the

---

[1] 但丁《神曲》中的地狱，亦作 Dante's Inferno。但丁在《神曲》一书中描绘了地狱（Inferno）、炼狱（Purgatorio）、天国（Paradiso）三层灵魂境界。地狱位于耶路撒冷的地下，从地面通往地心，是一个巨大无比的深渊，一共分为 9 层；很多神学家认为炼狱也在地下，但但丁从道德的意义上把它想象为一座耸立于海洋的高山；天国则位于第九重宗动天之上，永静不动。参见但丁著，田德望译：《神曲》（地狱篇），北京：人民文学出版社，2002 年，16—21 页。——编者注

biosphere, the reality of which is sometimes difficult to appreciate. In and out among the mass of the water and of the air, and upon and through the upper layers of the rocky surface of the earth, the thread of living matter is woven, now forming a solid fabric, penetrable only with force, and again assuming the character of a network, the meshes of which vary in size, but are never discontinuous, except where momentarily broken by the hand of man, or by an abrupt or cataclysmic disturbance, such as the sudden denudation of a mountain side by an avalanche, or the destruction of life by the spreading lava stream. But even then such interruption is but temporary, for the barest mountain side, and the most inhospitable lava fields, will again be covered by a mantle of vegetation when time has brought its kindly influence to bear to mellow once more the harsh surface, and reduce it to a state of hospitality to the eager host waiting on its borders.

While all these spheres form parts of the earth as a whole and their interrelation and inter-dependences form the ultimate study of the earth science, each of these spheres also calls for independent study and investigation. Disregarding the central spheres, and considering only those that are open to direct study, we have the following subdivision of the science of geology.

1. Lithology; The study of the lithosphere.

This implies not only the study of the rocks of which the earth's crust is made, a subject generally referred to as petrology and including the more special division of mineralogy, but also the structure or architecture of the crust of the earth, its original character and its deformation; the physical and chemical causes which have brought about these changes; their form and surface appearance and the influence they have had in determining the physical geography of the country under the modifying forces resident in the other spheres. This is the division of our subject to which the term geology is most often applied in the limited sense, and various subdivisions of it have been elevated into special sciences. Most noted of these are mineralogy, petrology, dynamic geology (physics and chemistry of the earth), structural or tectonic geology, stratigraphy in a narrower sense, physiography and palaeogeography.

2. Hydrology;[1] or the study of the hydrosphere.

In the field we investigate the oceans, their extent and interrelations, their chemical constitution and physical activities in calm and storm, their varying temperatures and salinity, their currents, the contours of their bottoms and their shores, the fluctuations of the sea-level, and the modification of the sea-bottom; the chemical deposits formed on their borders and the mechanical erosion by the waves on the shores; the distribution and nature of their sediments and the influence that all these factors have on the distribution of life.

Here too, we study the ponds and lakes and rivers, the underground waters and the springs to which they give rise, the chemical deposits formed by them and the erosion, transportation and deposition of sediments by the rivers; and once again, the influence that all these continental waters bring to bear on the distribution of the living organisms.

3. Atmology; or the study of the atmosphere.

This is a division more familiarly known by the name of *meteorology*, because of an ancient misconception that meteors, —star fragments that enter our atmosphere, —were somehow the causes of climate and of weather.

Among the many special fields of study and investigation in the atmosphere, its chemical composition and physical constitution, the water vapour and other gases which it holds, and its movements are of the first significance. The many kinds of storm winds and the laws which govern them, the monsoons and the planetary winds and their influence in creating deserts on the one hand and regions of plentiful rain supply on the other in conjunction with the topography of the land, form a fascinating phase of this subject and one of far reaching significance. Indeed, it is not going too far to say, that a comprehensive study of the winds will help to solve many problems in the past history of the earth, which now appear obscure. For it must not be forgotten that winds have always swept

---

1 原文无分号，为使上下文统一，此处加了分号。——编者注

across the surface of the earth, though they have not always followed their present courses, and that they have been one of the great agents which modified the surface, and influenced the distribution of life. And when we listen to the reconstituted sound, that in silence has travelled for hundreds and thousands of miles through the air, we realize that we are on the threshold of another great investigation in the science of Atmology.

4. Biology; or the study of the biosphere.

The sciences which deal with life include not only those that investigate their structure and anatomy, their physiology and pathology, but also those that deal with their psychology and the wide range of activities attempted by the human mind. Too long the old idea, that man is distinct from the rest of the organic world, has dominated the minds of the philosophers and led humanity astray. But now the scientific study of the mind forms the basis of the new philosophy, and if man is regarded as a unit of the biosphere, all his activities, all his thoughts and aspirations, are legitimate subjects of enquiry for the specialist of those particular phases of the science of biology. And as a phase of biology they are in turn a part of the grander science of geology, that all comprehensive science, whose field of study is the earth and all that appertains thereto.

But biology is not limited to the study of the animals and plants which live today, for these are but the latest comers on the surface of the earth, the descendants of long lines: of ancestors, whose roots go back to pre-Palaeozoic time. For more than 500,000,000 years life has existed on this earth, and the records which the ancient plants and animals have left behind, buried in the strata of the earth's crust, are the fossils, from the study of which we are slowly piecing together the history of life upon this earth. This is called a special science, Palaeontology, but it is after all only the biology of the past, in so far as it can be studied from the imperishable parts.

This then is Geology in its broader aspect; and if culture means enlightenment in its widest sense, as distinct from erudition or deep learning in a special field, surely no one can be considered truly cultured, to whom geology, as here defined, is a foreign field, no matter how erudite he may be in his own particular narrow subject. I believe the day will come when it will be held, that an outline of Earth history and an understanding of the agencies which have guided its development, will be given the same importance in the educational curriculum, that the outline of human history is given today. Nay, I hold that it will be regarded as far more fundamental, —for who is there that can understand human history, and indeed form a true conception of human evolution and the basis of human relationships, when he lacks the foundation on which alone he can build, the foundation created by an understanding of the evolution of the earth and all that implies.

While then geology, as a purely cultural subject is bound to become an integral part of general education, when those charged with such education have broken the shackles of the old tradition and themselves begin to glimpse the wider view, there is of course also the professional aspect of our science and to this we may devote a little attention.

The intensive cultivation of a science invariably implies specialization. No one could dream of mastering the whole subject of geology as we have here defined it, but no one should attempt specialization in any small portion of it, without acquiring and retaining a general view of the entire subject, for in that way alone can he realize where his own studies may be amplified and illuminated by those made in some other particular field. The best training for the attack of stratigraphic problems which I ever received was not from teachers in stratigraphy, but from a master in physiography. And yet physiographers will tell you that their subject is as far removed as can be, from stratigraphy and palaeontology.

The danger in our modern technical education lies, not so much in early specialization, but in the disregard of most of the other subjects, the immediate bearing of which upon the speciality may not be apparent. But I do not mean that the student should wait until he has covered the whole field of geology before he decides upon the portion to which he wishes to devote himself primarily. I once thought that this might be the most desirable method of procedure, and it may be so for those who have no special urge in any given direction. But I have now arrived at the conviction, that if a student feels especially attracted to one phase of the science, let him devote his energies chiefly to that field, but not exclusively so. And let him make a beginning in research work in his selected field, as soon as he has a sufficient grasp of the subject to have confidence in himself to carry out a piece of original investigation. And let him not be deterred by the fear of making mistakes. If we all waited until we felt we were incapable of making mistakes, no good work would ever be done. But the publication of an immature piece of work is another matter. There, careful advice should be sought from those capable of giving it. And while you make the attempt to solve a problem which has interested you, do not disregard or minimize the importance of careful training in other fields of science. Early in my own student career, I began research work on Devonian fossils, but while doing so, I also studied intensively the subjects of crystallography and mineralogy, sciences which one would regard as far removed from palaeontology. And yet I have never regretted devoting much time to these and other fields of study, for they have helped to widen the outlook in my own field.

And now one final word to young Chinese geologists. Remember that you are pioneers in the development of this science in your country, and that it depends upon the way you do your work, whether you add a permanent part to the structure which is building, or whether your contribution is a temporary one to be replaced in the future, because it cannot stand the test of time. Not broad generalizations, but intensive detailed work is needed. We must first mould the bricks of which the structure is to be built. And these metaphoric bricks should be moulded today as were those actual bricks in the early days of China, moulded to withstand the attack of time, so that after a thousand years they are today still superior to those that have been moulded since.

Make sure that what you contribute is sound and will stand the test of time. Give to your work the best that is in you and avoid superficiality, and do not worry if the bulk of your contribution is not very great. Quality not quantity counts, and in the careful moulding of your individual bricks, keep before you the vision of the great temple of science, that some day in the future will arise in this your country, and of which your individual bricks will form permanent and imperishable units. And with that bright vision in view, the intensive labour in detail will not seem irksome, because you know that without such intensive labour the bright vision cannot be realized.

——原载《国立北京大学地质学会会刊》（*Bulletin of the Geological Society of the National University of Peking*）1931 年第 5 期，

中文翻译见《科学周刊》1934 年 7 月 27 日

十年来中国地质研究工作之鸟瞰

葛利普著　胡伯素译

　　地质思想在中国虽萌芽甚早，但十年前中国人并无何种专门著述出现。至一九一九年中国地质调查所发行《地质汇报》（*Bulletin of the Geological Survey*），其所刊载之论文，虽属前此研究之结果，而地质研究之工作，要算以此刊为其嚆矢。

　　考中国官方进行地质工作，始于一九一二年，彼时南京临时政府实业部矿政司下设一地质科，后政府移至北京此科仍旧存在。

　　一九一三年地质调查所由丁文江博士组织成立，中国地质界遂开一新纪元。

　　初，地质教育已于一九一〇年在北京大学开始授习，主讲者为柏林梭尔格博士（Dr. F. Solger）。至一九一四年世界大战爆发，梭氏返国从军，此科遂尔停办。幸在此时前，调查所已经成立研究班，为造就人才之计，章鸿钊博士实主其事，丁文江、翁文灏二博士皆为当时之重要教师，梭尔格博士有一时期亦曾在彼中担任讲座也。

　　一九一六年第一期学生毕业，凡三十人[1]，留所服务者计十八人，地质调查所之成为独立机关自是年始。而一九一四年中国政府任命安特生博士（Dr. J. G. Andersson）为矿业顾问，实与地质学之科学研究以意外之有力激奋焉。

　　一九一六年北京大学恢复地质科，第一班学生计八人，即于一九二〇年卒业。自此以降先后毕业者至一百四十四人，今其中约有三分之一犹从事地质工作，未变初衷也。

　　十年前吾人所得之中国地质知识，大部分根据李希霍芬（von Richthofen）之名著《中国》五卷，洛川（von Locgy）[2]随 Szechenyü[3] 科学调查队来中国后

1　事实上，地质研究所录取 30 人，毕业 22 人，其中 18 人留在地质调查所工作。——编者注
2　匈牙利地质学家 Lajos Lóczy（1849—1920）。——编者注
3　匈牙利爵士 Béla Széchenyi（1837—1918），中文名塞切尼。——编者注

发表之论著，及康乃吉研究院¹委派来华之威里士（Bailey Willis）与布拉克瓦尔得（Black Welder）²二氏之著作三卷。

此外对于地质有极大贡献者亦颇有其人，最著者如佛采罗（Futterer）³、奥布鲁考（Obrutschew）⁴及一九〇七年至一九〇八年 Merzbacher⁵ Tianshan 考察队中诸地质家。而日本之地质学家与古生物学家亦有相当之研究，与吾人以不少之新知识。又来华之教士及游历家虽亦有采集标本，从事鉴定者，顾皆零碎不全，无整个著作可言。

属文至此，顾就所知，略述中国地质进行之概况，读者或亦感兴趣耶？

惟今所欲言者，拟将矿产沉积，泰山杂岩及造山作用等除外，而今吾人于中国地学智识之增进方面，三致其意。兹先从震旦纪（Sinian period）⁶述起：

震旦纪 初李希霍芬对震旦纪所下之定义，以寒武纪之岩层为其上部。威里士及布拉克瓦尔得则以震旦二字，指诸寒武奥陶纪岩层而言。此为徒感多事极不需要之更张也。此后吾等则力图恢复其原来之意义，使寒武纪以前之略受变质作用，而与元古界或太古界作不连续接触之水成岩层属之，并另成一系，为古生代之底层。在先此诸岩层与其下部元古界结晶岩曾被归在一类，及发现此二者之间有一大间期（Time Interval），并物理性质亦迥不相同，始行分开焉。

现则举一切寒武纪以前之古生代岩层，皆包括于此创用之震旦纪内，如北美之伯鲁特、由而塔、大峡谷（Belt, Uninta⁷, & Grand Canyon Series）诸岩层，苏格兰托里顿沙岩（Torridon Sandstone），及澳大利亚印度世界各处之寒纪前诸岩层统属之。

在此岩层中，吾人发见三新种，信为最古之藻类，暂归之于 *Collenia* 属内。

Collenia 在震旦纪灰岩内至丰富时，几乎触目皆是，俯拾即得，诚奇观也。

1　今译为卡耐基研究所（Carnegie Institution of Washington）。——编者注
2　应为美国古生物学家 Eliot Blackwelder（1880—1969），中文名白卫德。——编者注
3　应为德国地质学家、探险家 Karl Josef Futterer（1866—1906）。——编者注
4　现一般译为奥勃鲁切夫（Vladimir Afanasyevich Obruchev，1863—1956）。——编者注
5　德国地质学家 Gottfried Merzbacher（1843—1926），曾于 1902—1903 年在新疆天山地区考察。——编者注
6　关于 Sinian 一字之沿革，可参阅北大地质学会年刊第四期高振西君之 "Sinian 之意义在中国地质学上之变迁" 一文。
7　应为 Uinta，指美国犹他州的乌因塔盆地。——编者注

此种化石最早原自北美伯鲁特岩层内得之，继而大峡谷内亦有发现。再有一类阔翅类化石（Eurypterids）在伯鲁特层内，甚属丰富，在澳大利亚相当岩层中，亦所在多有，故在中国不无发现之可能也。

北京大学李四光教授及赵亚曾君对于扬子江流域之此类古老岩石曾尽力研究，证明宜昌峡冰碛层，确属真正之震旦纪，此六百至七百呎[1]之震旦纪石灰岩，介于冰碛层与含三叶虫 Redlichia 之下寒武纪岩层之间，盖即前此威里士等指为寒武奥陶纪者也。但吾人现在已深信其为震旦纪无疑矣，彼冰碛层者该震旦纪之底层也。

寒武纪 中国寒武纪之研究首推华尔特（Walcott）及默首（Mansay[2]）二氏。彼等曾各就布拉克瓦尔得搜集之材料，及云南方面之材料，加以整理，分别鉴定，皆为使寒武纪昌明之人，功不在少。

自后关于寒武纪之分布情形，分类及厚度，亦迭有发明。孙云铸博士之鉴定若干新种尚不计也。

在过去十年间尚有一重要事实，所不能不言者，即发现北方寒武纪与奥陶纪岩层间，一范围甚广之不连续（disconformity）存焉。此现象极为普遍，处处可以证明之。

再后孙云铸博士在寒武纪之最上部，找到直角石（Orthoceras），亦极有重大意义之事。

奥陶纪[3] 中国北部之奥陶纪，李希霍芬初归之为下石炭纪，直至发现标准化石后始证明其误。

威里士等虽于证实方面有所效力，顾于详细之分层，则鲜注意，故统名之曰济南灰岩（Tsinan limestone）。但在此十年内研究之结果，已证明此岩层至少可分为二部：其下层含 Proterocameroceras, Chihlioceras, Piloceras 与 Archaeocyphia 等动物群，相当于美国之毕克蒙坦（Beekmantown）[4]，上部代表中奥陶纪末期，中有 Actinoceras, Stereoplasmoceras 及腹足类等化石，此二部份为一不连续所分开，并其缺口[5]（hiatus）往南扩大，故下奥陶纪时海水退出，及中奥陶纪时复行侵入之事实，昭然若揭。北大一九二八级黄汲清君研

究西山地质时，已与此一最可征信之据矣。

又笛慈央（Chazyan）[1]时代之 Maclurea 层，亦经发现。

自李四光教授在扬子江发现含 Girvanellas 及 Archaeocyathid 动物群之下奥陶纪宜昌灰岩后，益与吾人对奥陶纪智识以莫大之长进，其影响于世界各部 Archaeocyathid 石灰岩之分类如何，已另于《北大自然科学季刊》[2]上讨论及之。

在此下奥陶纪之岩层上，为含 Orthoceras, Vaginoceras 与 Endoceras 甚富之晚期中奥陶纪宝塔石灰岩（Pagoda limestone），彼此作不连续之接触，其化石现在一九二四级俞建章君研究中。

此岩层即素负盛名之宝塔石所从出者，盖亦从前视为泥盆纪之物者。再此灰岩上随着，即为艾家山页岩（Neichia shale）含化石甚多。北大毕业生克拉罗佛博士[3]（Dr. Kolorova）鉴定之 Yangtzella poloi 允为其中之特出者。至在此区域内无上奥陶纪岩层，已为公认之事实，盖艾家山页岩上即直接为含笔石之最早泥盆纪页岩是也。

再后找到相当于低蒲克（Deepkill）[4]或下爱尔伦义（Lower Arenig）及罗曼斯克（Normanskill）或上兰得罗（Upper Llandeilo）年代之含笔石之岩层，要为过去数年内奥陶纪层序学上之一大事。至是吾人研究奥陶纪笔石之来源与分布方向者，更得充分之理由，与进一步之了解。

志留纪 吾人得于志留纪有进一步之认识，要由在扬子江及印度地质调查所布兰及乌里（Coggin Brown & Cowper Reed）[5]二君在云南发见早期志留纪笔石岩层是已。

关于志留纪动物化石，至此研究，亦略有进境。

后限于南方之志留纪，仅属志留纪下部之事实，亦渐昭著，在云南虽曾发现含特殊小形动物群之上志留纪，但其面积殊有限耳。

志留纪笔石亦发生于印度太平洋中，由亚洲向欧洲再转而向美洲逐渐分

---

1 即美国之中奥陶纪。

2 第一卷第二号。

3 美国人，女性。外国学生毕业于北大地质系者除彼外，尚有一日本人。该克氏因此化石，作一论文，遂荣膺博士学位焉。（此注中谓克拉罗佛为美国人，实为捷克人，全名 Františka Naděžda Kolářová。——编者注）

4 属下奥陶纪，相当于 Beekmantown。

5 John Coggin Brown（1884—1962），Frederick Richard Cowper Reed（1860—1946），均为英国地质学家并曾在印度进行地质学考察与研究。——编者注

布，则与奥陶纪者初无二致也。

泥盆纪　泥盆纪岩层层序方面，迄少显著之进步。但其动物化石，则因大宗采集之结果，已渐为人所注意。关于腕足类已有专志，在印刷中。

一九二三级王恭睦博士在南京附近乌桐山（Wutung）石英岩中，寻得中泥盆纪植物化石，此砂岩之年代遂始得与以古生物学上之证明。

下石炭纪　下石炭纪在中国则混然不明，维西（Viseen）层向无疑仅限于南方各地，但现则袁复礼先生见其在甘肃极形发达。久被认为代表下石炭纪之栖霞山石灰岩（Chihsia limestone），早扳一郎博士（Dr. Hayasaka）谓为属于下二叠纪，并得赵亚曾、乐森璕诸君之考证，乐君更在广西及他处之二叠纪底层发出 tetrapora 化石甚多。

太原系与二叠纪　北中国太原系之智识，吾人亦有相当之进步。其化石在十年前，固一无知者。虽然，在太原系于地质层序中之真正位置[1]，未得眉目而断定以前，自仍须最大之工作，庶能竟其全功。关于太原系化石，一九二三级赵亚曾君鉴定不少。同级田奇㻞君于海百合类亦有专著，同时哈雷博士（Halle）[2]鉴定之北中国二叠纪植物化石亦甚多。其著作流行颇广，盖为研究植物化石者所不可不读之书也。

中国地层史成就最大者或推二叠纪之一时代，因南中国二叠纪之发达较任何处为佳也。

前已言栖霞山石灰岩，现为下二叠纪之代表岩层，但中二叠纪则为乐平（Loping）或 Lpttonia 系所代表。上二叠纪在中国亦甚发达可观，广西之马平石灰岩（Maping limestone），盖颇富上二叠纪之新生物群者，为丁文江博士及乐森璕君采得不少。其专志不久亦可付印。现予等拟不特注意此亚洲二叠纪海产生物群，在二叠纪沉积历史中之重要；并其于欧洲二叠纪沉积历史关系如何，亦所欲推究者，由最近不断之工作，已与吾人不少之新见解也。[3]

中国十年来古生代地质研究工作经过情形，约略如此。兹再就中生代言之：

白垩纪　在此时期内白垩纪之研究，确有极大之收获，十年前中国盖亦未知有此岩层也。

1　属中奥陶纪，相当于 Chazy。
2　瑞典古植物学家 Thore Gustaf Halle（1884—1964），对中国植物化石多有研究。——编者注
3　顷葛氏反复陈述谓太原系为下二叠纪与南京栖霞山石灰岩及欧洲之 Uralian 相当。（此注原文位置不清楚——编者注）

自谭锡畴君在山东发现含鱼类昆虫及植物化石之下白垩纪后，继之赵亚曾、谢家荣二君研究之贵州层（Kweichow beds）之年代亦告决定为下白垩纪。而东北各省中含中生代鱼类及昆虫之白垩纪岩层，更继之先后闻于世。秉志博士对于北部白垩纪昆虫化石，已有重要专志刊行。

至张家口附近地层层序构造及地质史，巴布尔博士（Dr. George B. Barbour）于其专著中阐明详尽，适由地质调查所印就发行。更有进者，自中央亚细亚考察团及西北科学考察团去蒙古、新疆调查后，该二地中生代及新生代之地质，乃大放光明，增进吾人之智识尤属非尠。

第三纪　据近十年之考察，早期第三纪在中国北部极不发达，虽在数处曾发见始新统沉积（Eocene deposits），在满洲发现渐新统（Oligocene），中新统（Miocene）似全属阙如。但下上新统（Lower Pliocene）则又在北部异常发达，一提三趾马层（Hipparion beds）人盖多知之也。此种化石研究成绩甚佳，现已有多种专志可阅。一九二三级杨钟健博士，正悉心从事此种工作。

在第四纪中更新期（Polycene）尤受一般人之注意。非脊椎动物方面，已刊有专志一册，另一册在写作中。脊椎动物方面已有数种出版，而正在写作者尤多。

自一九二七级[1]裴文中君在周口店发现原人（*Sinanthropus Pekinensis*）之头颅骨后，研究脊椎动物化石者，更多奋然而起，感觉特殊兴趣，此惊人可纪之发现，遂以结束此十年来之古生物学工作焉，不可谓非可喜之事也。再裴君在发现北京人头颅骨前，亦曾获得同属原人之颚骨数枚及牙齿甚多，特附志于此。

故一回顾十年来往事，吾人可直言中国地层学上之发现，实甚伟大足纪，顾吾人对于中国地质之观察与估计，此盖其第一次也。吾人须知今日中国地质仍在萌芽时期，前之所成就者，仅为未竟工作之一眇小之小数，绝未可引以为足。后来之十年或数个十年内，自须有精明干练之本国地质学家出而继承未竟之大业，造成真实可靠之成绩，不特以解决头绪纷繁之地质学上诸难题，并使此各问题对于整个地球地质史之关系，亦日就阐明光大之途，斯诚非他人任也。

一九三〇年十月二十二日于国立北京大学

——原载《科学》1931年第15卷第6期

1　一九二七级或一九二一级等皆指在北大毕业时之年号而言。

Davidson Black: In Memoriam[1]

By Amadeus W. Grabau

The versatility of mind and breadth of outlook which was so characteristic of Dr. Black was strikingly shown in the keen interest he displayed in palaeogeography and the question of polar shifting.

By training and profession Dr. Black was an anatomist and his chosen field was the comparative study of ancient and modern man. But he soon realized that he would have to go farther afield than the limited area circumscribed by the tradition of his profession. He felt that man must be studied in the light of his environment and that to understand primitive hominids we must know something of the geography of their homeland and especially the vicissitudes of the climate which they had to endure.

Palaeogeography and the study of ancient climates is still in its infancy but this did not deter Dr. Black. With painstaking energy, he mastered the pertinent literature of the subject; and not content with that, he himself started lines of investigation suggested by his review of the present status of these problems.

First of all he felt the need of an adequate presentation of the various phases of Cenozoic and Psychozoic geography as determined in the variation of the position of the Earth's poles. To do this he devised special methods of projection, and in his laboratory he had a large globe, so mounted that it could be adjusted to the poles and make possible the determination of the new latitudes and longitudes. Some of the early results of his work have appeared in *Bulletin X of the Geological Society*[2] and they have already indicated the possibility of solving

---

1 此为步达生去世后，1934 年 5 月 11 日在北京举行的追思会上葛利普的发言，原无标题，此处以追思会录的标题代之。追思会录包括：翁文灏的追悼信（谢家荣代读）、许文生（Paul H. Stevenson）关于步达生的生平介绍，以及丁文江、葛利普、德日进、杨钟健、巴尔博、裴文中、顾临（Roger S. Greene）等的发言，最后附有步达生的著作目录。——编者注

2 Davidson Black, "Palaeogeography and Polar Shift: A Study of Hypothetical Projections," *Bulletin of Geological Society of China* (Grabau Anniversary Volume), 1931, Vol. 10, pp. 105–157. ——编者注

many perplexing questions, which have a bearing on the environment of early man.

Had Dr. Black lived, his contribution in this field would have done much to point the way to new investigations, and would have brought us a step nearer to an understanding of man's place in nature and the causes which have both forced and guided his progress toward an evermore complete humanity.

——原载 *Davidson Black (1884-1934): In Memoriam*, Geological Society of China/Peking Society of Natural History, 1934.

Dr. Henry Fairfield Osborn: An Appreciation

By Amadeus W. Grabau

In the death of Henry Fairfield Osborn America has not only lost one of its leading men of Science, but also one of its enthusiastic promotors of scientific exploration and research.

From 1877 to the year of his death, research was the keynote of his life work, and considerably over 800 memoirs, books and papers testify to his productivity.

At the very outset of his career, he learned the lesson "that research is work of the hardest kind, requiring persistence, intelligence and imagination".

This and the further lesson that the seeker after truth must expect no pecuniary rewards and that though recognition of his work may be long delayed, if it is conscientiously and well done, recognition will eventually come.

This was the keynote of his advice to students who proposed to enter the field of research in Palaeontology, and he had the gift, perhaps acquired from his great teacher Huxley, of inspiring his students with his own enthusiasm and the determination to devote the utmost of their powers to the task before them.

Osborn was a man of vision. He was not content with the mere gathering of facts, for facts to him had only values in their bearing on the wider problems of organic evolution and the origin, adaptations and distributions of organisms through the different periods of geological time.

And he was always ready to look at the problem from different angles, and willing to listen to arguments on opposing views. I have never known him to be dogmatic, and one of the pleasantest memories that I have of him as a colleague at Columbia, was of the weekly evening seminars where beer, pretzels and tobacco fostered an atmosphere of good fellowship and where the lively discussions of these major problems of biology and palaeontology and the keen but good-natured arguments of the opposing factions often continued far into the night.

Osborn had the ability to carry on several research problems side by side and in his tower room in the American Museum he had a separate table for each problem that he was engaged upon. And his staff of assistants, secretaries and librarians were so well trained that at the briefest notice they would supply all the material and literature necessary for the continuance of the research which he proposed to carry forward.

When he showed me over this room he had begun work on five great Palaeontological monographs, the researches on three of which, that is rhinoceros, horses and titanotheres, had been going forward for close on forty years.

For nearly half that time he had been engaged on two other monographs, one on the Sauropoda and the other on the Proboscidea, both of which were begun in 1900.

The monograph on the Titanotheres appeared in 1929, as monograph 55 of the United States Geological Survey, and in it he was able to include some of the important results of the discoveries in Mongolia.

And the other monographs were likewise enriched by the discoveries in this part of the world.

Osborn was always deeply interested in Asiatic Geology and Palaeontology and not only kept in touch with all the work that was being done in this part of the world, but was himself the prime mover in the organization and carrying out of the Palaeontological researches by the highly organized and well-equipped series of expeditions under the direction of Dr. Roy Chapman Andrews, which have opened up such promising fields for explorations in research in Asiatic Palaeontology.

He was an ardent advocate of the Asiatic origin of man and although the explorations made by the Central Asiatic Expedition brought to light little direct evidence, primarily perhaps because the work was carried on too far to the north, yet the indirect bearing which many of the discoveries had on the problem of the origin of the human race were of the utmost significance.

Partly in recognition of these results and because of his keen interest in scientific work in China, the Geological Society of China made him an honorary member in 1924 and the Peking Society of Natural History gave him the same recognition in 1926.

Although he had been honoured by all the leading scientific bodies of Europe and America, I have personal knowledge, that recognition bestowed on him by Chinese men of Science gave him the keenest pleasure and satisfaction.

While Professor Osborn will be remembered in scientific circles throughout the world for his many and manifold contributions, there are many of us who had the privilege of his near acquaintance and friendship, and to us, his death is a great personal loss.

——原载 *Peking Natural History Bulletin*, 1935, vol. 10, part 2

丁文江先生与中国科学之发展[1]

葛利普撰　高振西译

——是先锋，是热心工人。

建造中国地质学之基础，及擘划其发展之途径，丁文江博士[2]实具最大之功绩。博士之姓名，在地质学上所占之位置，恐较在其他任何学术方面更为重要。

丁博士心目中之地质学，极为广泛，范围所及，非只构成地球之材料，如矿物及岩石等，且包容形成及改动此种材料之种种动力，以及其渐渐演变之程序，进而对于地球之形状构造及经过历史等全体，作为研究之对象。于此，更涉及自亘古以来，地球陆面以上，及海水以内之生物焉。各种生物演进之程序，及足以影响其发展分布之各种因素，如关于地理气候及生物等，均在范围之中。在中国推行此等工作，需要经过高等训练之专门人才。造就此等专门人才之教育问题，在中国自属第一要图，而丁博士最早即献身于此。[3]

在欧洲科学思想发达以前，中国先哲对于地壳变动之基本性质，虽有明确之见解，而以后欧西竟超过远东，盖因能了解观察与实验之方法，足以改正哲学上之概说也。丁博士充分明了此种事实。发展东方科学，必须训练调查与实验之人才，且必须使此种人才在田野及实验室之内工作，而其所寻求必须是先寻求事实。

1　原著所述丁先生之事业与功绩，每引地质调查所为证。地质调查所之创设，为丁先生等所努力之结果，且任首任所长有年，多所擘划。近十余年来先生辞去所长职务，由翁文灏先生主持其事。而丁先生任该所出版之《古生物志》主编以至于今。且丁先生对于学术事业向具热诚，而与翁先生之交情又极密切，故即在翁先生任期以内，丁先生对该所之一切筹划与发展，随时均有极大之助力。原著云云，读者当不误会。——译者附注，下同。

2　丁文江 1911 年获格拉斯高大学理学学士学位，此后亦未有博士头衔，葛利普以博士称之，乃是尊称。下文《中国之古生物学》《十年来中国地质研究工作之鸟瞰》中亦有类似称呼。——编者注

3　民国初年，丁先生等创设地质调查所，惟工作人才缺乏，乃于民国三年北京大学地质系停办期间，借用该校之设备与校址，设立地质研究班，五年毕业，担任调查工作。其成绩优良者逐渐抽送留学。今日中国地质界之巨子，如谢家荣、王竹泉、叶良辅、李捷、谭锡畴、朱庭祜、李学清诸先生，均当时之学生也。

丁博士与其他曾受国外训练之领袖，均感觉此种教育工作之困难，丁博士乃运用其特有之能力以解决此科学教育问题。渠确认基本之科学训练，必须在本国讲授，于是需要适当之教师。渠自任相当之课程，其他课程，若不能在留学生中选得相当人才之时，则请外国人士相助。为求更高深及更专门之训练，渠确认必须将中国学生送出留学。但第一条件，必须淘汰成绩欠佳之学生，毫不姑息。惟其最适当者，方可予以留学之机会。

人才之训练，不过为事业之发端。研究之精神，必须确立。坚强之中心与重要之设备，必须创设。中国地质调查所之发展，在效能方面，能有今日之超越地位，实为丁博士纪功碑之一也。次为改组后之北京大学地质系[1]最初亦由丁博士之计议，其中一切设计，均曾予以密切之注意者也。

丁博士最初即感觉中国地质研究之困难在于地层内之化石知识之欠缺。此种化石，非特须搜集之，保存之而已，尤须予以科学之描述及说明。渠深觉此种工作之重要，因而筹画刊物，专门记载与解证中国生物之遗迹。伟大之《中国古生物志》刊行即为实现此计划。此四开本之专刊出版甚多，丁先生之意欲使此刊物较之其他国家之同类出版物有过之而无逊色。全志共分甲乙丙丁四种：甲种专载植物化石，乙种记无脊椎动物化石，丙种专述脊椎动物化石，丁种则专论中国原人[2]。第一册之出版，距今不及十五年，而今日之各别专集，已近一百巨册之多。此种大成绩实非他国所能表现。

化石必需科学的采集，方有最大价值。丁博士功绩之一，即为训练中国青年在地质学各方面从事实地调查工作。在大学中，渠均亲自领导学生作野外实习。且曾两次组织大规模之科学调查队，对中国西南部地质作有系统之研究，并采集化石。其一次为一九二八年广西调查，一次为一九三零年贵州之行。渠曾于一九一四年第一次调查云南，又加上述两次调查之结果，遂造

1　北京大学之地质系创设于光绪末年之京师大学时代，后因故停办。地质研究班毕业之后，先生等主张教育与调查研究事业应分工合作，因建议北大恢复地质系，任造就人才之责。调查所则专司调查研究工作。当承蔡元培先生之同意，于民国七年正式恢复地质系。民国九年丁先生为研究中国化石起见，聘请世界第一流学者，美国哥伦比亚大教授葛利普先生来华，在调查所领导古生物学之研究。但为训练青年计，同时复请葛先生在北大教书。今日之中国古生物学家，如孙云铸、杨钟健、斯行健、黄汲清、张席禔、乐森璕、田奇㻪、朱森、陈旭、许杰、计荣森等，直接为葛先生之高足，而间接为丁先生之培植。十九年蒋梦麟先生回长北大，二十年聘先生为地质学教授。五年来，课程改良，设备扩充，人数增多，及地质馆之建筑等，均丁先生与李四光等诸教授努力之结果也。

2　《中国古生物志》丁种主要涉及古人类及其文化，大多为对 20 世纪 20—30 年代周口店发掘出土的北京人化石的研究，亦涵盖部分辽宁沙锅屯、河南仰韶、西北地区半山和马厂等地的新石器文化。——编者注

成吾人对于中国西南部古生代地层知识之基础。《古生物志》之根据彼等所得之材料者，已出版十二巨册，计在两千页以上，附专图一百八十余版。而即将付印，及尚在编著之中者，为数尚多。

博士于亲身担任调查工作之外，常派遣多数有训练之中国青年调查中国各地，所获材料极其丰博。搜集所及，几包植物及动物两界之全部，惊人之发现"北京原人"[1]亦在其中也。

化石之研究，最初每托请外国专家。博士亦曾邀请数人来中国任此种工作[2]。此在最初为不得已。今者教育进步，中国青年对于此种研究多有优秀成就，其曾受欧美专门训练之人，已能应付更困难之工作。近年刊行之专集，大多数均出于中国古生物家之手。

丁博士与其他科学领袖人物均认为，欲科学在国家社会之利益上能有高能之应用，纯粹科学之研究实为其最重要之基础。然而丁先生对经济地质及国内矿产之富源方面，亦未尝忽视。记录地动现象之地震台是其一例；另一例则有"西园燃料研究室"（浙江金西园[3]氏及其后裔所捐建者）同为地质调查所之重要且兼顾之设备也。地质之测量作图，化石之采集，以及构造等变迁之推定，不过为调查工作之一面。土壤调查，及其于中国农业上之应用，亦为同等重要之事业也。在经济方面，如煤炭储量之调查及中国矿业之发展，而努力于周口店之开掘则属纯粹科学范围。二者均为地质调查所所兼顾之工作。

丁博士为中国科学界之最伟大人物之一，余所述不过其生平事业与功绩之一部。丁君之为人，非特具有过人之能力，且有远大之眼光，弘毅之魄力与勇气，识见所及，均能力行之而成事实！

丁博士以超众之才识与能力为其祖国努力，从来不为私图。其生平最热烈欣慰之事莫过于亲见某一个青年之中国地质学者成就某一件有价值之工作而能与欧美之同类工作比美之时。丁博士之遽尔长逝，科学界哀悼损失一个领袖，一个工作人员，一个主动之力量。博士之学生，博士之同事，与博士之朋友，又哀悼损失丁文江这个"人"！

——原载《独立评论》1936 年第 188 号

1　即周口店发掘出土的"北京人"。——编者注
2　请外人研究古生物，葛利普先生实唯一重要之大员，见上注。
3　此两处之"西园"当为"沁园"，指晚清著名浔商金焘（1856—1914）。——编者注

Nathaniel Gist Gee (April 20, 1876–Dec. 18, 1937)

By Amadeus W. Grabau

(Peking, February 28th, 1938)

To his many friends in Peking, and elsewhere in China, the news of the death of Dr. N. Gist Gee on Dec. 18, 1937 at his home in Greenwood, South Carolina, U. S. A., came as a great shock. Those of us, who were intimately associated with him during the infancy of our Society, will always remember him as the quiet, yet forceful, organizer to whose clear vision and energetic activities the Society owes its beginning and early growth.

When he, George Wilder, Sohtsu and Kungpah King[1], and I, met in Mr. King's house, in September 1925, to consider the question of the formation of a Natural History Society, it was he who had the practical ideas of procedure and we unanimously appointed him Organizing Secretary. It was he who wrote the constitution, which we adopted at the inaugural meeting on September 21st, 1925, almost without change, when the Society came into existence with 38 charter members.

That he should be elected its first Secretary-Treasurer was inevitable, and to his energetic activity it was chiefly due, that at the end of the first year the Society had a membership of 109, 2 honorary, 20 foreign corresponding members, 41 fellows and 46 members. From 1928–1929 he guided our Society as its president and continued to serve as councillor.

With this fostering of the young Society as a labor of love, and in addition to his exacting duties as Director of the China Medical Board in Peking, he yet found time and energy to contribute a number of scientific articles to scientific journals. Most notable among them are his descriptions of the fresh water sponges of

---

1　Sohtsu and Kungpah King 为金绍基和金绍城兄弟。——编者注

China, etc., published in the Bulletin of this Society.

Other writings include: Survey of Fauna and Flora of Soochow[1]; Birds of Yangtze Valley[2] (with Moffett); *Chinese Birds* (with Wilder and Hubbard)[3]; Chinese Sponges[4].

Dr. Gee first came to Soochow University, China, in 1901, from Columbia College (S. C.) where he was Professor of Natural Sciences. In Soochow, he occupied the chair of biology and was Head of the Department from 1901 to 1919, after which he returned home for a year to serve as Superintendent of Schools. Returning in 1921 he acted for a year as Far Eastern representative of the Spencer Lens Company and then became adviser on premedical education to the China Medical Board of the Rockefeller Foundation (1922–1925). From 1926–1928 he was assistant resident director, and concurrently field director of the division of medical education (1927–1928). During the years 1928–1932 he was adviser in natural science for China and Director of the China Medical Board in Peking. After his relinquishing the exacting duties of the Director's Office in 1932 when he returned to his native state, he became Vice-President of Yenching University, which post he held until 1935. After that he was Professor of Biology in Lander[5] College, Greenwood, South Carolina, from 1936 until his death.

He is survived by his wife and 4 children.

Prof. Gee received much recognition and many honors during his lifetime.

He was a Bachelor of Sciences and Master of Arts of Wofford College, and an honorary LL. D. He studied at Harvard, Chicago and Columbia Universities and was an honorary fellow of Phi Tau Phi, Chinese National Scholastic Honor

---

1 N. Gist Gee, "A Beginning of the Study of the Flora and Fauna of Soochow and Vicinity," *Journal of the North-China Branch of the Royal Asiatic Society*, 1919, vol. 50, pp. 170—184. ——编者注

2 N. Gist Gee and Lacy I. Moffett, *A Key to the Birds of the Lower Yangtse Valley, with Popular Descriptions of the Species Commonly Seen*, Shanghai: Shanghai Mercury Limited, 1917. ——编者注

3 N. Gist Gee, Lacy I. Moffett and George D. Wilder, "A Tentative List of Chinese Birds," *Bulletin of the Peking Society of Natural History*, 1926-1927, Vol. I, Part I & Part II+III. 此目录后来以 *Chinese Birds* 为名于 1948 年在上海出版。参见秦硕志著、孙承晟:《万卓志在中国的鸟类学研究（1894—1943）》,《中国科技史杂志》2023 年第 1 期，73—90 页。——编者注

4 祁天锡发表了不少关于中国淡水海绵的论文，发表于 *Lingnaam Agricultural Review*, *Peking Natural History Bulletin*, *China Journal of Science and Arts* 等期刊上。——编者注

5 原文为 Landon，应为 Lander。——编者注

Society, and Beta Beta Beta, Biological Honorary Society.

He was a member of the Board of Managers of the Fan Memorial Biological Institute in Peking. In 1926 he was a delegate to the Pan Pacific Science Congress in Japan. His membership in various scientific societies reflects the range of his interests and researches. He was a fellow of the Society of Parasitology and of the Society of Microbiology; corresponding fellow of the Ornithological Union. In China, besides the prominent part he took in the Peking Society of Natural History as Secretary, President and life member, he was a fellow of the Science Society of China and of the East China Education Association of which he was past president.

In scientific circles Gist Gee will be remembered as an enthusiastic lover of nature, a keen observer and logical reasoner, a careful and detailed worker in his chosen field, an organizer and a man of vision.

In education circles his memory will be treasured as a sympathetic and conscientious adviser, a helpful guide who could put himself in the student's place and look at his problems from the inside, —and withal, when necessity demanded, be a stern disciplinarian.

But his personal friends, those who knew him intimately, will love best to remember him for his warm handclasp, his kindly smile, the sympathy and understanding that showed in the modulated tones of his cultivated voice, his humor and shrewd understanding of his fellow men, to whom, nevertheless, he always imputed a desire for honesty and truth.

And so, we shall most cherish his memory for the lovable qualities, that were the inseparable attributes of Nathaniel Gist Gee—the man!

——原载 *Peking Natural History Bulletin*, 1937–1938, vol. 12, part 3

现代地层学者应具的勇气

葛利普撰　谢家荣译

此系第十二届地质学会年会专题讨论关于脉动学说前之引言，承葛先生允许将译文在本刊发表。——编者识

地史上学说和理解，现在已有许多是不适用的了，因为这都是依据局部的研究，而忽略了世界大体关系的缘故。大多数地质学家没有学逻辑，所以我要奉劝每个中国的地质学者，必须学习逻辑，才能作合理的推解，并且必须将其本身研究的问题和世界全局的关系，详细考量一番。

当二十五年前我的《地质学原理》（*Principle of Stratigraphy*）刚刚出版，总以为立论过僻，恐不能引起实地工作的地质家们的注意。现在呢，这本书虽还没有再版，其中有许多理论，或者已成过去，但国外的地质家，还不断的写信给我，誉为是一部名著。为什么我书的价值，直到二十五年之后，才为后起之秀的地质家所注意？倘若因服从旧说，足以阻止新理的发展，那末中国的地质学者倒可欣幸能自由发展，而不受旧说的束缚了。

一

因为渍于沉积学的旧说，和一切地层俱以海相解说的老话，直使流行的地层论文，写得肤浅可笑。中国的地质学者，应以在本国所见者为根据，而充分了解大陆沉积作用和大陆沉积物的重要。

伟大的黄河，是一个最大的教训，我们应当深切注意，用作推解地质现象的实例，因为黄河的沉积，并不是限于现代，而是普遍分布于地质时期之各时代的。

中国的震旦纪地层是一个好例，英国威尔斯的寒武纪和爱列臬地层 Arenig 都是标准实例。还有世界各处以及中国的志留纪地层，都可以黄河式的沉积来比较，来推解。

老红砂岩层在汉密勒（Hugh Miller）[1]、爱杰齐（Agassiz）[2]和盖开

1　休·米勒（Hugh Miller，1802—1856），苏格兰地质学家、作家。——编者注

2　路易·阿加西（Louis Agassiz，1807—1873），瑞士籍美国博物学家。——编者注

（Geikie）[1]等时代所认为系古湖沉积的名论，现在也因有中国的实例，渐渐要推翻了。诸如此类的例子，可谓不胜枚举。

这里是中国地质学者的职责；对于这伟大河流的沉积，应加深切研究，详列事实，阐发新理，以供将来的参考；这样才能对于古代沉积的成因，发生合理的推解。

二

其次是关于笔石的误解。最近在美国举行的地质学会会场上，还有一位著名美国古生物学者公然宣示在奥陶纪砂岩厚层内所含的笔石化石是属于深海沉积；他的惟一理由是因获得含放射虫化石的结核的缘故。现代的放射虫确乎聚积在深海之底，但这都是从漂浮地带沉降下去的。因为这放射虫的存在，遂将显然可与黄河式沉积相比的地层，硬指为深海沉积。但这样一来，大地槽（Geosyncline）的深度，不免要增加许多，并且这种极深的大地槽恐怕是为事实所不许罢！

这里显然是大地槽和前深渊（Fore deep）的分别的问题了，许多地质家尚还不能分清。事实上前深渊不能沉积较厚的粗砾，而海浪既能挟海草或笔石流入黄河冲积平原，那末也未尝不能同时挟运放射虫，这种简单的道理，反为我们的古生物家所不注意。

正因英国地质学的创造者承认笔石页岩是一种海洋沉积——这是彼国现知之惟一沉积——并因拉波华志（Lapworth）[2]氏未尝亲临中国，一览黄河平原的伟大，于是研究笔石化石的学者们，遂多泥于海成之说，而无法改进了。

三及四

多数美国地层学者泥于大陆迭动（Seesaw movement）或摇摆的学说，固执不化，遂为地层学进步的莫大阻力。同样的不幸，是他们对于陈旧的海陆固定说的崇仰。

他们未尝一问，"是否地层学能证明这种老话？"因为地球物理学者尝获得一二和地层学无大关系的材料，地层学者就居为异宝，来证明这种老话的真确（其实还有另一部分的地球物理家却能证明这许多老话是不对的），但究

1　阿奇博尔德·盖基（Archibald Geikie，1835—1924），苏格兰地质学家。——编者注
2　查尔斯·拉普沃思（Charles Lapworth，1842—1920），英国地质学家。——编者注

竟于本身问题，是否适用，却都未顾到，因此往往要委曲求全的来迁就这种老学说；倘能将本身问题，略一考量，就可发现这许多依据都靠不住了。当我们记起数学之磨只能磨其所加入的谷粉时，我们就可无疑的将这种相异的分子拒绝了。（其意即谓不要离题太远，妄为引证）

五

与之相似的还有地质系统分类问题，和以新事实与老分类谋相适合的一种不断的尝试。

老的分类法必须修改，因为地质系统分类最合理的根据，就是脉动学说。

积三十五年对于古生代地层的研究，使余深信此说之可靠。这里是中国地质学者能对科学全局作不灭贡献的新园地。

我并不希望消灭老的分类法，我所希望的是将许多事实，用不偏不倚的眼光，慎密考量，倘终于指示脉动说为真实的分类依据时，那末就请地质界同人竟无疑的采用之。

——原载《地质论评》1936 年第 1 卷第 2 期

The Development of the Natural Sciences in China[1]

By Amadeus W. Grabau
(The Geological Survey of China)

I am deeply conscious of the honor which you, the foremost body of scientific men of my country, have conferred upon me.

That I have been able to take an active part in the development of the natural sciences in China has been due, in the first place, to the fact that my coming to Peking has coincided with the awakening of interest in, and desire for scientific education along western lines, among the Chinese intellectuals.

The Geological Survey of China had been founded a few years before, and it had but recently completed its first task—the training of a number of men in the fundamentals of geological science.

It was still housed in an old adapted and inadequately equipped compound, with a library consisting of a few hundred books, and a few drawers of Chinese Paleozoic fossils. But, new buildings were under construction, and under the energetic guidance of Drs. Chang, Ting and Wong,[2] phenomenal progress was made in the equipment and acquisition of material and the investigation of Chinese geology. Dr. J. G. Andersson, foreign adviser to the Survey, had organized the scientific exploitation of the important deposits of fossil vertebrate remains and the study of these was undertaken by foreign paleontologists.

To me was assigned the study of the Chinese invertebrate fossils, while my task at the university has been the training of young Chinese paleontologists and stratigraphers.

Through the efforts of Dr. Andersson a fund had been provided to begin the

---

1 Remarks on receiving the Mary Clark Thompson Medal of the National Academy of Sciences. Read by Mrs. Grabau in the absence of Dr. Grabau in China.

2 Chang, Ting, Wong 分别指章鸿钊、丁文江、翁文灏。——编者注

publication of the *Palaeontologia Sinica* in four series: A. Fossil Plants; B. Fossil Invertebrates; C. Fossil Vertebrates, and D. Ancient Man. The support of this was subsequently taken over by the Survey.

The first two fascicles which I prepared appeared one in April and the other in September, 1922. Since then 95 fascicles have been issued with a total of 8,760 quarto pages and 844 plates. If we add the fascicles in press, or prepared, the total number of fascicles is well over a hundred, with more than 9,000 pages. The smallest fascicle comprises 14 pages and 1 plate, and the largest 441 pages and 31 plates.

The first paleontological memoir, written by a Chinese paleontologist, Dr. Y. C. Sun[1], appeared in 1924, and of the 43 fascicles of Series B, so far issued, 32 have been written by Chinese. Nearly all these are graduates of the National University.

At first graduate students had to be sent abroad for the study of vertebrates, but in recent years the preliminary training of workers in that field is carried on in Peking. This was made possible by the founding of the Cenozoic Laboratory of the Survey, which now takes care of all the work on fossil vertebrates, including the researches on the Peking Man (*Sinanthropus pekinensis*) under the direction of Drs. Weidenreich and Young and Pere Teilhard deChardin. Additional impetus had previously been given by the explorations of the Third Asiatic Expedition under Dr. Andrews and the coming to Peking of such internationally famous men as Granger, Matthew Nelson, Chaney, Berkey and the Swedish scientists under Sven Hedin, and his royal highness the Crown Prince of Sweden, an active patron of science.

The Survey has since acquired the Chaukoutien site and provided the necessary equipment for the extensive exploitation of these now famous deposits of the remains of ancient man.

The Soil Survey and Seismological Observatory in the Western Hills are other lines along which the Survey has branched out, in addition to its active pursuit of the study of the structural geology and economic deposits of China and

---

1 Y. C. Sun 指孙云铸。——编者注

the making of geological maps.

In 1920 the geological department of the university was reorganized and under the guidance of such leaders as Drs. J. S. Lee, C. Y. Hsieh[1] and the late V. K.Ting and others it developed rapidly. It is now housed in a building of its own, well equipped with lecture halls, laboratories, museum and library. The geological faculty consists of seven professors, two lecturers and four assistants.

Of the several hundred graduates of the department, the great majority is still in active geological work. Many of them are members of the national or of the various provincial surveys; others are teaching in various universities. The department now has resumed issuing its series of contributions, of which twelve have appeared, while others are in press.

In 1922, the Geological Society of China was organized with twenty-six charter members. At the first meeting (March 2) thirty-six new fellows and nine associates or student members were elected. Fifteen papers were presented, and the first volume of ninety-nine pages was published that year. Volume XV, totalling 574 pages, appeared in four parts during 1936.

At the annual meeting in February, 1937, sixty papers were presented and the membership was as follows:

Number of fellows..320

Number of student associates...........................66

Number of foreign corresponding members.....36

Number of living honorary members 1

Number of institutions, listed as members........6

Total...429

When I came to China, a Chinese fellow passenger, Dr. C. C. Ping[2], was returning after several years of study at Cornell University. It was an opportunity for discussing plans to develop research in natural history in China, for Dr. Ping was to take charge of the biological laboratory of the Science Society of China. This society was organized a few years previously for the promotion and diffusion of science. It publishes a monthly journal Science in Chinese, which is now in its

1　J. S. Lee 为李四光，C. Y. Hsieh 为谢家荣。——编者注
2　C. C. Ping 指秉志。——编者注

twenty-first year. Active work in the investigation of the fauna and flora of Central China was begun at once under Dr. Ping, the results appearing in English in numerous bulletins issued at irregular intervals. In Peking we organized the Peking Laboratory of Natural History, under the sponsorship of Mr. Sohtsu King, who has since become one of China's patrons of science. During two seasons he maintained a seaside laboratory at Peitaiho, where we collected the material for our illustrated guide to the shells of Peitaiho, in which 120 species were described and figured. Mr. King, who has since been elected to membership in several foreign malacological and conchological societies, has brought together the most complete library on conchology in China, and he and Dr. Ping are periodically issuing fascicles on the South Coast shells.

When the Fan Memorial Laboratory of Biology was organized, it took over most of the plans of the original Peking Laboratory, including the publication of the *Zoologia Sinica*, and extensively developed them, and since moving into its new and well-equipped quarters, it has become one of the leading biological research institutions of China.

Another direct outcome of the early activities of the Peking Laboratory of Natural History was the organization of the Peking Society of Natural History in 1925. The call was issued by Dr. N. Gist Gee, the ornithologist, as organizing secretary, and myself as convener, and the first meeting was held on September 21, with thirty-eight charter members.

Monthly meetings have been held ever since, with lectures and discussions on Chinese natural history. During the first year the membership rose to 101, including eighteen foreign correspondents and two honorary members. The society began at once to issue a bulletin, the first volume of 450 pages appearing in 1926. The present membership is 160 active members, 26 foreign correspondents and 4 honorary members, and the bulletin is now in its eleventh volume.

Besides this the society has undertaken the issue of the handbooks, of which four have appeared: (1) *Flowers of Peitaiho*, by R. D. Wickes; (2) *Shells of Peitaiho*, by Grabau and King, second edition; (3) *Hand-book of North China Amphibia and Reptiles*, by Drs. Boring, Liu and Chou, and (4) *Familiar Trees of Hopei*, by H. F.

Chow.

In addition the society has issued five monographs on Chinese medicinal plants and animal drugs, by Dr. Bernard Read, and one on minerals and stones used in medicine, by Drs. Read and Pak. It has also brought out a profusely illustrated manual of the dragon-flies of China, by Dr. J. G. Needham.

The scientific study of natural history is now a recognized intellectual pursuit in China, and those of us who were privileged to be present and in a measure give aid, during the early years of development, feel confident that in the years to come geological, paleontological, biological and archeological contributions by Chinese naturalists will become of increasing importance, not only to their home country, but to the world of science at large.

Chinese naturalists feel as I do, that in honoring me to-night you are giving recognition to the progress of the scientific work in China, and they take it as an encouragement for the unabated continuance of their endeavors.

With this interpretation of your award to me of the Thompson Medal and with my sincere personal thanks, I accept the honor.

——原载 *Science*, New Series, 1937, vol. 85, no. 2215

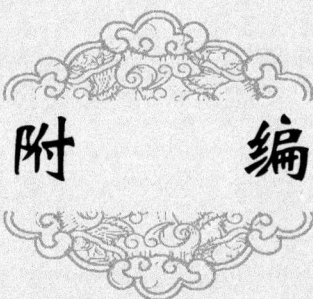

附　编

葛利普研究

葛利普与北京博物学会

孙承晟

摘 要 葛利普是 20 世纪具有重要影响的古生物学家和地质学家。他自 1920 年来华后，不仅对中国的古生物学和地质学教育与研究作出了不可磨灭的贡献，对生物学也十分关心。在他的号召下，1925 年成立了北京博物学会，吸引了很多在北京及部分外地重要的相关学者参加。该学会定期举办学术演讲、年会、野外考察，颁发"金氏奖章"，并出版很有影响的《北平博物杂志》，极大地促进了当时生物学的发展。在民国时期北京的科学界，形成了一个以葛利普为中心的科学团体，其中既有他的中国同事和学生，同时亦凝聚了当时北京内外的外国学者。通过对北京博物学会和《北平博物杂志》的分析，可窥见当时北京科学界的国际化程度和学术水准。

关键词 葛利普；金绍基；北京博物学会；北平博物学会；北平博物杂志；生物学；地质学

从 1914 年中国科学社成立起，民国时期的科学社团纷纷涌现，极大地促进了中国科学的发展，一些专业性社团，如中国地质学会（1922 年）、中国气象学会（1925 年）、中国生理学会（1926 年）、中国植物学会（1933 年）、中国动物学会（1934 年）等，对相关学科的发展影响尤为显著。纵观这个时期重要的科学社团，活跃其间的虽多为从国外归来的留学生或来华的外国科学家，但以外国人为主体的社团并不多见，由葛利普倡导下于 1925 年成立的北京博物学会（Peking Society of Natural History）[1]就是这样一个特别的学术团

1　Natural History 的中文翻译，民国早期多为"博物"，后逐渐被译为"自然历史""自然史"等，现亦有学者主张译为"自然志"，至今尚不统一。按当时报刊之记载，Peking Society of Natural History 最初名为北京博物学会。1928 年后国民政府南迁后，北京改名北平，该学会多被称为北平博物学会，但亦有北京博物学会之延用。为不致混乱，除引用原文外，本文一律作北京博物学会。下文所及的北京博物研究所类同（详下）。北平博物杂志（*Peking Natural History Bulletin*）则有所不同，因 1932 年 12 月时任燕京大学校长的吴雷川曾为该杂志题名《北平博物杂志》，影响较大。1949 年 12 月该刊中文名虽被改为《北京博物杂志》，但仅 3 年后即停刊。除特殊情况外，本文一般称之为《北平博物杂志》。
　　民国时期同类的学会尚有 1914 年成立的中华博物研究会（或中华博物学会，隶属于江苏省教育学会）、1916 年成立的北京高等师范学校博物学会和 1918 年左右成立的武昌高等师范学校博物学会，均有相应的杂志刊行。参见薛攀皋《中国最早的三种与生物学有关的博物学杂志》，《中国科技史料》1992 年第 1 期，90—95 页。

体。北京博物学会历时 27 年，是当时一个十分活跃的科学社团，对中国科学的发展有着较大的影响。然而，对此一重要的学会今人已鲜有提及[1]，其始末事迹多有不显。因此，本文将考其源流影响、组织结构，尤要说明葛利普对北京博物学会成立与发展的独特作用。

1 葛利普生平[1-13]

葛利普[2]（Amadeus William Grabau，1870—1946）1870 年 1 月 9 日生于美国威斯康星锡达堡（Cedarburg）一个德国血统的新教家庭。其祖父约翰（Johann Andreas August Grabau，1804—1879）出生于柏林东南郊（[12，1 页]），后为埃尔福特（Erfurth, Saxony）路德教牧师，[1,2]为反对当时腓特烈三世统一加尔文教派和路德教派的国家教会改革，1839 年移民到美国，在布法罗（Buffalo）创立三一路德会（Old Trinity Lutheran Church）和布法罗宗教会议、神学院（Buffalo Synod, College and Seminary）。其父威廉（Wilhelm Grabau，1836—1906）出生于埃尔福特，四岁时来到美国，在约翰的栽培下，亦成为一名路德教牧师。1863 年威廉与玛丽亚（Maria von Rohr，1845—1878）结婚，育有四子一女（另有两个夭折），葛利普是其中的第二个孩子。玛丽亚在葛利普 9 岁[3]时去世，但继母米娜（Mina Tobschall，1844—1928）对葛利普极为友善，以致后来葛氏在中国时还常为她庆祝生日。（[12，1—8 页]）

葛利普参加父亲的教区学校，学过管风琴，也学过木匠，但他对自然更有一种特别的爱好，经常外出收集岩石、昆虫、花朵，表现出异于同龄人的特殊兴趣。其继母对此着意培养，其父亦未以牧师之志而加以阻拦。1885 年，葛利普 15 岁时，举家迁回布法罗。他白天在一家装订公司作学徒，晚上参加夜校学习，业余时间则自己钻研植物学。在对布法罗西南的十八里溪（Eighteen-Mile Creek）等地的植物学考察后，他将兴趣转到地质学和古生物学。为加强地质学的学习，他参加了麻省理工学院克罗斯比（William O.

1 张孟闻在其《中国生物分类学史述论》（《中国科技史料》1987 年第 6 期，3—27 页）中对北京博物学会略有论及，并有较高的评价；薛攀皋在《20 世纪中国学术大典：生物学》（福州：福建教育出版社，2004 年，631 页）中撰有"北京博物学会"的词条，对该学会略有介绍；刘华杰则对《北平博物杂志》有过简要的介绍，见其《博物人生》（北京：北京大学出版社，2012 年，253—256 页）；罗桂环《北平博物杂志》亦有提及，见《中国近代生物学的发展》（北京：中国科学技术出版社，2014 年，360 页）。西文论及北京博物学会者，参见文献[12]第 295 页、文献[26]、文献[42]第 110—113 页。
2 葛利普初来华时，并有葛拉普、葛赉普、葛拉包、葛拉伯等中文名字，后统一为葛利普。本文涉及很多外国人，除部分暂未考之外，其中文名均据民国时期报刊或各相关档案查出。
3 丁文江在葛利普的"Biographical Note"中作 11 岁。参见文献[1]。

Crosby，1850—1925）[1]讲授的矿物学函授课程。克氏对他很赏识，给他提供了波士顿博物学会（Boston Society of Natural History）一个助理的职位，并于1890年让他作为特招生进入麻省理工学院地质学系学习。（[12，9—23页]）

在麻省理工学院，他选修了克罗斯比、巴顿（George H. Barton，1852—1933）[2]和奈尔斯（William H. Niles，1838—1910）等的课程，并多与波士顿博物学会、哈佛大学的亨肖（Samuel Henshaw，1852—1941）[3]、菲克斯（Jesse W. Fewkes，1850—1930）[4]、杰克逊（Robert T. Jackson，1861—1948）[5]和海厄特（Alpheus Hyatt，1838—1902）[6]等往来。尤其是海厄特，当时的波士顿博物学会博物馆馆长和哈佛大学比较动物学博物馆古生物部主任，在古生物学上对他影响甚深，并委以他波士顿博物学会博物馆公开讲师之职。同时，他还结识了当时美国地质学的一些领军人物，如亚历山大·阿加西（Alexander Agassiz，1835—1910）[7]、霍尔（James Hall，1811—1898）[8]、赖康忒（Joseph Le Conte，1823—1901）[9]、鲍威尔（John W. Powell，1834—

1　克罗斯比1871年入麻省理工学院，1873年在彭尼基斯（Penikese）岛跟随亚历山大·阿加西学习。1878年后一直在麻省理工学院地质学系任教。专长工程地质学，对波士顿地区的地质做了开创性的研究，先驱性地为大型工程（如水库大坝、地铁、隧道等）建设提供地质咨询。
2　巴顿长期在麻省理工学院教授地质学。
3　亨肖专长于昆虫学研究，长期作为海厄特的助手。1903—1927年任哈佛大学比较动物学博物馆馆长。美国人文与科学院院士、美国博物学会和美国动物学会会员，剑桥昆虫俱乐部（Cambridge Entomological Club）的发起人之一。
4　菲克斯1873年在彭尼基斯岛跟随亚历山大·阿加西学习，从事海洋动物学研究。长期在哈佛大学比较动物学博物馆工作。后亦研究印安人人种学、考古学。美国人文与科学院、美国国家科学院院士。
5　杰克逊1889年获哈佛大学博士学位。1892年入哈佛大学任教，兼职于比较动物学博物馆。专长棘皮动物化石的研究。美国人文与科学院院士，波士顿博物学会、美国地质学会、美国古生物学会会员。
6　海厄特师从路易·阿加西（Louis Agassiz，1807—1873），1862年毕业于哈佛大学劳伦斯科学学院。后为麻省理工学院动物学和古生物学教授，直至1888年。1870年起入波士顿博物学会。美国国家科学院院士、美国博物学者学会（American Society of Naturalists）的创立者之一。提倡新拉马克主义。
7　路易·阿加西之子。1862年哈佛大学毕业后，进入比较动物学博物馆任其父亲的助手，直至逝世一直在此工作，并长期担任馆长。同时还到多地开展调查研究，并在密歇根经营矿业公司。一生以动物学和海洋学研究而著称，亦是当时以工业资助科学的典范。继承其父，反对达尔文的进化论。1910年逝世于从英格兰到美国的大西洋考察途中。
8　霍尔早年从事纽约地区的地质调查，撰成很有影响的《纽约古生物志》（13卷）。与当时的著名学者有密切的交往，如路易·阿加西、赖尔（Charles Lyell，1797—1875）等，并培养了很多著名的学生（如沃尔科特），在当时的科学界深有影响。曾任美国科学促进会主席（1856）、美国地质学会首任会长（1889）、美国国家科学院50名创始会员之一。
9　赖康忒1850年弃医赴哈佛大学师从路易·阿加西学习地质学和动物学。1869年入加利福尼亚州大学，教授地质学、生物学等课程，直至逝世。一生兴趣广泛，涉猎医学、地质学、化学、心理学、光学、生物学等领域。曾任美国科学促进会和美国地质学会主席。参见杨丽娟、韩琦《晚清英美地质学教科书的引进——以商务印书馆〈最新中学教科书·地质学〉为例》，《中国科技史杂志》2014年第3期，316—331页。

1902）[1]、沃尔科特（Charles D. Walcott，1850—1927）[2]、钱伯林（Thomas C. Chamberlin，1843—1928）[3]等，均对他的学术生涯有直接或间接的影响。[1]

1896 年，葛利普以对十八里溪的动物学研究获麻省理工学院学士学位。次年获得哈佛大学奖学金进入地质学系，主要师从杰克逊、谢勒（Nathaniel S. Shaler，1841—1906）[4]和戴维斯（William M. Davis，1850—1934）[5]。1900 年以《纺锤螺及其同源动物之系统演化》（*Phylogeny of Fusus and Its Allies*）[6]一文获博士学位。在哈佛期间，他即被伦斯勒理工学院（Rensselaer Polytechnic）聘为讲师，毕业后为该学院教授。1901 年因哥伦比亚大学的邀请，他放弃了此一职位，前往哥大任古生物学讲师，1903 年任副教授，1905 年即升任教授。一战期间，美国具有排德思潮，但葛利普继续宣扬德国文化，后因人事纷争和一战中继续支持德国而与校方产生龃龉，于 1919 年被哥伦比亚大学解聘。[1, 12]

1918 年底，丁文江随梁启超等前往欧洲考察，并出席巴黎和会。考察结束后，1919 年丁文江赴美访问，其中一个任务就是受北大校长蔡元培的委托，物色地质学家到中国工作。在美国地质调查所怀特（David White，1862—1935）[7]

1　鲍威尔专长于地质学、人种学。19 世纪晚期长期担任美国人种学局（Bureau of American Ethnology）主任和美国地质学会主席，是当时最有影响的科学家之一。

2　沃尔科特以对寒武纪动物化石和岩石的研究而著称。活跃于许多重要学术机构，曾担任美国地质调查所主任、史密森学会秘书长、国家科学院院长等，国家研究委员会的创立者之一。

3　钱伯林早年在亚历山大·温切尔（Alexander Winchell，1824—1891）指导下研究地质学。后参与威斯康星地质调查，开始研究冰川，进而探讨冰川气候与地球历史等根本问题，强调地球大气（尤其是二氧化碳）构成变化对地球的重要性，并提出星子假说（planetesimal hypothesis）模型。1887—1891 任威斯康星大学校长，将威斯康星大学从组织和精神上变成了一所真正的大学。1891 年，出任芝加哥大学新成立的地质系主任，作出重要贡献，并创办了著名的《地质学杂志》（*Journal of Geology*）。美国国家科学院、美国人文与科学院、美国哲学学会院士。曾任美国地质学会主席、美国科学促进会主席。1909 年受洛克菲勒基金会委托，和他的儿子及另外两人组成一个四人小组，前往中国进行考查。

4　谢勒 1862 年毕业于哈佛大学劳伦斯科学学院，路易·阿加西最欣赏的美国学生。1864 年开始在哈佛大学任教直至去世。多次参与美国海岸调查局的工作，对填海地质学（reclamation geology）作出重要贡献，进而深入思考人地关系等重大问题，撰写了大量地质学科普著作。1895 年任美国地质学会主席。

5　戴维斯 1869 年于哈佛大学获科学学士学位，1870 年获工程学硕士学位。曾作为庞培利（Raphael Pumpelly，1837—1923）的助手到北太平洋进行地质调查。1877 年后长期在哈佛大学任教。一生从事气象学、地质学、地形学领域，均有重要建树。"侵蚀循环"（cycle of erosion）是戴维斯最重要的地理学理论，影响深远。美国地质学会创立者之一，1911 年任主席；美国地理学家协会创立者之一，并于 1904 年、1905 年、1909 年任主席。美国人文与科学学院院士、美国国家科学院院士。

6　该文 1904 年作为史密森学会（Smithsonian Institution）的一种专刊出版。

7　怀特 1886 年以对伊萨卡附近的泥盆纪植物化石研究获康奈尔大学学士学位。后一直任职于美国地质调查所，从 1903 年起亦担任史密森学会古植物馆主任，以对古生代的植物化石研究而著称。美国国家科学院院士，还担任过美国国家科学院副院长、美国古生物学会副会长、美国地质学会主席等职。比利时地质学会和中国地质学会荣誉会员。

的引荐下，丁文江聘请到刚被哥大解聘的葛利普，任北京大学地质学系教授和地质调查所古生物学室主任。[11]1920 年 10 月，葛氏抵达中国，随即开始在北京大学地质学系授课，并在地质调查所开展研究工作，长期住于西城豆芽菜胡同 5 号。[1]杨钟健后曾回忆"多年前北平豆芽菜胡同的葛先生寓所，无形中成了北平一个文化中心。"[14]

来华后，葛利普不仅全身心致力于古生物学和地质学研究，亦极为热心教学，深受中国学生爱戴。中国第一代、第二代古生物学家，大部分都受到他的直接影响，如孙云铸、赵亚曾、杨钟健、黄汲清、斯行健、计荣森、许杰、尹赞勋、赵金科、田奇瓗、乐森璕、俞建章、朱森、陈旭、张席禔、丁道衡、卢衍豪、王鸿祯等。此外，秉志的《中国白垩纪之昆虫化石》、李四光《中国北部之蟆科》也都是在他的鼓励下完成的。[3,5]葛利普还积极参与创立中国早期的地质学、生物学学术团体，如北京大学地质研究会（1920 年）、中国地质学会（1922 年）、北京博物学会（1925 年）、中国古生物学会（1929 年）等。

1937 年，全面抗战开始，北京大学和地质调查所南迁，葛利普因风湿腿疾滞留在京。他痛恨日本人的侵略行径，拒绝与伪北大合作。珍珠港事件后，太平洋战争爆发，他经济来源断绝，生活极为艰苦，且不久被送到英国大使馆集中，身心受到极大摧残，神志亦日渐不清。抗战胜利后，中华文化基金会恢复葛利普薪俸，虽有朱家骅等的慰问和地质界同人的悉心照料，但终因胃部出血，于 1946 年 3 月 20 日在北平逝世。遵其遗愿，在其逝世一周年，他被安葬于北京大学地质馆（沙滩旧址）前，并由汤用彤代胡适主祭。[5,15]1982年 8 月 13 日，在中国地质学会成立 60 周年之际，葛利普墓被迁至北京大学（燕园）西门附近。[16]

葛利普在来华前，已是著名的古生物学家，发表了很多有影响的著作，大部头者如《北美标准化石》（*North American Index Fossils*, Vol. I—II，1909—1910）、《地层学原理》（*Principles of Stratigraphy*，1913）、《硅酸盐以外的非金属矿床地质》（*Geology of the Non-metallic Mineral Deposits other than Silicates*，1920）、《地质学教程》（二卷）（*Text Book of Geology*，1920—1922）。来华后，则以中国地质材料致力于古生物学和地质学的研究，著述不断，尤为重要者，如协助丁文江创办《中国古生物志》（*Palaeontologia Sinica*），卷帙浩繁（他自

---

1 葛利普使用信笺地址为"5 Tou Ya Tsai Hutung, Peiping, West City"。豆芽菜胡同位于丰盛胡同附近，1965年改名为民强胡同，20 世纪 90 年代被拆除建为住宅小区。参见严肃编：《北京市街巷名称录（附城区街巷图)》，北京：群众出版社，1986 年，116 页，附图 39。

己撰著的就有 7 种），深受国内外学者重视。另外，他所撰写的《中国地质史》（二卷）（*Stratigraphy of China*，1924—1928）、《蒙古之二叠纪》（*The Permian of Mongolia*，1931）、《脉动理论下的古生界地层》（四卷）（*Palaeozoic Formations in the Light of the Pulsation Theory*，1934—1938）、《年代的节律：从脉动理论和极控理论看地球的历史》（*The Rhythm of the Ages: Earth History in the Light of the Pulsation and Polar Control Theories*，1940）、《我们居住的世界：地球历史新论》（*The World We Live In: A New Interpretation of Earth History*，1944, published in 1961）等，更是为他赢得了世界性的声誉。[3,9,12]终其一生，他著作 300 余种，累计 20 000 多页，在现代科学史上亦不多见。[11]葛利普的科学贡献主要在于古生物学和地质学，进而归于全球性的脉动和极控理论，是 20 世纪最重要的古生物学家和地质学家之一。

由于葛利普在古生物学和地质学上的重要贡献，他于 1925 年获得中国地质学会首届以其名字命名（王宠佑捐资设立）的葛利普奖章，1934 年获北京博物学会颁发的金氏奖章（详下），以及 1936 年美国国家科学院的汤普森（Thompson）奖章。为纪念他在科学上的卓越成就，1976 年月球上的一个山脊以葛利普命名。（[12]，459 页）

葛利普 1901 年与著名的作家、社会活动家玛丽·安亭（Mary Antin，1881—1949）[1]结婚，育有一女约瑟芬（Josephine Esther Grabau）。他 1920 年来华，妻女均未随行。1933 年赴美参加第 16 届国际地质大会，与其家人进行了团聚，这是他到中国后唯一一次离开中国之行。

葛利普对中国极为友好，常呼吁中国人要自强，尤其要在科学上取得进步。1930 年，在其六十岁时，以丁文江和美国驻华大使为证人立下遗嘱，将其全部图书赠予中国地质学会，现部分仍存中国地质图书馆。[3]

葛利普逝世后一个月，中国地质学会与中央研究院地质研究所、中央地质调查所于 1946 年 4 月 20 日在重庆联合举行了隆重的追悼会，由俞建章（国立重庆大学地质学系主任）代表时任中央研究院地质研究所的所长李四光主祭，李春昱报告生平事略，尹赞勋报告学术贡献，均对其学术成就及在中国

1　安亭 1881 年出生于俄国波洛茨克（Polotzk）一个犹太家庭，1894 年随家人移民到波士顿。著有《从波洛茨克到波士顿》（*From Plotzk to Boston*，1899）、《应许之地》（*The Promised Land*，1912）、《敲门者》（*They Who Knock at Our Gates*，1914），反映了当时犹太移民如何融入美国社会与文化的历程，受到广泛欢迎，尤其是后两书共售出 10 万多册。安亭因此声名鹊起，并成为一个活跃的作家和社会活动家。葛利普 1920 年来华前，两人即已分居。安亭晚年生活消沉，1949 年 5 月 15 日逝世于纽约州罗克兰。

的地质教育工作作出了高度评价。[17]翁文灏亦撰文悼念，称"近数十年来世界知名的外籍科学家，在中国服务最久而贡献最多的，要算一位地质学大师葛利普先生了。"[18]可谓公允之论。

2　北京博物研究所

在谈北京博物学会之前，须提及稍前成立的北京博物研究所（Peking Laboratory of Natural History）。

葛利普不仅致力于古生物学和地质学的研究，对生物学也一直保持着浓厚的兴趣。1920 年从美国来华时，刚好与秉志同船，旅途中即商讨如何在中国开展生物学研究。[1]秉志归国后，创建了中国第一个生物系和生物研究机构，即 1921 年成立的东南大学（原南京高等师范学校）生物系和 1922 年成立的中国科学社生物研究所。[19]与此相呼应，葛利普则于 1925 年创建北京博物研究所和北京博物学会。1928 年 10 月 1 日，秉志、胡先骕等筹建的静生生物调查所在北京正式成立，北京博物学会多人参加，祁天锡（Nathaniel Gist Gee，1876—1937）、金绍基（1886—1949）[2]、葛利普先后致辞。任鸿隽曾有过这样的记述："最后古生物学家葛利普先生手挟两杖，扶掖而至，为极诚挚之演说，力陈研究纯粹科学之重要，并言中国科学虽属幼稚，然世界科学仍同在幼稚之境，望中国学者不必气馁，听者掌声雷动。"[20]以上各机构的建立，相互促进与支持，为中国生物学的发展做出了奠基性的贡献。

1925 年在葛利普和金绍基的努力下创建的北京博物研究所，其意似与中国科学社生物研究所相类。北京博物研究所位于嘎嘎胡同（现东城区协作胡同）11 号[3]，所址和设备全部由金绍基提供，是一个关于中国动植物研究的私立组织，致力于对中国动植物的系统调查和相关著作的出版。研究所隶属于

1　Amadeus W. Grabau, "The Development of the Natural Sciences in China," *Science*, New Series, 1937, 85 (2215): 551-553；重熙：《葛利普博士之荣誉》，《科学》，1937，21（9/10）：744-748。感谢胡宗刚先生提示此条材料。

2　金绍基，字叔初，浙江南浔人，1902—1905 年随其兄金绍城赴英国伦敦大学国王学院游学，归国曾在南洋公学、邮传部等任职，后从事实业，是民国时期重要的实业家，曾担任中华教育文化基金会、协和医学院董事，对当时的科学事业多有资助。1925 年与葛利普共同组织成立北京博物研究所、北京博物学会，1930 年资助成立沁园燃料研究室。主要研究在贝类，曾与葛利普合著《北戴河的贝类》等。后移居香港。参见陆剑：《南浔金家》，杭州：浙江人民出版社，2006 年，40—45 页，136—138 页；孙承晟：《在商业与科学之间：金绍基的科学活动及其身份转型》，《科学文化评论》2020 年第 1 期，56—72 页。

3　该所中文名字、地址、所属研究室及与地质调查所的关系可见于瑞典斯文·赫定档案中葛利普使用的信笺。感谢业师韩琦研究员提供该材料。

地质调查所，研究则主要集中于对更新世软体动物（作为地质调查所古生物研究室的补充）和现代动植物。

研究所的成员是自愿的，没有薪酬，更像是一个自由的学术团体。葛利普为主任，金绍基则担任监护人（Custodian）和荣誉秘书（Honorary Secretary）。研究所分为 7 个研究室，并有相应的规制和设想：[21]

（1）软体动物研究室，由金绍基负责。葛利普、金绍基和科拉罗娃（F. N. Kolarova，1901—1945？）女士在河北北部（北直隶）海湾收集了大量的海洋软体动物，多布森（R. J. Dobson，燕京大学最早的生物系教授[1]）则收集了一些海洋无脊椎动物。为此项研究，还在北戴河建立了一个海洋生物实验室。金绍基通过广泛渠道收集全国各地海岸的软体动物、贝壳及其他海洋动物，并积极与世界各地图书馆或书商联系，系统搜罗关于软体动物研究的文献，试图建立一个全面的图书馆。

（2）斧足类动物研究室，由科拉罗娃负责。[2]

（3）淡水和陆地无脊椎动物研究室，由祁天锡负责。祁天锡在中国收集了不少轮虫、淡水海绵动物和水蛭，并发表不少研究成果，他还正准备一部关于淡水海绵动物的专著。

（4）鱼、两栖和爬行动物研究室，由秉志负责。秉志在南京已对中国的鱼进行了大量的研究，他的一些同事则开展两栖动物、爬行动物及昆虫的研究。因与中国的食物供应相关，希望不久的将来此一部门发展成为一个大型的研究所。

（5）鸟类研究室，由万卓志（George D. Wilder，1869—1946）负责。万卓志与祁天锡、慕维德（Lacy I. Moffet，1878—1957）及其他国外鸟类学家对中国北方鸟类进行了多年的研究，相关的研究成果或鸟类目录将发表于北京博物学会杂志上。万卓志还正准备一本关于鸦科的专著，将发表于《中国动物志》（*Fauna Sinensis*）第一辑。

（6）植物研究室，由伊博恩（Bernard E. Read，1887—1949）负责。伊博恩多年来收集并研究中国医用植物。工作主要在北平协和医学院开展，其专著近期将作为《中国植物志》A 辑（*Flora Sinensis*，Series A）的第 1 卷由北京

1　"A Brief History of the Department of Biology of Yenching University," 燕京大学档案 YJ1931018，藏北京大学档案馆。

2　科拉罗娃关于 "Triplecia" Poloi 的研究，参见 F. N. Kolarova, "The Generic Status of 'Triplecia' Poloi," *Bulletin of the Geological Society of China*, 1925, 4(3&4): 215-219.

博物研究所出版。

（7）绘画研究室，由金绍城（金绍基长兄）负责。他是当时著名的画家，着手关于中国鸟类的插图，纯粹是自己的爱好。他还免费提供对一些年轻艺术家的培训，尤其注重动植物的画法，为北京博物研究所及地质调查所出版的著作提供插图，可谓当时将艺术用于科学的先驱。此外，他还承诺赞助关于中国动植物著作的出版，希望能激发人们对中国科学事业的支持。[22]

此外，葛利普亦表示，研究所暂无人开展昆虫方面的研究，但希望不久的将来能与北京其他研究机构合作建立一个昆虫学研究所。

研究机构的成果主要体现在出版物上。按照葛利普等的设想，北京博物研究所将致力于中国动植物研究专著的出版，研究文章（如物种目录、新种属的初步描述，以及关于中国动植物的博物观察等）则将刊登于北京博物学会杂志上。研究专著将分两类：《中国动物志》和《中国植物志》。两类之下，又分为若干系列，或以生物种属分，或以地理区域分，由各研究室主任负责并编辑。每一系列不定期出版，每卷约 300—400 页，容量、行文及图版等基本与由地质调查所刊行的《中国古生物志》（*Palaeontologia Sinica*）相类，图版约 30 幅，尽可能为彩色。葛利普还对论著的撰写作了详细的说明。

这些设想后来并未完全实现，除《北平博物杂志》定期出版外，《中国动物志》和《中国植物志》并未如预期刊行，但还是出版了 6 种博物手册（Handbook）和 11 种专书，成为北京博物研究所的一项重要工作，经费主要由金绍基赞助。6 种博物手册中包括葛利普和金绍基合著的《北戴河的贝类》（*Shells of Peitaiho*）、万卓志和胡本德（Hugh W. Hubbard，1887—1975）合著的《中国东北部的鸟类》（*Birds of Northeastern China*）等。11 种专书则多为发表在《北平博物杂志》上的抽印本，伊博恩即占 9 种之多。

应该看到，北京博物研究所的人员较为匮乏，研究工作也难以展开，植物部分仅有伊博恩着手研究。或许正是因为如此以及经费的有限，研究所建制难以延续，才有 1925 年底成立的北京博物学会。1930 年，因《北平博物杂志》的重组（由北京博物学会和燕京大学生物系共同筹办），北京博物研究所正式解体，并将其所存有的出版物和现金转赠《北平博物杂志》，以资助杂志的出版运行。[23]

3　北京博物学会

3.1　成立[24]

在发起人（Convener）葛利普和组织秘书（Organizing Secretary）祁天锡

的筹划下，北京博物学会于 1925 年 9 月 21 在协和医学院解剖楼举行成立大会。首先由临时主席葛利普发表热情洋溢的开场白，接着由翁文灏、安得思（Roy Chapman Andrews，1884—1960）和谷兰阶（Walter Granger，1872—1941）先后致辞，均表示成立这样一个学会的必要性，希望国内外同好对中国的动物、植物进行全面深入的调查和研究；学会不仅面向专家，对学生或业余爱好者也敞开大门，将是热爱自然的人们交流的一个平台。然后宣布创始会员（Charter Members）38 人（其中中国人 12 人）：

Bert G. Anderson[1]、安得思、巴尔博（George B. Barbour）、步达生（Davidson Black）、博爱理（Alice M. Boring）、Rachel E. Burner (Miss)、E. D. Congdon、Ernst de Vries、多布森（R. J. Dobson）、H. K. Fung、祁天锡、葛利普、谷兰阶、Gladys M. Jefferis (Miss)、John F. Kessel、金绍城、金绍基、科拉罗娃（F. N. Kolarova）、李顺卿、李济、林可胜、林同曜、刘汝强、慕维德、Heinrich Necheles、毛里士（Frederick K. Morris）、N. C. Nelson、伊博恩、许文生（Paul H. Stevenson）、孙云铸、C. H. Taine、Roy C. Tasker、E. Tschepourkovsky、钟观光、万卓志、翁文灏、胡经甫、温森德（H. S. Vincent）[2]。

创始会员逐条讨论由祁天锡拟定的章程（Constitution）草案，未作太多修改，即形成正式章程共 14 条，主要内容包括：

（1）确定学会名称为"北京博物学会"（Peking Society of Natural History，图 1）[3]。

（2）会员包括六类：常驻会员（resident-members）、研究员（fellows）、会友（associates）、非常驻会员（non-resident members）、通讯会员（corresponding members）、荣誉会员（honorary members），均须选举产生；研究员须为致力于中国动植物研究的会员，由全体理事一致同意方能获得资格。

（3）会费：常驻会员和研究员每年鹰洋（Mexican Dollar, Mex.）[4]5 元，非常驻会员 3 元，会友 2 元，通讯会员和荣誉会员免收会费；一次性捐助 75 元

1　协和医学院口腔外科教授，见协和档案。

2　燕京大学制革系主任。

3　林可胜曾提议名为中国科学院（Chinese Academy of Science），但会员们均认为中国科学院应该由中国学者自己在政府许可下创办，故未获通过。参见文献[29]。

4　鹰洋即墨西哥银元，1842 年墨西哥铸造流通于世，因正面镌有一只鹰衔蛇的图形，故亦称鹰洋。晚清进入中国，最初流通于广东地区，后至全国，成为在华流通最广的外币。参见王利中：《民国前期（1912 年—1927 年）中国货币制度研究》，新疆大学 2003 年硕士学位论文。

者即可获得终身会员资格。

（4）学会行政事务由一名会长、两名副会长、一名秘书和一名司库（秘书和司库可合一）负责，一年一选。学会并有理事会（Council），由不少于六名理事组成，负责学会的发展、会员选举、筹备会议等事宜。

（5）学会出版杂志《北京博物学会会志》（*The Bulletin of the Peking Society of Natural History*）。

图 1　北京博物学会钢印（感谢程正庆先生惠允）

此外，章程还对会员和理事的选举、任期、组织学术活动等作了详细的规定。讨论完章程之后，选举了未来一年的学会负责人。

会长：万卓志；

第一副会长：翁文灏；

第二副会长：金绍基；

秘书兼司库：祁天锡。

并选举了以下理事（Councilors）：

葛利普、李顺卿（三年期）；

林可胜、谭熙鸿（二年期）；

步达生、钟观光（一年期）。

最后由会长万卓志发表"北京一些常见鸟类"（Some Common Birds of Peking）的演讲；金绍城则展示了一些中国古代关于鸟类的绘画。金绍城为北京博物研究所绘画研究室的负责人，随着北京博物学会的成立，他希望能在更大范围之内为各类相关著作提供插图。但他不幸于 1926 年去世，故祁天锡、慕维德、万卓志的《中国鸟类目录试编》（*A Tentative List of Chinese Birds*）的下卷（刊于《北平博物杂志》第 2—3 期）便是献给金绍城的，以示对他的敬意。

在葛利普的召集下，1925 年 9 月 30 日在协和医学院举行理事会，会长万卓志被推选为当年理事会主席。关于学会成立的简讯刊于当时国内外著名的报刊杂志，如 *Nature*、*Science*、*Journal of Science and Arts*、*Peking Leader*、*Shanghai Times*、*China Press* 等。

理事会会议不定期举行，讨论关于学会的各种相关事宜及增选会员和通讯会员。如 1925 年 11 月 27 日下午，在葛利普豆芽菜胡同的家中举行理事会议，讨论学会出版物及与其他学会交流等事。1925 年 12 月 11 日，在葛利普演讲结束后举行理事会会议，讨论北京博物研究所与北京博物学会的合并。1926 年 2 月 26 日，理事会讨论年会的召开，并决定斯坦福大学的创始校长佐敦（David Starr Jordan，1851—1931）和美国自然史博物馆的馆长奥斯朋（Henry Fairfield Osborn，1857—1935）为荣誉会员，通讯会员若干。此外，理事会也会特邀其他人员参加讨论，如 1926 年 9 月邀请安特生（Johan Gunnar Andersson，1874—1960）、德日进（Pierre Teilhard de Chardin，1881—1955）、毕士博（Carl Whiting Bishop，1881—1942）等参加，安特生就邀请瑞典王子访华（由北京博物学会、地质调查所、协和医学院联合举办）通报了相关事宜，毕士博则表示希望与史密森学会（Smithsonian Institution）等机构协商对北京博物学会提供资助。

1926 年 3 月 12 日，北京博物学会举行年会，万卓志发表了"北京博物学会的功能"的会长演说，其中强调了葛利普对学会的特别贡献，称他为学会真正的奠基者（"the real father of this Society"）。[25]祁天锡也指出成立北京博物学会是葛利普的设想。[26]葛利普不仅是发起人，后并成为荣誉会员、终身会员，及唯一的终身理事（Life Councillor），足见葛氏对学会的重要影响。

葛利普发起成立北京博物学会很可能是受到了波士顿博物学会的影响。葛利普早年曾在波士顿博物学会学习、工作多年，自言受益很多。北京博物学会很大程度上可看成是北京的波士顿博物学会。葛利普还主张应在各地设立博物学会，且这些博物学会必须是"非专门的"，以激发普通人的博物兴趣为宗旨，但应秉持科学的基础与方法。[27]1929 年 3 月 22 日，葛利普在北京博物学会第四次年会上致辞时曾引"波士顿老人"（the grand old man of Boston）

爱德华·埃弗里特·黑尔（Edward Everett Hale，1822—1909）的名言而指出他发起成立北京博物学会的初衷："每个人，除了他自己的职业，应该有一个业余爱好（avocation）。"[28]

按照学会最初的设想，会长每年一选，分由中国和外国学者轮流担任，副会长两人，中外各一，[29]理事亦大体照顾中外学者的比例（表1）。但是，从会员成员来看，大多数还是外国学者。如刚成立后一年的 1926 年，共有会员 111 名，其中中国人仅 24 名[30]。至1929 年，会员增至 148 名，中国人也仅 40 名而已，[31]外国会员多来自美国、英国、德国、法国、日本等。这在当时中国的学术团体中，不能不说是很特别的。至 1933 年，会员达到最多的 193 名。[32]

表 1　北京博物学会历届会长、秘书、司库及理事

年度	会长	副会长	秘书	司库	理事
1925—1926	万卓志	翁文灏、金绍基	祁天锡	祁天锡	葛利普、李顺卿（1925—1928）1 林可胜、谭熙鸿（1925—1927） 步达生、钟观光（1925—1926）
1926—1927	翁文灏	金绍基、步达生	伊博恩	伊博恩	祁天锡、金绍城2
1927—1928	金绍基	祁天锡、胡经甫	伊博恩	伊博恩	步达生、林可胜 博爱理、Henry Meleney（代替祁天锡）3
1928—1929	祁天锡	林可胜、伊博恩	刘汝强	刘汝强	葛利普（终身理事）4 博爱理、福敦（A.B.D. Fortuyn）5（1926—1929） E. de Vries、步达生（1927—1930） 丁文江、胡经甫（1928—1931）
1929—1930	林可胜	丁文江、伊博恩	许文生	许文生	葛利普（终身理事） E. de Vires、步达生（1927—1930） 丁文江（后为金韵梅）、胡经甫（1928—1931） 刘崇乐、金韵梅（后为顾临）（1929—1932）
1930—1931	伊博恩	丁文江、步达生	许文生	许文生	葛利普（终身理事） 金韵梅、胡经甫（1928—1931） 刘崇乐、顾临（1929—1932） 胡先骕、福敦（1930—1933）

1　起止时间指任职时间，下同。

2　*Bulletin of The Peking Society of Natural History*, 1928, 2(4): i.

3　*Bulletin of The Peking Society of Natural History*, 1928, 2(4): vii.

4　由于对北京博物学会的卓越贡献，理事期满的葛利普 1928 年被选为终身理事，并在第三届年会上获得通过。参见 "The Third Annual Meeting and Banquet," *Bulletin of The Peking Society of Natural History*, 1928, 3(1): lvii-lviii.

5　福敦，协和医学院解剖科教授。

续表

年度	会长	副会长	秘书	司库	理事
1931—1932	丁文江	步达生、胡经甫	福敦	福敦	葛利普（终身理事） 许文生、顾临（1929—1932） 胡先骕、韦尔巽（S. D. Wilson）[1]（1930—1933） 秉志、博爱理（1931—1934）
1932—1933	何博礼	胡经甫、胡先骕	福敦	福敦	葛利普（终身理事） 韦尔巽、侯祥川（1930—1933） 秉志、博爱理（1931—1934） 窦维廉[2]、李汝祺（1932—1935）
1933—1934	胡先骕				
1934—1935	许文生				
1935—1936	胡经甫				
1936—1937	福敦				
1937—1938	许雨阶				
1938—1939	许雨阶	魏登萃、李汝祺	冯兰洲	马文藻	博爱理、杨维义、张景钺、郭毓彬[4]
1939—1940	李汝祺	博爱理、冯兰洲	钟惠澜	武兆发	葛利普（终身理事） 张景钺、袁贻瑾、张锡钧、冯兰洲、寿振黄、John Cameron[5]
1940—1941	博爱理	冯兰洲、德日进	钟惠澜	武兆发	葛利普（终身理事） 福敦、吴宪、寿振黄、万卓志、张锡钧[6]
1941—1947	冯兰洲				
1947—1948	钟惠澜				
1948—1949	林宗扬	娄克思、武兆发	谢少文	谢少文	胡经甫（1947—1949）、郭毓彬（1948—1949） 博爱理、胡先骕（1948—1950） 冯兰洲、汤佩松（1948—1951）
1949—1950	武兆发	汤佩松、郭毓彬	冯兰洲	冯兰洲	博爱理、胡先骕（1947—1950） 冯兰洲、汤佩松（1948—1951） 胡经甫、钟惠澜（1949—1952）
1950—1951	汤佩松	谢少文、胡先骕	张宗炳	张宗炳	冯兰洲、刘崇乐（1948—1951） 胡经甫、钟惠澜（1949—1952） 张景钺、裴文中（1950—1953）

注：据北京博物学会历次会议纪要、《北平博物杂志》及其他杂志刊登学会消息编制，资料不全待补

1　中文名字见 Yenching University Directory (1927-1928), vol. X, no. 25, p. 15. 藏北大档案馆。

2　关于窦维廉，参见《燕京大学人物志》（第一辑），252—253 页。

3　或即魏敦瑞。

4　《北京博物学会年会》（《科学》，1937 年，21 卷 6 期，487 页）中写为郭育宾，疑误。——编者注。

5　《北京博物学会第十四届年会》，《科学》，1939 年，23 卷 5/6 期，327—328 页。

6　《北京博物学会第十五届年会》，《科学》，1940 年，24 卷 7 期，580—581 页。

会员包括当时很多活跃于北京科学界的外国学者，除葛利普、祁天锡外，尚有如安特生、德日进、安得思、谷兰阶、步达生、巴尔博、顾临（Roger S. Greene，1881—1947）、胡恒德（Henry S. Houghton，1880—1975）、新常富（Erik Nyström，1879—1963）等，中国地质学的开创者丁文江、翁文灏、孙云铸、袁复礼等亦参与其事。

北京博物学会每月召开一次公开的学术演讲（一些还被辑录刊于《北平博物杂志》），每年举行年会（图 2）并组织一次野外考察，抗战前从未间断，此外还有不定期的学术演讲，[33]北京博物学会组织之力可见一斑。1941 年，太平洋战争爆发，日本在北京实行高压统治，北京博物学会亦被迫陷于停顿，直至 1947 年方才恢复。

图 2 北京博物学会十周年纪念（1935 年）
前排左三至左五分别为：胡先骕、葛利普、胡经甫（1935—1936 年度会长），
来源：《北平博物杂志》1935 年第 10 卷第 2 期

3.2 金氏奖章

为促进中国的博物学研究，金绍基出资于 1929 年设立了金氏奖章（The King Senior Medal）和金氏青年奖（The King Junior Prize），均每年颁发一次。前者为一枚金质奖章，用于奖励在博物学领域做出原创性贡献的学者，以纪念金绍基亦深爱博物学的父母金焘（字沁园）夫妇。后者为一枚铜质奖章和 20 元奖金，奖励 20 岁以下的最佳标本收集者，以纪念其兄金绍城。[34]

历年金氏奖章获奖者及成就如下：[32,35]

1930 年，尼登（James G. Needham）（关于中国昆虫的研究）；

1931 年，胡先骕（关于中国植物图谱的研究）；

1932 年，步达生（关于"北京人"的研究，图 3）；

1933 年，秉志（关于古动物学的研究）；

1934 年，葛利普（关于中国地质学的研究）；

1935 年，翁文灏（关于大地构造学和地震学的研究）；

1936 年，伊博恩（关于中国古代医用药物的研究）；

1937 年，胡经甫（关于中国昆虫的研究）；

1938 年，魏敦瑞（关于北京猿人齿系的研究）；

1939 年，朱元鼎（关于中国鱼类的研究）；

1940 年，万卓志（关于中国鸟类的研究）；

1941 年，张春霖（关于中国淡水鱼类的研究）。

1942 年后，学会陷于停顿，金氏奖章及青年奖亦随之停颁，后未再继续。我们可看到，与学会会长分由中外学者担任相呼应，金氏奖章亦隔年分别颁给中外杰出的研究人员，反映了学会的国际化程度。

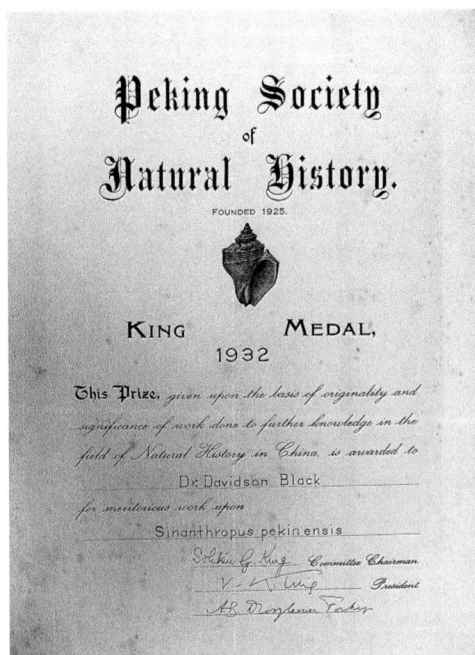

图 3　步达生获 1932 年度"金氏奖章"证书（多伦多大学图书馆藏）

4 《北平博物杂志》

北京博物学会甫一成立，发起者便计划创办一份属于学会的杂志，并设想将此一杂志分作学术（Technical Series）和教育（Educational Series）两个系列。此外，尚有仿照《中国古生物志》之体例，由北京博物研究所出版《中国动物志》（*Fauna Sinensis*）和《中国植物志》（*Flora Sinensis*）两种专著系列的计划，[36]但这些设想后来并未完全实现。杂志的学术系列后即为连续刊行的《北平博物杂志》，由于研究和稿源有限，教育和专著系列则合为北京博物研究所（暨北京博物学会）不定期出版的专刊，很多是从《北平博物杂志》上的文章抽印而成，并未达到每年出版一部的计划。

学会杂志 1926 年创刊，每年一卷。第 1 卷第 1 期为祁天锡、慕维德、万卓志合撰《中国鸟类目录试编》之上部，第 2 和第 3 期合为下部，上下部共370 页，第 4 期为学会纪要、演讲论文汇编和会员录。

至第 4 卷的出版，《北平博物杂志》即走上规范发展的道路。因刊行这样一份杂志超乎学会本身的力量，从 1930 年第 5 卷起，北京博物学会和燕京大学生物学系达成协议，共同经营这份杂志，杂志之名略有变动，从 *Bulletin of The Peking Society of Natural History* 变为 *Peking Natural History Bulletin*，并成立一个由 4 人组成的独立编委会（Board on Publications），[1]两人代表北京博物学会，两人代表燕京大学生物学系，具有较大的自主性。[23]新的编委会由祁天锡、伊博恩、胡经甫、博爱理组成，前两者代表北京博物学会，后两者代表燕京大学生物学系。至于出版经费，北京博物学会号召每位会员捐助 4 元以支持杂志的刊行，另一部分则由燕京大学生物学系筹措；此外，尚有售卖出版品（专书和抽印本）所得和洛克菲勒基金会的不定期捐款以作为资助。

杂志自第 7 卷第 2 期（1932 年 12 月号）起，配有时任燕京大学校长吴雷川（1870—1944）题写的中文刊名"北平博物杂志"（图 4）。1941 年 12 月，第 16 卷第 2 期刚一出版，燕京大学和出版社即被日军占领，杂志未获发行。[2]自此，北京博物学会陷于停顿，《北平博物杂志》亦被迫停刊。战争结束后，学会恢复，《北平博物杂志》于 1948 年复刊，第 16 卷第 2 期则被重印发行。[32,37]1949 年以后，在中国的外国人纷纷返回，中国的文化事业也发生

---

1 早期 4 卷的编辑委员会主要由翁文灏、林可胜、步达生、祁天锡、伊博恩、福敦、博爱理等担任。
2 Lucius Porter 曾设法取出 10 份给博爱理，后被分发给一些见到的会员。参见文献[32]。

了根本性的变化，会员及杂志发表文章者均基本为中国学者。此外，从第18卷第2期（1949年12月）起，期刊中文名即被改为《北京博物杂志》。从第19卷第1期（1950年9月）起，期刊论文加有中文摘要。但此举并未持续多长时间，1952年6月，随着第20卷第2—4期合刊发行，此后再未有出版。

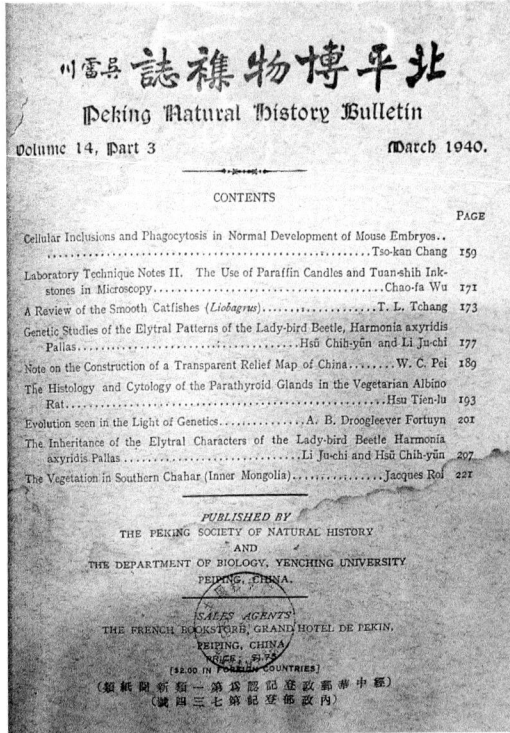

图4 《北平博物杂志》书影

《北平博物杂志》可谓北京博物学会最大的贡献之一。26年时间里，除日据时期外，《北平博物杂志》从未中断，共刊行了20卷，内容基本保持稳定，反映了学会成立20多年来博物学的研究水平。20卷《北平博物杂志》，共发表有568篇文章，可大体分为植物学、动物学、生理和医学、地质学、综合类，以及关于学会活动和相关人物的介绍，其中最多的是关于动物学、生理和医学类的文章，分别达292和173篇（表2）。《北平博物杂志》也刊有少量的地质学文章（部分是辑录在北京博物学会所作的学术演讲），如葛利普关于周口店古人类化石、中国北部地层和脉动理论的报告或论文，翁文灏关于地震和李四光关于地质力学的论文，以及新常富在山西地质考察的成果等。

表2　《北平博物杂志》刊发文章分类表

作者分布＼文章类别	植物学	动物学	生理与医学	地质学	综合	学会活动和人物介绍
外国学者	11	95	30	4	31	
中国学者	17	197	143	2	12	
总计	28	292	173	6	43	26

注：论文（及演讲报告）无论长短，均以篇计；合作撰写者，每个作者以人均计；少量微生物学论文，归入综合类或生理与医学类。

《北平博物杂志》早期发表文章者以外国人居多，但自1932年第7卷起，随着越来越多的外国留学生返回以及本土生物学者的成长，中国学者发表论文开始赶超（这与吴雷川1932年12月题写"北平博物杂志"之中文刊名亦很巧合），并不断持续下去（图5）。这从一个侧面反映了生物学在中国逐渐获得本土化的过程。1949年以后，北京博物学会转变为一个完全由中国人自主的学术团体，在杂志发表文章者也基本都是中国人了。

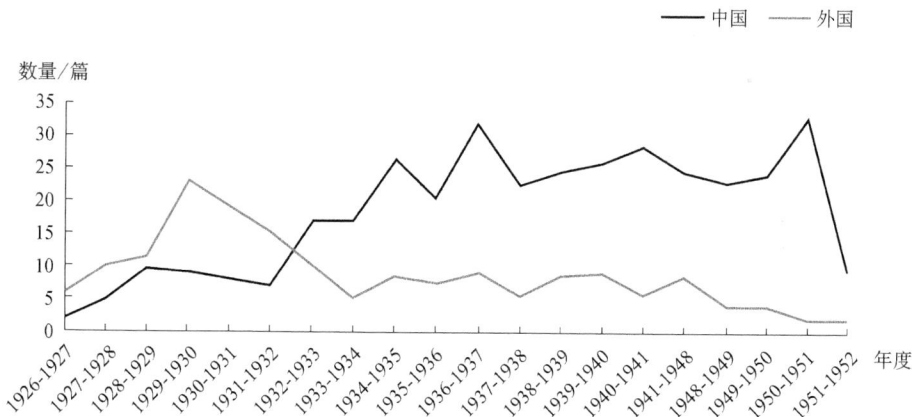

图5　《北平博物杂志》中外学者历年发表文章示意图

纵观《北平博物杂志》之编辑和论文发表，祁天锡、伊博恩、胡经甫、博爱理、李汝祺、胡梅基等贡献最大，他们或长期担任杂志编辑，或经常在上面发表研究成果。作为发起人的葛利普，尽管他的研究主要在古生物学和地质学，也在《北平博物杂志》上发表了7篇文章，其中2篇关于软体动物，2篇关于地质学，1篇关于古生物学，另2篇分别为关于奥斯朋和祁天锡的纪念文章。

《北平博物杂志》主要用英文发表文章，间有少量德文或法文。北京博物学会与国外多家学术机构交换杂志，1935 年达 170 余种[38]。其中多为美国、英国，并有加拿大、澳大利亚、德国、印度的研究机构，重要者如美国国会图书馆、美国自然史博物馆、美国农业部、费城科学院、伦敦科学博物馆、柏林大学动物博物馆、悉尼林奈学会、加拿大农业部等。[39]

5 尾声

随着新中国的成立，1949 年 10 月 12 日，新的北京博物学会在协和医学院举行第一次理事会，出席者有武兆发、汤佩松、郭毓彬、胡先骕、胡经甫、林宗扬、冯兰洲，讨论每月学术报告、会员会费、采购纸张以备杂志出版之需等事宜。1950 年 5 月 27 日在协和医学院举行第二次理事会，出席者与前次同，讨论了以下事项：

（1）决定冯兰洲为《北京博物杂志》的主编，胡经甫、博爱理和林宗扬为副主编。尽管资金困难，但杂志仍坚持定期出版，并讨论在杂志论文附加中文摘要的提议，议决自第 19 卷起执行。此外，1950 年 4 月已将杂志的所有过刊及其他博物手册或专刊从燕京大学移至协和医学院。

（2）任命 1950—1951 年学会行政及理事会成员：汤佩松为会长，谢少文[1]和胡先骕为副会长，张宗炳为秘书兼司库；并公布了理事成员。[2]

（3）决定 1950 年 6 月 18 日举行学会年会。

1950 年 6 月 18 日在北京基督教女青年会礼堂举行年会，共有 80 名会员及部分友人参加。此次年会的特别之处，在于恰逢学会成立 25 周年，并且是新中国成立之后的第一次年会。会上分别报告了学会现状、出版情况，公布 1950—1951 年学会领导和理事。前任会长武兆发作题为"蝗虫雄性生殖细胞中染色体与中心体之关系"（Chromosome-centriolar relationship as illustrated in the male germ cell of the grasshopper）的报告。

博爱理因要返回美国，在年会上作答谢词。[40]抗战结束后，博爱理虽亲国民党，但随着抗战胜利，越来越多的人支持共产党，她亦惊奇地发现"尽管我过去反对共产党，但我现在亦对未来充满希望！"太平洋战争爆发后她曾被日军关押，并于 1943 年 11 月作为美日之间交换人员遣返美国。但在二战胜利一年后的 1946 年秋，她又返回燕京大学。1950 年 8 月，并非由于年龄或政

1　关于谢少文，参见《谢少文论文选辑》中"我的治学之路"。

2　这些理事成员与之后不久召开年会上所宣布的最终人选有所区别。

治原因而是因为她姐姐的健康状况，她返回了美国，从此直至 1955 年去世再未来过中国。[41]

冯兰洲作学会现状报告，指出 1949—1950 年共有 68 名会员，北京 61 名，外地 5 名，国外 2 名。他希望随着局势稳定，会员会有较大的增长。尽管财务困难，但《北京博物杂志》已定期出版，无论数量还是质量都将有提升。报告还列出未来半年的学术报告 6 次，只有第 4 次为外国人 Alfred Koehn，其余均为中国学者演讲。[40]

通过此次年会，我们可清楚地看到，随着新中国政权的稳定和外国人的纷纷撤离，新的北京博物学会也发生了很大的变化，外籍会员仅剩 2 名，总的会员数亦降至 68 名，成为一个中国人自主的学术团体。《北京博物杂志》亦有明显的中国化，自第 19 卷（1950 年 9 月）起，每篇论文均附中文摘要；编辑部则从行将消失的燕京大学搬到协和医学院。

自 1950 年年会之后，就很难看到关于学会的活动。若以 1952 年 6 月《北平博物杂志》的停刊算，学会共走过 27 年。其最终去向应是并入其他相关学会，进而汇入全国自然科学专门学会联合会或全国科学技术普及协会。[1]在 1926—1952 年的 26 年中，除 1942—1948 年被迫停刊，《北平博物杂志》共出版了 20 卷。杂志上发表了不少高水平的学术论文，促进了中国生物学的繁荣与发展，但终因生命太短而令人惋惜。

6 结语

作为 20 世纪上半叶在中国最有影响的外国科学家，葛利普通过研究古生物学和地质学、培养学生、筹划《中国古生物志》等，对中国的地质科学作出了不可磨灭的贡献。因对中国生物学的关切，在他的发起下于 1925 年成立了北京博物学会，吸引了当时在北京内外很多重要的生物学、医学和部分地质学学者，尤以外国学者为多，形成了一个以葛利普为中心的科学团体，足见葛利普在当时的巨大影响。至 1952 年"解体"，北京博物学会共存在了 27 年，除 1942—1948 年的动荡 7 年，该学会实际运行了 20 年。期间通过组织定期的学术演讲、年会、野外考察，颁发"金绍基奖章"，极大地促进了当时生物学的发展。北京博物学会还出版了很有影响的《北平博物杂志》20 卷，许多著名的学者在此发表他们的论著，很大程度可反映当时生物学研究的成果

---

1 关于北京博物学会的最后归宿，现尚未见到相关档案材料以资说明。

和水准。《北平博物杂志》刊发文章共计 568 篇，可大体分为植物学、动物学、生理和医学、地质学、综合几类，其中最多的是关于动物学、生理和医学类的文章。早期发表论文者以外国人居多，自 1932 年第 7 卷起，中国学者发表论文开始赶超，并不断持续下去，从一个侧面反映了生物学在中国逐渐获得本土化的过程。通过北京博物学会和《北平博物杂志》，我们亦可窥见当时北京科学界的国际化程度和学术水准。葛利普及其倡导成立的北京博物学会对于中国生物学的发展功不可没。

　　致　谢：本文撰写过程中，承业师韩琦研究员提供档案资料并悉心指点，熊卫民博士、胡大年博士、姜丽婧博士及两位匿名评审人给予宝贵建言，谨致谢忱。本文初稿还于 2014 年 11 月 14—15 日在布拉格查理大学举行的 Science as Knowledge, Ideal and Practice in 20th Century China 研讨会上报告，笔者感谢罗然（Olga Lomová）教授、胡吉瑞（Jiří Hudeček）博士的邀请及与会者富有启发性的评论与建议。

参 考 文 献

[1] Ting V K. Biographical Note [J]. *Bulletin of the Geological Society of China*, 1931, 10: iii-xviii.
[2] Shimer H W. Memorial to Amadeus William Grabau [J]. *Proceedings Volume of the Geological Society of America Annual Report for 1946*, 1947: 155-166.
[3] Sun Y C. Professor Amadeus William Grabau: Biographical Note [J]. *Bulletin of the Geological Society of China*, 1947, 27: 1-26.
[4] Gregory W. Minute on the Life and Scientific Labors of Amadeus William Grabau (1870-1946) [J]. *Bulletin of the Geological Society of China*, 1947, 27: 31-34.
[5] 孙云铸. 葛利普教授[J]. 科学，1948，30（3）：70-72.
[6] 寿振黄. 古生物学大师葛利普教授年表[J]. 科学，1948，30（3）：73-75.
[7] Kay M. Grabau, Amadeus William[C]//Gillispie C C. *Dictionary of Scientific Biography*, Vol. 5. New York: Charles Scribner's Sons, 1981: 486-488.
[8] 潘云唐. 葛利普——中国地质科学工作者的良师益友[J]. 中国科技史料，1982，（3）：22-30.
[9] 杨静一. 葛利普传略[J]. 自然科学史研究，1984，3（1）：83-89.
[10] Johnson M E. A.W. Grabau and the Fruition of a New Life in China [J]. *Journal of Geological Education*, 1985, 33 (2): 106-111.
[11] 王鸿祯. 葛利普教授——中国地质学界的良师益友[C]//王鸿祯主编. 中国地质事业早期史. 北京：北京大学出版社，1990：81-93.
[12] Mazur A. *A Romance in Natural History: The Lives and Works of Amadeus Grabau and Mary Antin* [M]. Syracuse: Garret, 2004.
[13] 于洸. 葛利普教授——忠于中国地质事业的美国人[C]//郭建荣，杨慕学等. 北大的大

师们. 北京：中国经济出版社，2005：18-30.

[14] 杨钟健. 科学家是怎样成长的？——纪念葛利普先生逝世二周纪念作[J]. 科学，1948，30（3）：65-69.

[15] 佚名. 平市地质学家昨公葬葛利普[N]. 申报，1947-3-21：5.

[16] 佚名. 葛利普教授之墓迁入北京大学[J]. 地质论评，1982，28（6）：618-619.

[17] 朱夏. 葛利普教授悼会记[J]. 地质论评，1946，11（1/2）：1-6.

[18] 翁文灏. 悼地质学大师葛利普先生[N]. 重庆大公报，1946-03-28：2.

[19] 薛攀皋. 中国科学社生物研究所——中国最早的生物学研究机构[J]. 中国科技史料，1992，13（2）：47-57.

[20] 任鸿隽. 静生生物调查所开幕记[J]. 科学，1929，13（9）：1263-1264.

[21] Grabau A W. The Peking Laboratory of Natural History [J]. *China Journal of Science and Arts*, 1926, 4: 193-196.

[22] Anonymous. In Memoriam of Mr. Kungpah King [J]. *Bulletin of the Peking Society of Natural History*, 1928, 2 (4): iii-iv.

[23] Fortuyn A B D. Concluding the Tenth Volume [J]. *Peking Natural History Bulletin*, 1936, 10 (4): 361-363.

[24] Anonymous. Proceedings of the Society for 1925–26 [J]. *Bulletin of the Peking Society of Natural History*, 1927, 1 (4): i-xxxiv.

[25] Wilder G D. The Natural History Society's Functions [J]. *Bulletin of the Peking Society of Natural History*, 1927, 1 (4): xx-xxxiii.

[26] Haas W J. *China Voyager: Gist Gee's Life in Science* [M]. New York: M. E. Sharpe, Inc., 1996: 179-182.

[27] 葛利普. 中国科学的前途：葛利普先生在民国十八年八月二十五日中国科学社年会公宴的演说词[J]. 任鸿隽，译. 科学，1930，14（6）：759-777.

[28] Anonymous. Minutes of Council Meetings [J]. *Bulletin of the Peking Society of Natural History*, 1929, 4 (1): 91-92.

[29] Anonymous. Dr A. W. Grabau's Address in the Tenth Anniversary [J]. *Peking Natural History Bulletin*, 1935, 10(2): 90-92.

[30] Anonymous. Membership List 1925-26 [J]. *Bulletin of the Peking Society of Natural History*, 1927, 1 (4): lxviii-lxxv.

[31] Anonymous. Membership List 1928-29 [J]. *Bulletin of the Peking Society of Natural History*, 1929, 4 (1): 103-115.

[32] Anonymous. Boring A M. Early Days of the Peking Natural History Society [J]. *Peking Natural History Bulletin*, 1950, 19 (1): 77-79.

[33] Anonymous. The Tenth Anniversary [J]. *Peking Natural History Bulletin*, 1935, 10 (2): 89-92.

[34] Anonymous. Regulations for the King Medals [J]. *Bulletin of the Peking Society of Natural History*, 1929, 4 (1): 95-97.

[35] Anonymous. Peking Society of Natural History Proceedings [J]. *Peking Natural History Bulletin*, 1948, 16 (3&4): i-ii.

[36] Wong W H. A Prefatory Note [J]. *Bulletin of the Peking Society of Natural History,* 1926, 1 (1): i-ii.

[37] Anonymous. Announcement [J]. *Peking Natural History Bulletin*, 1941, 16 (2): back cover.

[38] 北京博物学会十周年纪念[J]. 科学，1935，19（10）：1665.

[39] Anonymous. Scientific and other Learned Bodies, Exchange List [J]. *Bulletin of the Peking*

Society of Natural History, 1928, 3 (1): lxvii.

[40] Anonymous. News of the Society [J]. *Peking Natural History Bulletin*, 1950, 19 (1): 74-76.

[41] Ogilvie M B, Choquette C J. *A Dame Full of Vim and Vigor: A Biography of Alice Middleton Boring; Biologist in China* [M]. Amsterdam: Harwood Academic Publishers, 1999, 158: 170-181.

Amadeus W. Grabau and the Peking Society of Natural History

SUN Chengsheng

Abstract   Amadeus W. Grabau (1870–1946) was a world famous paleontologist, geologist in the 20th century. After coming to China in 1920, he made indispensable contributions to the development of Chinese geological sciences. He was also concerned with biological sciences in China. As a result of an appeal by him and his encouragement, the Peking Society of Natural History (PSNH) was founded in 1925. This Society attracted almost all the famous scholars interested in the biological scieneces in Peking (Beijing) and beyond, and issued *Peking Natural History Bulletin*, a very important journal at the time. In addition, the Society also vigorously promoted biological studies in China by organizing regular lectures, annual meetings, excursions, and natural history awards (the King Medal). A scientific community formed around Grabau, which included not only his Chinese colleagues and students, but also many foreign scientists. The study of PSNH and its *Bulletin* can help us to understand the internationalization and academic standards of the scientific community in Peking at that time.

Keywords   Amadeus W. Grabau; Sohtsu G. King; Peking Society of Natural History; *Peking Natural History Bulletin*; biology; geology

——原载《自然科学史研究》2015 年第 2 期

海进海退和大陆漂移之地球"沧桑"史

——葛利普的脉动和极控理论

孙承晟

　　摘　要　葛利普的科学贡献主要在于古生物学和地层学，以及在此基础上所提出的全球性脉动和极控理论。因其着眼点在于整个地球之历史，故后者更受其重视。本文对葛氏脉动理论的产生、发展、影响及相关背景进行了新的探讨，指出该理论在 1913 年的《地层学原理》一书中即有发端，至 1933 年发表于《时事日报》（*The Peiping Chronicle*）的文章奠定基本框架，同年在美国的第 16 届国际地质大会正式提出。葛氏自美国返回中国后，在与国内学者深入讨论的基础上，对该理论进行了修正、发展，并与极控理论相结合，1940 年以《年代的节律：从脉动理论和极控理论看地球的历史》（*The Rhythm of the Ages: Earth History in the Light of the Pulsation and Polar Control Theories*）一书进行系统的总结。脉动理论以全球性的海进和海退解释规律性地层的形成，并对地层（尤其是古生界）作了新的划分；极控理论则以魏格纳大陆漂移说为基础，用来说明硅铝圈在地球自转的带动下围绕两极的运动，并作为脉动理论的补充，说明规律沉积中的不规则性。这两个理论虽引起不小的反响，但揆诸实际，还是因太过理想主义而未能得到进一步的发展。

　　关键词　葛利普；脉动理论；极控理论；地质学；古生物学；海进；海退；大陆漂移

1　"沧桑"新解

　　晋代葛洪（284—364）在《神仙传》中记载王远与麻姑的对话："麻姑自说：'接侍以来，已见东海三为桑田，向到蓬莱，水又浅于往者，会时略半也，岂将复还为陵陆乎？'方平笑曰：'圣人皆言，海中行复扬尘也。'"[1]这可谓世界上最早关于海陆变迁的论述。后颜真卿（709—785）任抚州刺史时撰

《有唐抚州南城县麻姑山仙坛记》，记述了此事，对抚州南城县麻姑山赞誉有加，并说麻姑山山顶古坛之"高石中犹有螺蚌壳，或以为桑田所变"，注意到山石中的螺蚌壳，并与葛洪的"沧海桑田"相呼应，堪为化石观念的最早萌芽。[2-3]

宋代沈括（1031—1095）在其巨著《梦溪笔谈》亦记载北太行山崖之间的螺蚌壳之遗迹，并以此表明陆地常为河海所淹没。[4]朱熹（1130—1200）则采纳邵雍（1011—1077）十二万九千六百年为宇宙演化周期之说，进而指出："常见高山有螺蚌壳，或生石中，此石即旧日之土，螺蚌即水中之物。下者却变而为高，柔者变而为刚，此事思之至深，有可验者。"[5]其中关于宇宙之演化虽多为思辨，但以沧海桑田的现象加以说明，不能不说是难能可贵的。[6]

这些或朴素或思辨的论述，反映了中国古人关于海陆变迁的思想，在世界上亦属地质学之萌芽。19 世纪上半叶，以赖尔的《地质学原理》为标志，地质学作为一个现代学科在西方得以建立。经过几代学者的探索与争论，所谓"沧海桑田"逐渐成为地质学中一个显然的事实。因此李四光 1918 年在其伯明翰大学的硕士学位论文[1]中开篇即以中国古代的"沧海桑田"（blue seas change into mulberry fields）一语引申，进而展开关于中国地质学的论述，后发展为地质力学。

"沧海桑田"更可用来反映葛利普（Amadeus W. Grabau，1870—1946）[7-14]的主要地质学思想。1924 年，葛利普在其《中国地质史》（*Stratigraphy of China*）第一卷卷首以《神仙传》和《朱子语类》中关于沧海变桑田的论述作为题记，引出是书主题，[2]书中以海进海退理论解释地层的形成。后来，他的多篇关于脉动理论的文章均以"沧桑"一语作为中文的翻译，如 1933 年在华盛顿第 16 届国际地质大会上发表的 *Oscillation or Pulsation*？即被翻译为《沧桑论中之轩轾》，1933 年 11 月以"沧桑论及古生代地层之分法"宣读于中国地质学会第十次年会，1934—1935 年在《国立北京大学自然科学季刊》（*The Science Quarterly of the National University of Peking*）连续发表的长文 *Palaeozoic*

1　感谢刘晓博士提供该论文。

2　这两段引文由章鸿钊帮助摘引并翻译。葛利普还在"中国之古生物学"（《科学》，1931 年，15 卷 8 期，1207—1211 页）一文中对朱熹的该段文字进行阐释。1928 年，《中国地质史》下卷则引《庄子·寓言篇》中的"万物皆种也，以不同形相禅，始卒若环，莫得其伦，是谓天均。天均者，天倪也。"作为题记（英文翻译采用英国驻华使馆领事倭纳（Edward Chalmers Werner，1864—1954）转译法国传教士戴遂良（Léon Wieger，1856—1933）的法译本。

Formations in the Light of the Pulsation Theory（后刊行成书四卷）则被译为《古生代地层之沧桑观》[1]。地球的"沧桑"史通过脉动和极控理论获得了新的解释。

葛利普综合中西方地层、古生物资料提出的脉动和极控理论是 20 世纪重要的地质学理论之一。脉动理论发端于 1913 年的《地层学原理》（*Principles of Stratigraphy*），极控理论则主要源自魏格纳的大陆漂移说，1933 年正式将两者结合起来，至 1940 年在《年代的节律：从脉动理论和极控理论看地球的历史》（*The Rhythm of the Ages: Earth History in the Light of the Pulsation and Polar Control Theories*）一书中进行系统的总结，并不断用来诠释地球的历史，发表了很多种论著。作为葛氏自认为最重要的科学贡献，脉动理论当时在国内外即引起广泛的关注，如计荣森和田奇瑭随即分别发表长篇评论，对海进海退机制及其对地层的重新划分进行了分析。[15-17]西方学者亦多有评论。[2]美国著名古生物学家舒克特（Charles Schuchert，1858—1942）很大程度赞同规律性的海进海退，但对葛利普的地层新划分却颇多批评。[18]孙云铸则结合地壳运动、沉积相和古生物群等方面对脉动理论提出了修正。[19-20]

阮维周从地层纪录和动力来源两方面正确地指出葛利普的脉动理论是后来兴起的海底扩张和板块构造学说的先声。[21]王鸿祯曾对葛利普的脉动和极控理论有过简要的论述。[12]约翰逊（Markes Johnson）指出葛利普是在综合北美、欧洲和亚洲的地层信息提出脉动理论的。[22]马文（Ursula B. Marvin）将葛氏的脉动和极控理论作为一个全球的框架进行阐述，作了高度且中肯的评价。[23]澳大利亚科学史家奥尔德罗伊德（David R. Oldroyd）将葛利普的脉动学说置于世界地质学史的背景中进行评述，但表示该理论是在与西方隔绝的情况下提出来的，则与事实不符。（[24]，251—256 页）梅热（Allan Mazur）为葛利普及其夫人玛丽·安亭（Mary Antin，1881—1949）撰写了精彩的传记，其中对脉动和极控理论也作了适当的述评，指出葛氏坚持脉动理论很大程度上是基于美学上的信念。（[13]，353—355 页、361—364 页）然而，关于脉动理论的发展脉络、葛氏古生物研究与脉动理论之间的关系、脉动理论与极控

---

1 以上中文译名见《国立北京大学研究教授工作报告》（1933 年），23—26 页。感谢郭金海研究员提供此材料。民国时期的地质学者似乎复活了"沧桑"的地质学含义，多加采用，如莫柱孙 1944 年就曾在《新学生》撰有长篇连载的《大地沧桑史话》，名曰"新石头记"。

2 如 G. B. Barbour 于 1938 年在 *The Geographical Journal*，A. A. M 于 1939 年在 *The Geographical Journal*，Raymond E. Janssen 于 1939 年在 *The Journal of Geology*，H. J. F. 于 1942 年在 *Geography* 分别对 *Palaeozoic Formations in the Light of the Pulsation Theory* 一书相关卷次发表书评。

理论之间的关系，以及极控理论与魏格纳的大陆漂移说之间的关系等，尚有不少讨论的必要。因此，本文将从科学史的角度对该理论的渊源、发展及其影响进行系统的梳理，以深入理解葛利普在地质学史和 20 世纪地质学在中国本土化过程中的地位。

2 海进海退：地球之脉动

葛利普的脉动理论可分为两个部分：一是规律性海进（transgression）和海退（regression）的脉动机制，二是以此对地层的重新划分。规律性的海进海退可谓脉动理论的核心和基础。

关于海进海退的现象很早就有人注意。最早对海平面升降进行系统研究的是法国外交官德马耶（Benoît de Maillet，1656—1738）。他综合笛卡尔地表聚集水和尘埃的理论，以及阿拉伯关于山崖海生动物贝壳的观察等，从而收集有关海平面逐渐下降的详尽证据，并提议建立水文观测站严密监测这种变化。（[22]；[24]，123—124 页）

拉瓦锡（Lavoisier，1743—1794）在从事化学研究之前，曾对地质学有过兴趣。受其化学老师鲁埃勒（G. -F. Rouelle，1703—1770）的启发，认识到深海沉积和沿海沉积的不同，进而有过海进和海退的猜想。（[24]，102—103 页）

法国著名科学家居维叶（Georges Cuvier，1769—1832）和矿物学家布龙尼亚（Alexander Brongniart，1770—1847）合作，通过对巴黎盆地沉积环境的考察，发现那些看起来相似的地层，有的能看到贝壳化石，有的却全无化石的痕迹，他们因此得出巴黎地区曾有过海进和海退的结论。（[24]，158 页）

奥地利著名地质学家修斯（Eduard Suess，1831—1914）进一步从地球整体的角度来看待海进和海退。在其著名的《地球的面貌》（*Das Antlitz der Erde*）一书中，他将构造理论与地层记录中所揭示的全球地史联系起来，指出海底的塌陷将会造成世界范围的海退，随后出露的陆地被侵蚀，使得沉积物不断增加，填入洋盆，导致海水复侵入陆地。他将大陆与大洋盆地的消长看作一次次的循环，从而引致周而复始的海进海退，最终使地球再一次被一个统一的大洋淹没。[23]他还创造"海面升降运动"（eustatic movement）这一术语，以解释大地构造与地层形成之间的关系，此即我们今天所常用的"海面升降"

（eustasy）概念之前身。（[24]，239—242 页；[25]）

以全球的视角研究海面升降、地壳运动和地层对比的还有美国著名的地质学家钱伯林（Thomas C. Chamberlin，1843—1928）。他相信地球历史中存在着"脉动"现象，强弱的地质构造相互间隔，形成一次次的脉动，因此引发海进和海退的循环。周期性的海进和海退，使得地层柱上的主要地层系统得以形成。（[24]，248—249 页）

作为一位百科全书式的地质学家，葛利普深谙上述各种学说，尤其是修斯和钱伯林[1]的理论，此外他对当时世界各地的地层资料亦极为熟稔。结合这些地层资料（尤其是北美、欧洲和中国），葛利普研究了全球性的海进海退现象，由此提出解释全球地层系统和地球变迁的脉动和极控理论，并通过规律性的海进和海退将古生代重新划分为 14 个脉动纪和 14 个间脉动纪。（[26]，27 页）但他并未认同欧洲流行的地球收缩说，而是采用爱尔兰地质学家乔利（John Joly，1857—1933）地壳岩石放射性热源周期性生成并传导的理论，指出地球以此能源而发生不断的地壳隆起与塌陷，持有一种美国式的"均衡"说。

葛利普说："周期性的增热通过冷的大洋底传导而消失。膨胀之后是收缩，大洋随着其洋底的变化而升降；海进之后是海退，一个脉动纪就完成了。"他深信，岩石放射性引发的脉动为地球带来永恒的生命力，使之永葆青春。（[26]，15 页）但他依据乔利的理论估算，海进和海退大概各需 3000 万年（30Ma），间脉动纪又约 3000 万年，这样以 14 个脉动纪和 14 个间脉动纪组成的古生代，即要经过漫长的 12.6 亿年（1260Ma），以现在约 2.9 亿年（541—252Ma）的古生代分期，实在太过漫长了。[23]

3　脉动和极控理论：一种新地球观

3.1　脉动理论的提出

一般认为，葛利普的脉动理论发端于 1933 年 7 月在华盛顿召开的第 16 届国际地质大会。事实上，早在 1913 年的巨著《地层学原理》（*Principles of Stratigraphy*）一书中，即已出现脉动思想的端倪。他以地球作为一个整体（The Earth as a Whole）来看待古生物和地质现象，并将地球从上到下分为气圈（atmosphere）、水圈（hydrosphere）、岩石圈（lithosphere）、岩浆圈（pyrosphere）、

1　葛利普在来华前，即与钱伯林等美国著名地质学家熟悉。参见文献[14]。

重圈（barysphere）、地核（centrosphere），以及分布于上三圈层的生物圈（biosphere）。他在水圈部分专门解释了海进和海退及其对地层沉积的影响。（[27]，1—16 页）他还明确表示该书的目的是"把我们努力获得的全部事实和原理汇集起来，通过遗留在岩石中的地质记录来解释地球的历史"。（[27]，vii）

1920 年丁文江延揽葛利普来华，那时葛氏尚对被哥伦比亚大学解聘一事颇有纠结，但或许是希望建立一个全球性的地层理论，他还是被召唤来到大洋彼岸。1920 年 10 月刚到中国，葛利普即投入到对中国古生物和地层的研究和在北京大学地质学系的教学工作，并随即于 1920 年在中国或美国的期刊上发表多篇关于中国古生物的论文，尤其是 1922 年为新创刊的《中国古生物志》连续刊发新著，并在《中国地质学会志》上发表《震旦系》（"The Sinian System"）的长文，深受学界重视。他以百科全书式的地质学知识，一方面迅速掌握中国的古生物学、地层学资料，另一方面以此完善他的知识结构，试图构建一个会通中西的地球理论。

通过对中国地层学的深入探索，他于 1924 年出版《中国地质史》（*Stratigraphy of China*）上卷（古生代及之前），下卷（中生代）于 1928 年问世，堪为中国地层学的里程碑著作。他在书中系统探讨了中国的地层，奠定中国地层学的基础。同时，他一直强调地质学应将地球作为一个整体来研究[28]，并开始用海侵海退理论来解释地层之形成[29]。但他那时关于中国古生代地层的划分仍属于传统的范畴。

前人殊少注意到，关于脉动和极控理论以及地球历史的整体框架其实早在 1933 年即已建立。1933 年 5 月 28 日和 30 日葛利普在《时事日报》（*The Peiping Chronicle*）上发表《年代的节律——脉动论，地球历史的新观点》（"The Rhythm of the Ages: The Pulsation Theory, A New Aspect of Earth History, Lecture at Yenching University"）一文（葛氏在燕京大学的演讲），首次提出了"脉动理论"，并将之与魏格纳的大陆漂移学说相提并论。

该文除未提及"极控理论"之名，实际已奠定葛氏后来的全部理论框架。作者以地外星体的假设以及大陆漂移说解释了泛大陆的产生、海陆格局的形成，以及生物的出现和演化等地球的恢宏历史。文中着重阐述其脉动理论，指出规律性的海进海退是全球现象，海进即正脉动（positive pulse-beat）将海水带入所有地槽（geosyncline），海平面上升，海退即负脉动（negative pulse-beat）使得海水撤退，海平面下降。如此一正一负脉动构成一次脉动纪。海洋的进退可用以解释地层的形成和生物的生灭。文中还以海进海退阐述了古生代各纪地层的三分法，即每个系的下部代表海进，中部为海退，上部又代表海进，下一系的下部则为海退，依次类推。[30]

3.2 第 16 届国际地质大会

脉动理论是葛利普在第 16 届国际地质大会上正式提出来的。1933 年 7 月 22—29 日，第 16 届国际地质大会在华盛顿召开，共有 54 个国家的 1182 人注册了会议（实际参会 665 人），34 个国家派遣了 141 名官方代表参会。中国共注册 8 位代表[31]，但只有 6 人参加：巴尔博（George B. Barbour，1890—1977）夫妇、步达生（Davidson Black，1884—1934）、葛利普、德日进（Pierre Teilhard de Chardin，1881—1955）、丁文江，[1]阵容强大。

1933 年 6 月 23 日，葛利普同丁文江、德日进、赫那（Nils Horner）[2]、葛利普的秘书伍德兰（Alice Woodland）女士等从上海乘船前往美国，参加第 16 届国际地质大会。（[13]，348—360 页）因美国地质学会提供部分经费，葛利普方能参加此次会议。[32]这是葛氏自 1920 年来华后唯一一次离开中国之行，那时他的腿疾已经十分严重，且已基本不出门进行地质考察，可见他对此次会议的重视。此行的主要目是在该会议上宣扬其脉动理论，同时亦借此机会返美探亲访友。

在此次会议上，他共发表了 3 篇论文，一篇即是关于脉动理论的，另外两篇则为与丁文江合作关于中国石炭纪和二叠纪的论文。1933 年 7 月 25 日下午，葛利普在"中古生代"（Middle Paleozoic）分会上作了《沧桑论中之轩轾》（"Oscillation or Pulsation"）的报告，对哈尔曼（Erich Haarmann）的颤动理论（oscillation theory）提出质疑，进而提出脉动理论。哈尔曼 1930 年在其《颤动理论》一书中，指出大陆的垂直运动是构造地质的主要动力，一地上升，必有另一地下沉，上升之地因不断侵蚀而沉积于低地，因此伴随有海进海退之现象，成为地质构造的次要动力。[33-35]

1　翁文灏原也计划参会，但后因事未成行，参见李学通：《翁文灏年谱》，济南：山东教育出版社，2005 年，90 页。但翁文灏 1933 年 4 月曾为中国化石人的研究致信组委会，表示会上将公布相关研究报告，后步达生在会上散发了该报告并作总结，参见文献[30]第 1171 页。李四光可能是因筹建地质研究所办公楼而未参会，但递交了论文，参见马胜云、马兰编著：《李四光年谱》，北京：地质出版社，1999 年，111—112 页。中国代表提交论文为：1）葛利普《沧桑论中之轩轾》；2）丁文江、葛利普《中国之二叠纪及其对二叠纪分层的意义》；3）丁文江、葛利普《中国之石炭纪及其与密西西比纪和宾夕法尼亚纪的关系》；4）李四光《东亚构造的框架》；5）德日进《大陆地质学中山麓砂砾之意义》；6）步达生《中国化石人》；7）巴尔博《中国之黄土》。参见文献[31]和《国立北京大学研究教授工作报告》（1933 年）23—26 页。

2　赫那是中瑞西北考查团（The Sino—Swedish Scientific Expedition to the North Western Provinces of China）成员，在青海、新疆做了大量的考察工作，此次是一起乘船返回瑞典。

与哈氏相反，葛利普认为海进与海退为地质构造的主要动力，因此带来的陆地升降运动则居于次要的地位。他在文中论述了海平面有节奏的脉动，正脉动代表全球性的海进，广遍的代表性构造沉积形成，大量海洋生物同时侵入地槽和陆缘浅海；接下来的全球性海退（或称为负脉动）则导致陆相沉积的形成或广泛的侵蚀。下一次脉动又是全球性的海进。因此两次海进之间则为沉积间断（hiatus）或陆相沉积。这正好形成了每一个系的三重划分，如某一系的上下层为海进所形成，中间为海退，下一系的上下层则为海退，中间为海进，造山运动与海进海退的关系亦可因此获得解释。葛利普因此对古生代地层（寒武系、志留系、泥盆系、狄南阶、宾夕法尼亚系、二叠系）作了重新划分（图1）。[36]

报告后，德国地质学家施蒂勒（Hans Stille，1876—1966）和美国地质学家伊莱亚斯（Maxim K. Elias，1889—1982）分别与葛利普进行了交流与讨论。前者赞成葛利普提出的海进海退确有全球性的意义，但其原因更多地在于构造运动，因而构造运动也是地质现象的主要动力。葛利普对施蒂勒的评论表示感谢，但不赞同他认为构造运动是主要动力的看法，虽然在其报告中引用了施蒂勒关于构造运动的研究，作为脉动间歇期的地质作用。[36]

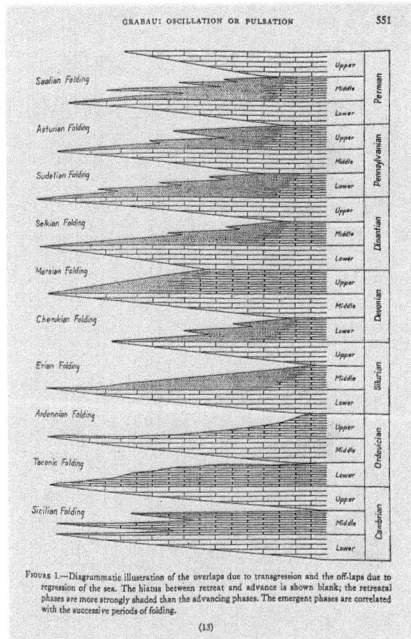

图1 海进海退与构造运动关系示意图

在 166 篇所提交的论文中，葛利普的这篇报告被认为是此次大会上 11 篇最为重要的文章之一。此外，戴维斯（W. M. Davis，1850—1934）的《山地沙漠地貌学》（"Geomorphology of Mountainous Deserts"）、李四光的《东亚构造的框架》（"The Framework of Eastern Asia"）亦名列其中。[37]至于葛利普与丁文江合作的两篇论文，反响并不大，为此丁文江还显得比较沮丧。[38]

葛利普会后与伍德兰女士前往波士顿、布法罗会见师友、亲人。两个月后，1933 年 9 月初，葛利普抵达加利福尼亚州，在加利福尼亚大学伯克利分校作了关于脉动理论的演讲[39]。9 月 8 日，葛利普、伍德兰和德日进等从旧金山乘船返回中国。

3.3　脉动理论下的古生代地层

葛利普自美国返回中国后，即全力发展和完善其脉动理论，至 1940 年，以《年代的节律：从脉动理论和极控理论看地球的历史》（*The Rhythm of the Ages: Earth History in the Light of the Pulsation and Polar Control Theories*）为名的洋洋巨著诞生，脉动和极控理论及一部新地球史最终形成。在此期间，葛利普经常撰文或报告，与同行交流讨论。如他多次在中国地质学会年会上报告其关于脉动理论的论文。1933 年 10 月于中国地质学会与北京博物学会的联合会议上报告。1933 年 11 月以"沧桑论及古生代地层之分法"宣读于中国地质学会第十次年会，引起热烈讨论。[15, 40]1936 年，在南京举行的中国地质学会第十二次年会上发表论文《依据脉动学说来划分古生代系统并各系统名称之订定》，因未亲自出席，由计荣森代为宣读，并以图示说明。报告后亦有热烈讨论，如翁文灏指出脉动说属全球性的理论，而葛利普所拟的新名多用中国地名，且有新旧混用者，似有不妥；此外，据葛氏理论，同一脉动纪的造山运动应大体相当，但为何太平洋沿岸的中生代运动甚为剧烈，而大西洋沿岸的则轻微不著，对脉动说提出了质疑。中央大学的贝克（Hans Becker）则表示很赞赏提出脉动说的理论勇气，但在此一理论尚未获得证实之前，即用于全球颇有危险。黄汲清在赞赏葛氏的理论勇气之余，亦表示需厘清所谓正负脉动的显著区别，以及它们是以海的范围还是沉积的类型而确定。葛利普对这些质疑都作了相应的答复。[41-43]

葛利普系统综合世界各地的古生物和地层信息，对上述论文或报告扩充发表。1934—1935 年，在《国立北京大学自然科学季刊》连续发表长文《古生代地层之沧桑观》（*Palaeozoic Formations in the Light of the Pulsation Theory*），根据脉动理论将古生代地层分为：下寒武脉动系、中寒武或艾伯塔（Albertian）脉动系、寒奥脉动系。这些文章修订后于 1936—1938 年分为 4 卷出版。惜后续各卷未能问世。

在多方交流讨论的基础上，葛利普对其理论作了修正，即将原来每一系以下、中、上的三分法改为二分法，下部代表海进，上部代表海退，共同成为一个独立的脉动单位。[15]传统的三分法是美国地质学家威廉斯（Henry S. Williams，1847—1918）提出来的，每一个系分为下、中、上三个部分，用以表示古（Eo-）、中（Meso-）、新（Neo-）之意，代表动物或植物群之兴、盛、衰三个连续的时期。葛利普刚提出脉动理论时，亦采用三分法，但其意却与威廉斯略异，所谓下、中、上分别代表连续的海进或海退层序，标志则是海相动物群。但由于这种三分法与脉动理论不相协调，他将之改为两分法，如寒武脉动系分为海进的下部和海退的上部，成为一独立的脉动周期，但这仅占传统寒武系的一部分。1936 年，葛利普根据脉动理论对古生代地层重新作了厘定，在《脉动理论下的古生代地层分类修订》（*Revised Classification of Palaeozoic Systems in the Light of the Pulsation Theory*）一文中将古生代分为 12 个脉动周期。[41]在 1940 年出版的《年代的节律：从脉动理论和极控理论看地球的历史》一书中则被划分为 14 个脉动。（[26]，27 页）

在地球的脉动历史中，间脉动纪（interpulsation period）代表海退之后以陆相沉积为主的地质构造时期。1938 年 2 月 26—28 日，中国地质学会在长沙举行年会，葛利普在会上发表《间脉动期在中国地层学上之意义》一文，由王钰代读。[44]在重新划分 14 个脉动纪的基础上，葛利普以世界各地地质现象为例，对 14 个间脉动纪中主要的地质活动：侵蚀（erosion）作用、陆相沉积（continental sedimentation）、造山变形（orogenic deformation）、火山活动（volcanicity）进行了系统的论述。[45]葛氏认为地槽是"山链的孵化器"，地槽消耗殆尽的时候就是发生褶皱之时，山脉因此生长形成，这就是间脉动纪。（[26]，51 页）

规律性的海进海退是脉动理论的基础。葛利普认为地球的脉动能用保存于地层记录的海进和海退现象加以说明，全球性规律的海进海退就像人的脉搏一样，导致各地区地层建造、古生物演化及古地理变迁具有一定的节律。他正是以此为原则并以丰富的化石证据对古生代地层作了新的划分。但这个划分同他的脉动理论一样，也经历了不断的修正与发展。现以《中国地质史》、《沧桑论中之轩轾》、《脉动理论下的古生代地层分类修订》（简称《修订》）、《年

代的节律：从脉动理论和极控理论看地球的历史》（简称《年代的节律》）为
例说明葛氏对于古生代地层划分之发展（表1）。[1]

表1 葛利普古生代地层划分之演变

序号	《年代的节律》（1940）	《修订》（1936）	《沧桑论中之轩轾》（1933）		《中国地质史》（1924—1928）	
14	二叠脉动系	二叠脉动系	Supra	二叠系	上	二叠系
			上			
13	乌拉尔丁斯克脉动系（Uralinskian）	乌拉尔脉动系	中		中	
			下		下	
12	顿巴斯脉动系	顿巴斯脉动系	上	宾夕法尼亚系	上	石炭系
		狄南脉动系	中		中	
11	维宪脉动系		下		下	
10	丰宁脉动系	丰宁脉动系	上	狄南系	上	狄南系
			中		中	
		泥盆脉动系	下		下	
9	泥盆脉动系		上	泥盆系	上	泥盆系
8	志泥脉动系（Siluronian）	志泥脉动系（Siluronian）	中		中	
		志留脉动系	下		下	
7	志留脉动系		上	志留系	上	志留系
		奥陶脉动系	中		中	
6	奥陶脉动系		下		下	
		寒奥脉动系	上	奥陶系	上	奥陶系
5	斯基达夫脉动系		中		中	
			下		下	
4	寒奥脉动系	寒武脉动系	上	寒武系	上	寒武系
3	寒武脉动系	塔康脉动系	中		中	
2	塔康脉动系		下		下	
1	震旦脉动系					

　　葛利普1933年提出脉动理论时，他对于古生代地层的划分还采取传统的
方式，分为寒武、奥陶、志留、泥盆、狄南、宾夕法尼亚（石炭）、二叠7个
系，每个系又分为上、中、下三个统。至1936年则分为12个脉动系（pulsation
system），但方案尚不甚清晰。1940年则明确分为14个脉动系，每个系分上下
两部，下部代表海进，上部为海退，共同组成一个脉动沉积地层。每两个脉

---

[1] 葛利普生前亦从事中生代、新生代的地层新划分，但未能完成问世。

动系之间代表一个间脉动纪，如震旦脉动系和塔康脉动系之间为比尔森纪（Pilsenerian），以此类推从下至上分别为：普里布拉姆纪（Pribramian）、阿克通纪（Arctomian）、若贝尔纪（Rhobellian）、彼得洛夫纪（Petrovian）、普林利蒙纪（Plynlimonian）、沙林纪（Salinan）、奥里斯坎尼纪（Oriskanian）、蒙特普勒桑特纪（Monteplaisantian）、毛赫丘克纪（Mauch-chunkian）、拉纳克纪（Lanarkian or Hintonian）、洛钦瓦尔纪（Lochinvarian）、后昆古尔纪（Post Kungurian or Lebachian）、阿帕拉契纪（Appalachian）。（[26]，27 页）

3.4 极控理论

在葛利普的地球理论体系中，脉动理论是其核心和重点，极控理论则用来配合前者一起说明地球的演变与历史。与脉动理论一样，关于极控理论的整体框架早在 1933 年即已形成。在《年代的节律——脉动论，地球历史的新观点》一文中，除未提及"极控理论"之名，实际上葛利普已经以地外星体的假设、大陆漂移说以及乔利的岩石放射性理论解释了泛大陆的产生、海陆格局的形成，以及生物的出现和演化等宏大的地球历史。

1940 年，葛利普出版《年代的节律：从脉动理论和极控理论看地球的历史》一书，是其对脉动理论和极控理论的总结。书中用这两个理论系统阐述了从原始地球到人类出现的历史，被他看作是最重要的贡献，并将此书献给他的妻子安亭、女儿约瑟芬和两个外孙女。

葛利普指出在众多的地质现象中，有两类是最为重要的。其一是海洋的脉动，通过节律性的全球海进和海退，影响了古生代以后地层的形成和生物的迁移。其二是极控（polar control），即硅铝圈（sial-sphere）在地球自转离心力的作用下绕极轴的周期运动。（[26]，vii—ix）两者合在一起即可解释地层的形成以及整个地球的历史。葛利普采纳当时的地球结构理论，将原始地球（primitive earth）从外至内分为六个圈层：最外为气圈（atmo-sphere），主要由二氧化碳组成；下面是包裹整个地球的水圈（hydro-sphere），厚度约 2.64 公里；以下依次为富含硅和铝的酸性岩层硅铝圈（sial-sphere），厚度约 40 公里；富含硅、镁、钙、氧化铁的硅镁圈（sima-sphere），厚 1200 公里；其下或为富铁橄榄岩圈（pallassite sphere），厚 1700 公里；最中心为由镍、铁及其他重金属元素组成的镍铁圈（nife-sphere），厚 3500 公里。各圈的比重从地表至地心累增，地表为 2.75，硅铝圈和硅镁圈交接处约 3.1，硅镁圈和橄榄岩圈交接处 4.75，橄榄岩圈和镍铁圈交接处约 5.0，地心达 11.0，整个地球的平均比

重则为 5.52。（[26]，6—8 页）

原始地球何以转变成为现在的样子？在当时板块学说尚未兴起，天体动力学尚未成熟的情况下，葛利普设想一个外来的客星体（stellar visitor）靠近地球南极，在适当的位置将地球北部的硅铝圈吸引至南极周围，形成魏格纳所谓的泛大陆（Pangaea），硅铝圈在被拖拽的过程中即形成山脉和地槽；水圈则被迫向北方撤退，淹没硅铝圈移出的硅镁圈，形成最初的泛大洋（Panthalassa）。（[26]，11—13 页、18—19 页）此后，出露的硅铝圈（即泛大陆）就在地球自转离心力（impetus）的驱动下周期性地移动，各大陆分裂形成。大陆边缘遭遇硅镁圈的阻碍便会形成地槽，另一端则因牵拉张力产生断裂或火山。因地球自转轴基本恒定，两极和气候带保持不变，因此大陆在移动的过程中会遭遇不同的环境与气候带，从而产生化石链条的断裂以及冰川的迁移。（[26], viii）但在一些火成力量（如岩脉、岩柱、岩基）入侵极盖（sial-cap）附近时，因周围力量或比重的变化，从而导致两极顺时针或逆时针的移动，进一步会影响到硅铝圈的漂移。（[26]，13 页）这就是所谓的极控理论。在《年代的节律：从脉动理论和极控理论看地球的历史》一书的扉页葛利普以一块桌布（代表硅铝圈）包裹着一个地球仪，来模拟泛大陆的形成（图 2）。在该书的最后，作者则以 24 幅彩色插图表示从泛大陆到第四纪详细的海陆变迁，尤其是地槽和陆表海位置的变化。此外，葛利普还猜想地球的生命来自于地球之外的星体。（[26]，16 页）

极控理论是在魏格纳（Alfred Wegener，1880—1930）大陆漂移说的基础上发展起来的。魏格纳 1912 年在法兰克福会议上最早提出大陆漂移的假说。1915 年出版著名的《海陆的起源》，从地球物理学、古生物学、古气候学、大地测量学等方面论证了现在的海陆格局是如何从"泛大陆"漂移形成的。其核心思想是，现在地球上所有的陆地在石炭纪之前是连接在一起的泛大陆，周围则是泛大洋；从中生代开始，泛大陆在潮汐力和地球自转产生的离心力的共同作用下，产生向西和向赤道的分裂、漂移，形成现在的海陆格局。[46]但早期魏格纳这种"犁地式前进"的大陆漂移理论，因其驱动力实在太过微小，再加上他作为一名气象学家的身份，故支持者寥寥。1929 年《海陆的起源》第四版才把大陆漂移的动力归于地幔对流。[23]

葛利普对魏格纳大陆漂移说极为赞赏。早在 1933 年的那篇《年代的节律——脉动论，地球历史的新观点》一文中即支持该学说，并将自己的脉动理论及极控理论与之相提并论。这在当时对大陆漂移说普遍不看好的情况下是很罕见的。很明显，葛利普的极控理论与魏格纳大陆漂移说十分相似，不同的是葛利普设想了一个外来的星体以解释泛大陆的形成，且他的大陆漂移

图2　《年代的节律：从脉动理论和极控理论看地球的历史》（1940年）
中泛大陆形成示意图

并没有魏格纳所说的潮汐力。可以设想，倘若他能活到20世纪后半叶，一定会是板块构造理论的积极推动者。此外值得一提的是，黄汲清在瑞士的老师阿尔冈（Emile Argand，1879—1940）亦是当时为数不多大陆漂移说的支持者之一，曾以该理论对阿尔卑斯山和亚洲的地质构造进行解释。

极控理论不仅用来解释大陆的起源及相关地质现象，而且还可说明脉动理论下一些不规则的地层沉积。根据脉动理论，地层的累积形成因海进海退的规律性而应该较为规则，但实际情况并非如此。在葛利普看来，极控理论下的硅铝圈漂移则可解释地层沉积的不规则现象，如地壳之间的碰撞可产生地层不整合，硅铝圈从南到北或从北到南的移动可出现化石证据的误植等，这些均使原本规律的地层沉积产生改变。（[26]，viii—ix）

4　脉动理论的反响

自葛利普提出其脉动理论后，随即引起国内外学者的重视和评论。国内

如田奇瑛和计荣森分别发表长篇评论[15-17]，翁文灏、黄汲清等亦提出相应的看法与质疑。德日进亦相信地球表面确实存在节律性的运动，但对葛利普将脉动与沉积、造山等联系起来却不认同，当然他也指出脉动理论在不少地方仍具有启发性，对传统的地层理论带来了新的看法，并在给朋友的信中调侃葛利普因"脉动"而变得年轻起来。[47]1西方学者对葛利普的著述亦给予了及时的评价，其中作为乌尔里克（Edward Oscar Ulrich，1857—1944）同盟的舒克特虽对脉动理论颇为欢迎，尤其对其著作中所给出的丰富的古生物信息和系统的世界地层对比高度赞赏，但对葛氏所作的古生代地层新划分并不满意，很多都不赞成。[18]总之，中外学者对脉动理论的一个共同点是，他们在赞扬葛氏广博的知识和非凡的理论勇气之外，均提出了相应的质疑。

的确，尽管规律性的海进和海退确实存在，但单以脉动理论来解释地层的划界显然不够。对此，正如舒克特所指出的：海进是否为全球同时发生的现象；如何区分海进与颤动；海进后可能没有海退但也可能伴随着多次海退，而且其中有些海退造成很厚的沉积，这些都是葛氏脉动理论所无法解释的。[18]因此葛利普在中国的第一个助手孙云铸以更全面和实际的态度对葛利普的脉动理论提出了修正。他承认葛氏所提出的规律性海进和海退确乎存在，但单以此和化石证据来对地层进行划分显然不够，因此提出古生代地层应按沉积旋回（cycle of sedimentation）、地壳运动（diastrophism）、动物群组合（faunal assemblage）三个条件分类，从而把古生代地层划分为 8 个系：塔康系（Taconian）、寒武系、奥陶系、志留系、泥盆系、狄南系、石炭系、二叠系，其中寒武系、石炭系和二叠系可各分为两个统，其他五个系则分为三个统。[19, 20]

1949 年后，西方科学家在中国遭到不同程度的批判。或许是由于葛利普的巨大影响以及他对中国人极为友好的人格魅力，对他的批评没有太上纲上线，但与 1949 年前相比也有了微妙的反差。[48]自称受葛利普影响最深的杨钟健即在不同场合对葛氏提出批评。1953 年，在学习《实践论》一次会议上，杨钟健说："1. 科学工作者搜集事实是唯物的，但要通过实践。……Grabau 葛利普脉动学说把这理由找到唯心一方面去。吾道一以贯之，没有经过实践。2. 列宁所说，要经过实践，成果才能达到。3. 发现是偶然的还是有系统的？"可见杨钟健乃是基于马列主义和毛泽东的实践论而指出葛利普的脉动理论没有经过实践。杨钟健接着对奥斯朋提出了批评："Osborne 晚年有许多

1　感谢业师韩琦研究员提示该文献。

惟心著作，但少年时说哺乳动物是始于中亚，但以后以为人类始于蒙古，那是错了。朱子以山上的化石认定山谷可以在海中，是惟物的，但是许多理论是惟心的。科学家有所谓 Pet theory。"[49]1 接续上述引言，杨钟健实际将葛利普的脉动理论和奥斯朋人类起源于蒙古的猜想都归为所谓的"宠物理论"（Pet theory）2，意指他们都太偏执自己的理论而忽略了其客观性。杨钟健的批判虽带有意识形态的情绪，但将葛利普的脉动理论比作"宠物理论"却也不无合理性。

现在来看，正如舒克特和孙云铸所评论的那样，葛利普虽具有百科全书式的古生物学和地层学知识，但他所创建的脉动理论及古生代地层新划分却因理想化而与实际有较大出入。这是该理论未能得以继续传承和发展的原因。至于在美国，有学者指出，葛利普的海平面升降和脉动理论并没有得到恰当的反应和发展，可能与他在美国没有学术继承人有关。[22]

一个花絮，据葛利普夫人战时从曾长期担任协和医学院校长顾临（Roger S. Green，1881—1947）处获得的一份重庆的时报（1942 年 11 月 10 日）记载，日本人出于对葛利普在脉动理论贡献的认可，给予他每月 6 美元的资助。[50] 此时，他应已被日本人囚禁于东交民巷的集中营，该消息的真实性存疑。即使属实，以他对日本人的憎恨，恐怕也不会接受。

5 余论

葛利普虽以古生物学和地层学见长，但其着眼点却在于整个地球的历史。因此，他在 1913 年的《地层学原理》一书中即表达了通过地质纪录还原地球漫长历史的宏愿，并且采用海进海退来解释地层的沉积。他毕生均以此为目标，尤其是 1920 年入华后，在综合中西方地质资料和各家之说的基础之上，于 1933 年提出了脉动和极控理论，至 1940 年以《年代的节律：从脉动理论和极控理论看地球的历史》一书进行全面的总结，系统地解释了整个地球之"沧桑"史。

脉动和极控理论是 20 世纪重要的地质学理论之一。脉动理论以丰富的古生物、地层知识，指出由于海底的升降及所造成的全球规律性的海进和海退，导致地层的规律性累积，阐释了沉积、侵蚀连续不断的循环思想，并对传统的地层分类作出新的调整。动力方面，葛利普采用了乔利地壳岩石因放射性

---

1 感谢上海科技教育出版社殷晓岚博士帮助核查此条资料。
2 "宠物理论"是指理论创建者对自己理论如对宠物一般极度偏爱，而对其他理论有抗拒心理，从而有失客观性。

产生热效应的理论，指出地球因此而具有驱动海进海退的"永动"能源。同时，葛利普在魏格纳大陆漂移说的基础上，提出大陆在地球自转离心力的驱动下绕极轴周期性移动的极控理论，"泛大陆"慢慢分裂成大陆碎块，最终呈现出现在的海陆格局；并作为脉动理论的补充，说明规律沉积中的不规则性。

即使以现在的海平面升降和板块构造学说而言，这亦不失为一个宏大而富有启发性的理论。但以实际情况而言，还是因太过理想主义而未能得到进一步的发展。正如孙云铸、舒克特等所指出的那样，葛利普单以规律性海进海退来解释地层的形成失之偏颇；且葛氏诉诸"特设性"的星体来解释泛大陆的形成、将地球自转的离心力作为硅铝圈漂移的动力，以及假设生命来自地球之外，现均不被接受。然而，他通过脉动理论对世界地层的对比、硅铝圈绕极轴的漂移、大洋底部的热能转换在当时都是极具前瞻性的，他广阔的视野和创新精神仍是人类的宝贵财富，更不用说他在古生物学、地层学上的卓越贡献了。

葛利普生前曾说过，他的脉动理论时人未必能理解，但三十年后必得世人之公认。[51]其自信可见一斑。这不由让人想起开普勒1619年在其《世界的和谐》第五卷序言所说的："……总之书是写成了，骰子已经掷下去了，人们是现在读它，还是将来子孙后代读它，这都无关紧要。既然上帝为了他的研究者已经等了6000年，那就让它为读者等上100年吧。"[52]两者之言论颇有异曲同工之处。此处开普勒兴奋溢于言表的正是他终生探索而得出的宇宙的秘密："天体的运动只不过是某种永恒的复调音乐。"现在看来，开普勒最重要的贡献是他发现的行星运动三定律（为牛顿提出万有引力定律铺平了道路），而其所谓的"世界和谐"仅具有一种形而上学的审美价值。葛利普也一样，他自认为最重要的脉动和极控理论很大程度上也是出于美学的信仰，如今也仅具有历史的意义，但他对理论的自觉追求及所作的思考仍具有鲜活的价值。

葛利普曾指出，科学的历史分为思辨、实证、理性三个阶段。思辨即古代对自然的玄想，实证就是通过收集科学事实认识自然，最后的理性阶段则是通过广泛的科学事实提出科学理论。他还多次说过，科学不仅在于收集科学事实，更重要的是通过科学事实提出科学理论。具体到地质学，地质资料的积累固然重要，但对这些资料的分析进而解释地球的历史和各种地质现象是一项更重要的工作。[30, 53]因此不难理解他为何更看重其脉动和极控理论。葛利普一生不仅皓首穷经积累世界各地的地质学知识，对科学理论的总结更是不遗余力。他所掌握的浩瀚的古生物、地层学知识令人惊叹，其老而弥坚的科学精神更让人钦佩。田奇㻞曾表示"他这种'老而弥坚'、'为学不倦'的

精神，实值得我们十二万分的佩服！"[16]胡适 1937 年 1 月 9 日在参加葛利普
67 岁生日宴会后在日记中称"此公半生残废，而努力作学问，至死方休，真
是我们的模范！"[54]可谓这个伟大学者的真实写照。

　　致　　谢：承潘云唐教授、张九辰研究员、颜茂都博士、魏荣强博士、王
光旭博士给予宝贵建议并审阅初稿，使我获益很多，特致谢忱！本文部分初稿
分别在第 40 届国际地质科学史会议（2015 年 6 月，北京）和"赛先生在中国：
中国科学社成立百年纪念暨国际学术研讨会"（2015 年 10 月，上海）报告。

参 考 文 献

[1] 葛洪. 神仙传校释[M]. 胡守为校释. 卷 3. 北京：中华书局，2010：94.

[2] 章鸿钊. 中国研究地质学之历史[J]. 中国地质学会志，1922，（1）：27-31.

[3] 尹赞勋. 中国古生物学之根苗[J]. 地质评论，1947，12（1-2）：63-69.

[4] 沈括. 新校正梦溪笔谈[M]. 胡道静校注. 卷 24. 香港：中华书局，1975：237.

[5] 黎靖德，王星贤. 朱子语类[M]. 卷 94. 北京：中华书局，1986：2367.

[6] 李仲均. 我国古代关于"海陆变迁"地质思想资料考辨[J]. 科学史集刊，1982，（10）：16-21.

[7] Ting V K. Biographical Note [J]. *Bulletin of the Geological Society of China*, 1931, 10: iii-xviii.

[8] Sun Y C. Professor Amadeus William Grabau: Biographical Note[J]. *Bulletin of the Geological Society of China*, 1947, 27: 1-26.

[9] Kay M. Grabau, Amadeus William[C] // Gillispie C C. *Dictionary of Scientific Biography*, Vol. 5. New York: Charles Scribner's Sons, 1981: 486-488.

[10] 潘云唐. 葛利普——中国地质科学工作者的良师益友[J]. 中国科技史料，1982，（3）：22-30.

[11] 杨静一. 葛利普传略[J]. 自然科学史研究，1984，3（1）：83-89.

[12] 王鸿祯. 葛利普教授——中国地质学界的良师益友[C]//王鸿祯主编. 中国地质事业早期史. 北京：北京大学出版社，1990：81-93.

[13] Mazur A. *A Romance in Natural History: The Lives and Works of Amadeus Grabau and Mary Antin*[M]. Garret: Syracuse, 2004.

[14] 孙承晟. 葛利普与北京博物学会[J]. 自然科学史研究，2015，34（2）：182-200.

[15] 计荣森. 葛利普氏之脉动学说[J]. 科学，1935，19（8）：1186-1210.

[16] 田奇㻪. 对葛利普氏脉动学说之我见[J]. 地质论评，1936，1（5）：523-530.

[17] 田奇㻪. 对葛利普氏脉动学说之我见（续）[J]. 地质论评，1937，2（6）：515-532.

[18] Schuchert C. What is the Basis of Stratigraphic Chronology?[J]. *American Journal of Science*, 1937, 34 (204): 475-479.

[19] Sun Y C. Bases of the Chronological Classification with Special Reference to the Palaeozoic Stratigraphy of China[J]. *Bulletin of the Geological Society of China*, 1943, 23(1-2): 35-56.

[20] 孙云铸. 海侵的基本概念和问题——着重讨论中国古生代各纪动物群及其分区[J]. 地

质学报, 1963, 43 (2): 99-115.

[21] 阮维周. 葛氏脉动学说与海底扩张[C]//阮维周教授文集. 1992: 100-108.

[22] Johnson M. A. W. Grabau's Embryonic Sequence Stratigraphy and Eustatic Curve[C]// Dott R H (ed). *Eustasy: The Historical Ups and Downs of a Major Geological Concept*. Boulder, Colorado: Geological Society of America Memoir 180, 1992: 43-54.

[23] Marvin U B. The Global Theories of Amadeus W. Grabau (1870-1946): A Retrospective View[C]//Wang H Z, et al. *Comparative Planetology, Geological Education, History of Geology* (Proceedings of the 30th International Geological Congress, Vol. 26), Utrecht (Netherlands): VSP, 1997: 165-175.

[24] 戴维·R. 奥尔德罗伊德. 地球探赜索隐录: 地质学思想史[M]. 杨静一译. 上海: 上海科技教育出版社, 2006.

[25] Wegmann E. Suess, Eudard[C]// Gillispie C C. *Dictionary of Scientific Biography*, Vol. 13. New York: Charles Scribner's Sons, 1981:143-149.

[26] Grabau A W. *The Rhythm of the Ages: Earth History in the Light of the Pulsation and Polar Control Theories*[M]. Peking: Henri Vetch, 1940.

[27] Grabau A W. *Principles of Stratigraphy*[M]. New York: A. G. Seller and Company, 1913.

[28] Grabau A W. *Stratigraphy of China* (Part I: Palaeozoic and Older)[M]. Peking: Geological Survey of China, 1924: 1-3.

[29] Grabau A W. *Stratigraphy of China* (Part II: Mesozoic)[M]. Peking: Geological Survey of China, 1928: 773-774.

[30] Grabau A W. The Rhythm of the Ages. The Pulsation Theory, a New Aspect of Earth History. A Lecture Delivered at Yenching University[N]. *The Peiping Chronicle*, 1933-5-28: 6; 1933-5-30: 5 & 8.

[31] Anonymous. *International Geological Congress: Report of the XVI Session, United States of America* (1933)[M]. Washington, 1936: 16-17.

[32] 丁文江. 苏俄旅行记 (1934—1935) [C]//欧阳哲生主编. 丁文江文集 (第七卷). 长沙: 湖南教育出版社, 2008: 107-117.

[33] Haarmann E. *Die Oszillations-Theorie: Eine Erklärung der Krustenbewegungen von Erde und Mond*[M]. Stuttgart: Ferdinand Enke, 1930.

[34] Holmes A. The Mechanism of Earth Movements (Review on *Die Oszillations-Theorie*)[J]. *The Geographical Journal*, 1931, 77 (2): 164-165.

[35] Melton F A. Review on *Die Oszillations-Theorie*[J]. *The Journal of Geology*, 1931, 39 (3): 296-299.

[36] Grabau A W. Oscillation or Pulsation[C]//*International Geological Congress: Report of the XVI Session, United States of America (1933)*, Washington, 1936: 539-553.

[37] Nolan T B & Siegrist M. Recalling the 16th IGC, Washington 1933[J]. *Episodes*, 1987, 10 (4): 329-331.

[38] 黄汲清. 我的回忆: 黄汲清回忆录摘编[M]. 北京: 地质出版社, 2004: 113.

[39] Grabau A W. Palaeozoic Formations in the Light of Pulsation Theory (Pt. 1, The Lower Cambrian Pulsation)[J]. *The Science Quarterly of the National University of Peking*, 1934, 4 (1): 27-184.

[40] 中国地质学会第十届年会纪略[J]. 湖北教育月刊, 1933, (3): 163-167.

[41] Grabau A W. Revised Classification of the Palaeozoic Systems in the Light of the Pulsation Theory[J]. *Bulletin of Geological Society of China*, 1936, 15 (1): 23-51.

[42] 中国地质学会第十二次年会记事[J]. 地质论评. 1936, 1 (1): 81-86.

[43] 翁文灏. 翁文灏日记[M]. 李学通等整理. 北京：中华书局，2010：11.

[44] 地质学会年会首次在湘举行[N]. 申报（汉口版），1938-02-27：2.

[45] Grabau A W. The Significance of the Interpulsation Periods in Chinese Stratigraphy[J]. *Bulletin of Geological Society of China*, 1938, 18 (2): 115-120.

[46] Wegener A. *The Origin of Continents and Oceans*[M]. Skerl J G A (trans). London: Methuen & Co. Ltd., 1924.

[47] Cuénot C. *Teilhard de Chardin: A Biographical Study*[M]. Colimore V (trans). London: Burns & Oates, 1965: 156-157.

[48] 张九辰. 科学史事的时代解读：对中国地质学史的案例分析[J]. 自然科学史研究，2015，34（1）：74-87.

[49] 竺可桢. 竺可桢全集[M]. 第13卷. 上海：上海科技教育出版社，2007：109.

[50] Shimer H W. Dr. A. W. Grabau in China[J]. *Science* (New Series), 1943, 97 (2529): 555-556.

[51] 杨钟健. 地质学家葛利普[J]. 人物杂志，1947，2（2）：60-62.

[52] （德）开普勒. 世界的和谐[M]. 张卜天译. 北京：北京大学出版社，2011：3-4.

[53] Grabau A W. A New Interpretation of Earth History[J]. *Peking Natural History Bulletin*, 1940, 15 (1): 1-12.

[54] 胡适. 胡适全集[M]. 第32卷. 季羡林主编. 合肥：安徽教育出版社，2003：605.

The Vicissitudes of the Earth in the Light of Trans- & Regression and Sial-crust Shifting: Amadeus W. Grabau's Pulsation and Polar Control Theories

SUN Chengsheng

Abstract   Amadeus W. Grabau (1870–1946), a world famous paleontologist and geologist of the 20th century, made outstanding contributions to paleontology, stratigraphy, and advanced global pulsation theory and polar control theory. He focused his attention on the latter topic. This paper, a new investigation of the development and influence of his pulsation theory, argues that this theory can be traced back to the *Principles of Paleontology* in 1913. Its framework was then laid out in an article published in *The Peiping Chronicle* in 1933, and was formally presented at the 16th International Geological Congress (Washington, USA) in the same year. After returning to China, Grabau revised and developed the theory based on communications with Chinese scholars, systematically summarizing it in *The Rhythm of the Ages* in 1940, along with the polar control theory. With the pulsation theory, Grabau explained and gave a new (Paleozoic) stratigraphic classification based on global transgressions and regressions; while with the polar control theory, mainly based on Alfred Wegener's theory of continental drift, he demonstrated sial-crust shifting around the polar axis owing to the impetus of the Earth's rotation. Here he also tried to explain the irregularity of sedimentation, which should be rhythmic according to his pulsation theory. Though these two theories had worldwide influence at the time, they could not, however, on the basis of fact, be further developed due to being over-idealistic.

Keywords Amadeus W. Grabau; pulsation theory; polar control theory; geology; paleontology; transgression; regression; sial-crust shifting

——原载《自然科学史研究》2015 年第 4 期

"他乡桃李发新枝"：葛利普与北京大学地质学系

孙承晟

摘　要　葛利普1920年应丁文江之邀来华，出任北京大学地质学系教授和地质调查所古生物室主任。本文以中西档案及民国报刊文献为基础，还原葛利普在北大的授课、演讲活动，以及被聘为北大和中华教育文化基金会"研究教授"之经过，揭示他对北大乃至中国古生物学教育的独特贡献。葛利普在北大培养了一批古生物学家，奠定了中国的古生物学基础，堪称"中国古生物学之父"；其科学活动及所扮演的角色与路易·阿加西对美国科学的贡献极为相似，故亦可称为"中国的阿加西"。此外，葛利普于1920—1921年在北大开设的"地球与其生物之进化"系列演讲，系统介绍了当时西方最新的古生物学知识、生物进化论和遗传学理论，代表了新文化运动中科学传播的一面。

关键词　葛利普；北京大学；中基会；研究教授；古生物学；地质学；进化论；遗传学

作为"百日维新"的成果之一，京师大学堂自1898年创立以来，因倡导科学教育，且有京师同文馆的并入，便不断聘任洋教习授课。1916年蔡元培执掌北大以后，更有"兼容并包"之理念，此举愈得大力提倡。据统计，1924年北大共有西方教职人员19人，1925年为18人，1926年16人。[1-3]早期的洋教习以日本人为主，如岩谷孙藏（1867—1918）、服部宇之吉（1867—1939）、氏家谦曹（1866—1939）、杉荣三郎（1873—1965）、矢部吉祯（1876—1931）等，[4]另有"西学总教习"丁韪良（William A. P. Martin，1827—1916）[1]和德国人梭尔格（Friedrich Solger，1877—1965）；[5]后来以欧美人士居多，如毕善功（Louis R. O. Bevan，1874—1975）[2]、葛利普（Amadeus W. Grabau，1870—

1　丁韪良因教学懒散于1902年被张百熙免职。关于北大的洋教习，另可参见北京大学国际合作部编：《北大洋先生》，北京：北京大学出版社，2012年。

2　毕善功早年在山西大学任教，1911年被北京大学聘任，参见文献[1]。其兄Rev. H. L. W. Bevan亦在中国传教，参见http://gutenberg.net.au/ebooks15/1500721h/0-dict-biogBe-Bo.html。

1946）、柯劳文（Grover Clark，1891—1938）夫妇、钢和泰（Alexander von Staël-Holstein，1877—1937）[1]、洪涛生（Vincenz Hundhausen，1878—1955）[2] 等。他们均为当时北大乃至中国的现代学术做出了不可磨灭的贡献，其中葛利普的贡献尤为显著。

葛利普 1870 年出生于美国威斯康星锡达堡（Cedarburg）一个德国血统的新教家庭，1896 年于麻省理工学院地质学系获学士学位，1900 年以《纺锤螺及其同源动物之系统演化》（*Phylogeny of Fusus and Its Allies*）一文获哈佛大学博士学位。1901 年前往哥伦比亚大学任教，1919 年因故被校方解聘。受丁文江邀请，1920 年来到中国，任北京大学地质学系教授和地质调查所古生物室主任。1937 年因腿疾未随北大南迁，太平洋战争爆发后，被囚禁于东交民巷集中营。抗战胜利后获释，1946 年 3 月 20 日因病逝世于北京。[6-15]

1930 年葛利普六十寿辰时，章鸿钊曾撰有《葛利普教授六秩之庆》一诗，云："老眼看从开辟时，小周花甲似婴儿。藏山事业书千卷，望古情怀酒一卮。故国莼鲈添晚思，他乡桃李发新枝。东西地史因君重，灿烂勋名奕叶期。"[16] 第一句表明葛氏之六十寿辰，第二句说其著作等身，第三句说他后半生在中国桃李满天下，最后一句则表达了他在中西方地质学史上的崇高地位，概括了葛利普传奇的一生。

纵观葛利普一生的活动，他在中国长达 26 年的生活可能更具传奇色彩，无论是教书育人还是科学研究都较他在美国时更受人瞩目。中国近代的古生物学事业很大程度上奠基于葛氏之功。因此，翁文灏说："近数十年来世界知名的外籍科学家，在中国服务最久而贡献最多的，要算一位地质学大师葛利普（Amadeus W. Grabau）先生了。"[17] 本文依据中西档案及民国报刊文献，以葛利普在北京大学地质学系的教育活动为中心，分析他对中国古生物学的发展所作出的独特贡献，从中亦可窥见民国科学教育的一个侧面。

1　中国古生物学之父

京师大学堂于 1909 年设立地质学门，这是中国地质教育的开端。但当时人们对于何为地质几无认知，极少有人报考，地质学门遂于 1913 年第一届毕

---

1 参见王启龙编著：《钢和泰学术年谱简编》，北京：中华书局，2008 年；王启龙、邓小咏：《钢和泰学术评传》，北京：北京大学出版社，2009 年。

2 参见吴晓樵："洪涛生与中国古典戏曲的德译与搬演"，《德国研究》2013 年第 1 期，84—95 页。

业生后停办。时任工商部矿政司地质科长的丁文江因觉得训练地质人才和地质调查之重要，于 1913 年创立了地质调查所和地质研究所，均隶矿政司。其中地质研究所乃是假北京大学地质学门停办的机会，借用北大的校舍、图书和标本，甚至延聘了北大地质学门的德籍教授梭尔格。后从比利时留学回来的翁文灏则在研究所担任专任教授。丁文江一方面在研究所教古生物学，同时还肩负调查所之职前往正太铁路、西南地区进行了开创性的地质调查工作。地质研究所 1916 年共毕业学生 22 名，其中大多成为中国地质学的基石。[18-19]

　　1916 年后北大重新以丁文江的理念开办地质学系，毕业生则可供地质调查所用人之需。但早期的学生到地质调查所求职，面试的结果令丁文江很不满意。丁文江拿着许多带有零分的成绩单找到胡适，两人一同前往北大校长蔡元培处反映。蔡元培虚心听取了丁文江的意见，并委托他物色地质人才到北大任教。[20]1919 年丁文江等陪同梁启超赴欧参加巴黎和会，会后丁又前往美国，得以聘请葛利普来华工作；同时得知李四光刚从英国伯明翰大学地质学系获硕士学位，遂亦向蔡元培推荐。因葛利普和李四光的加入，北大地质学系才有了飞速的发展，不仅成为国内地质教育的翘楚，在世界上也颇为引人瞩目。

1.1　授课

　　虽然丁文江早在 1914 年即在地质研究所开设古生物学课程，成为中国第一个教古生物学的人。[21]但中国真正的古生物学教育，却是始于葛利普。1920 年 10 月底葛利普从美国抵达北京，《北京大学日刊》11 月 3 日即发布通告，他将于该日为地质学系三年级学生开设古生物学实验课，并于每周二、周三下午讲授高等地史学和地层学课程。[22]这是中国报刊最早介绍葛利普行程的记载。

　　随后，葛利普在北大开设了系统全面的古生物学课程。据 1926 年的北京大学外籍教员档案，葛利普在北大地质学系共开设过 8 门课程，每门课程每周 1—8 学时不等，计有：进化论（1 个学时）、高等地层学（2 个学时）、高等地层学实验（3 个学时）、古生物及标准化石（2 个学时）、古生物及标准化石实验（3 个学时）、地史学（2 个学时）、地史学实习（2 个学时）、中国古生物学实验（8 个学时），如图 1 所示。授课门数和课时与洪涛生相当，均属最多之列。[3]

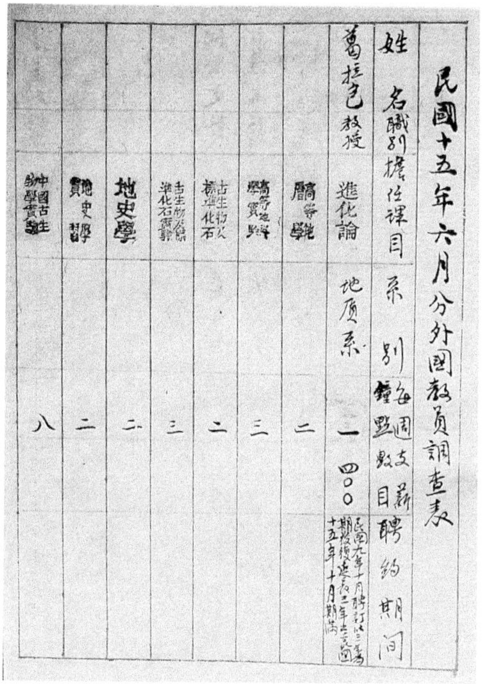

民國十五年六月分外國教員調查表

姓名	職別	擔任課目	系別	每週鐘點數	支薪目	聘約期間
葛利普	教授	進化論	地質系	一	四〇〇	[illegible]
		高等地層學		二		
		高等地層學實驗		三		
		古生物及標準化石		二		
		古生物及標準化石實驗		三		
		地史學		二		
		地史學實習		二		
		中國古生物學實驗		八		

图 1 葛利普开设课程（北京大学档案馆）

除进化论外，葛利普的这些课程在北京大学地质学系历年的课程指导书或课程表中均有反映，[1]因此进化论一科当指他 1920—1921 年在北大开设的系列演讲（详下）。按葛利普的设计，这些课程的进阶顺序如下：地史学及实习、古生物及实验、高等地层学及实验、中国古生物学。地史学及实习的内容为：地史概论、地史之分段（太古界、元古界、古生界、中生界、新生界、灵生界）、生物进化与地层年代之关系、地质图之用法、地层年代之鉴别法、中国地史概论，教材则为其所著《地质学教科书》（*Text Book of Geology*, part II）。古生物学及实验一科首先介绍化石之由来及其保护法、古今生物界之比较，接着讲授无脊椎动物化石、脊椎动物化石和植物化石，均涉及相应的分类、年代、鉴别和兴替。高等地层学及实验为一门更高阶的课程，在地层学概论的基础之上，重点讲授美洲、欧洲和亚洲（注重中国）古生界地层之对比，参考书为其所著《地层学原理》（*Principle of Stratigraphy*）。中国古生物学专论

1 如《国立北京大学地质学系课程指导书》（1923—1924 年度），载《北京大学日刊》，1923-09-28，第三、四版；《国立北京大学地质学系课程指导书》（1924—1925 年度），载《北京大学日刊》，1924-09-26，第二至六版；《国立北京大学地质学系课程》，载《北京大学日刊》，1929-09-23，第一、第二版；《地质系课程表》（1931—1932 年度），载《北京大学日刊》，1931-09-10，第四版。

中国各地质年代尤其是寒武纪和奥陶纪之化石，是一门专题研究课程，参考资料多采自《中国古生物志》和《中国地质学会志》上发表的论著。第四学年，还要指导古生物学门[1]的学生撰写古生物学论文[2]。除了葛利普之外，尚有孙云铸、赵亚曾、徐光熙、杨钟健等或协助或独立开设相应的古生物学课程。

图 2　葛利普在课堂上（《国立北京大学地质学会会刊》第 4 期，1930）

葛利普的课程很受学生欢迎（如他 1921 年的古生物学课有 50 名学生，地史学 40 名，比较地层学 60 名）[23]，并赢得了广泛赞誉，被认为可与欧美大学地质学系相比美（图 2）。如黄汲清说：

> 第三年的主课是中国地层学、古生物学、光性岩石学及岩石分析和中国矿床，分别由葛利普、李学清、谢家荣担任，中国地层学是按葛老师主编的 *Stratigraphy of China*（书名曾译为《中国地质史》）讲授，内容大半取材于中国，讲了两年，我们受益最多。葛老师是古生物学专家，教的古生物学也多取材于中国，而且有讲师孙云铸先生辅导，学生们真正学到好东西。
>
> ……

---

1 随着教师阵容的强大和开设课程的系统化，北大地质学系至少从 1923 年开始，即有专业化的设置：第一、二学年为基础课程，第三、四年则细分为矿物岩石学门、经济地质学门、古生物学门。古生物学独树一帜，而且往往成为其他学门课程的基础，这不能不说是葛利普的功劳。

2 参见《国立北京大学地质学系课程指导书》（1924—1925 年度），载《北京大学日刊》，1924-09-26，第二至六版，其中的少数课程名称与前述北京大学档案记载略有差别（如指导书中的古生物学及实验当为古生物及标准化石、古生物及标准化石实验），但并不影响其实质和内容。

　　　总起来讲，北大地质系水平可以和当时外国大学之地质系比美，特
别是葛利普的讲课最为突出。（[24]，28—29 页）

胡伯素亦有过生动的评论：

　　　综观上表，知北大地质系课程之完备，固足顾盼自豪，而各教授又
皆硕学鸿儒，一时上选，尤令人啧啧称道，至其经验宏富，教法优良，
更有足纪者：如葛利普先生胸藏万卷，每发为议论，必滔滔不绝，如长
江大河，一泻千里，如山洪暴发，溃堤决岸，莫可收拾，至得意之处且
眉飞色舞，声重如播鼓，此时学生子而不为此老引入胜景者，未之有
也。……[25]1

　　为利于教学，葛利普还编写了翔实的《北京大学理本科三年级古生物学》
《北京大学理本科三年级古生物学实习》《北京大学地质系三四年级高等古生物
学实习》等讲义，均为英文。翻阅这些厚重且已发黄的讲义（部分现仍存于中
国科学院南京地质古生物研究所），我们可感受到葛氏当年所付出的心血。

　　葛利普不仅教学有方，特殊情况下还将课堂搬到家中。1920—1930 年，
北平教育经费不稳定，欠薪时有发生，因而导致罢课、教员离校等情形，但
葛利普总是请学生到他家上课研讨。[26]黄汲清说："当有的时候，北平各大学
教授普遍欠薪罢课时，葛先生不罢课，还把学生带到自己家里上课。真是好
样的！"（[24]，78 页）

　　丁文江说，在北大为教潮罢课之际，作为一个外国人，葛利普非但没有
抱怨薪水拖欠，反而还请学生到家中去授课，其精神可感！[27]还说："他不但
是工作极勤，而且是热心教育青年的人。当北京大学屡次索薪罢课的时候，
他总是把地质系的学生叫到他家里去上课。他因为'风湿'病的原故，两腿
不能走动，手指也肿胀，然而他的工作比任何人要多。"[28]

　　蒋梦麟后来亦回忆道："这个外聘的洋教授（指葛利普）虽然近半年没拿
到薪水……可见到我不但没有怨言，还一个劲地催我快开课呢。"2

　　当然，除了葛利普之外，北京大学的地质学系还拥有何杰、王烈、丁文
江、李四光、王绍瀛、朱家骅等名家，也不断增添谭锡畴、谢家荣、孙云铸、

1　胡伯素此文据他发表于《国立北京大学卅一周年纪念刊》（1929 年，72—77 页）上的同名文章增改而成，
　关于地质学系各位老师的风格描写均为后来所加。
2　转引自陈军：《北大之父蔡元培》，北京：人民文学出版社，1999 年，436—437 页。蒋梦麟的这段话多有
　转述，但其原始出处不确，待考。

赵亚曾、何作霖、杨钟健、徐光熙、斯行健等新生力量，还从外面邀请翁文灏、钟观光、巴尔博（George B. Barbour，1890—1977）等来给学生开课。这使得北大地质学系在 20 世纪上半叶一直独占鳌头。[25]据章鸿钊统计，至 1936 年，全国各高校地质学系共有毕业生 264 人，北大即占 188 人。[29]

1.2　中国的阿加西

在 1921 年给美国的朋友巴斯勒（Ray Bassler，1878—1961）的一封长信中（图 3），葛利普详细叙述了他刚到中国一年所开展的地质调查、教育与研究活动，是了解他早期在华活动的难得史料。其中很大篇幅用来说明当时中国的地质学教育状况，并希望美国的科学家努力帮助训练中国的年轻学生，以使中国的地质学能有一个坚实的基础，并巩固中国与美国之间的友谊。信中说：

> 我们在大学的目标是培养更多的中国学生，以使他们能探寻这个幅员辽阔的国家的地质结构和历史，这些学生需要接受全面的训练。美国的地质学家和古生物学家能为此做出自己的贡献，中国人将对他们的所作所为心存感激，这同时也能极大地加强两个国家之间的友谊。中国人对美国和美国人极为钦慕，他们需要美国科学家的帮助以获得精神上的独立。尽管这里正遭受动荡和不安，但这个年轻的共和国有着伟大的未来，我们必须尽可能以最合适的方法来训练这些为未来而工作的学生。就我个人而言，我需要所有美国朋友们的同情、建议和帮助。从这些朋友的积极回应来看，我对此目标的实现怀有极大的信心。……[23]1

葛利普不仅是世界著名的地质学家，其境界之崇高和人格之伟大亦罕有人能及，信中我们还能深切感受到他对中国怀有的特殊情感。他常在不同场合强调科学是无国界的，并号召中国人要自强，努力发展科学。这与 1949 年后几十年间很多中国人所说的外国科学家形象大相径庭。[30]

在促进中国科学发展的信念下，葛利普不遗余力在北大开展古生物学教育，每周花去大量的时间。面对与美国不一样的学生和文化，他采取"阿加西法"（Agassiz method）来训练中国学生：

> 因中国学生崇尚权威，偏好书本，因此对他们的训练必须采取与美国学生不一样的方式，至少在最开始应该如此。阿加西法将是最有效的

1　Allan Mazur 最早发现并引用了该信，参见文献[14]，248—250 页。Evan Erickson 先生耐心帮助辨认书信，特致谢意。

方式。我给每位学生一些各种各样的标本，让他们将所有的书本知识抛诸脑后，分类出有共同特征的标本，并指出这些特征。尽管这种训练很耗时间，但确是最为成功的方式。……他们必须学会观察、推理，进而形成独立判断和依靠自我的能力，而不是过度诉诸权威。[23]

可见，葛利普对中国地质学教育不仅怀有满腔的热忱，同时亦有清醒的认识。所谓"阿加西法"，就是瑞士籍美国著名博物学家路易·阿加西（Louis Agassiz，1807—1873）创立的一种科学研究方法，即将学生置于实验室或野外，阅读"自然之书"，方法就是不带任何成见地观察、比较，从而得出结论。这个方法在当时的哈佛大学影响很大，[31] 19 世纪下半叶美国的很多科学家都在波士顿接受阿加西的训练。[32-33]

正是在葛利普的努力下，20 世纪上半叶中国尤其是北大地质学系在古生物学上取得了举世瞩目的成绩。中国早期的古生物学家及地质学家，大部分都受到他的直接影响，如孙云铸、赵亚曾、杨钟健、黄汲清、斯行健、计荣森、许杰、尹赞勋、赵金科、田奇瑪、乐森璕、俞建章、朱森、陈旭、张席禔、丁道衡、卢衍豪、王鸿祯等。秉志的《中国白垩纪之昆虫化石》、李四光《中国北部之蠖科》也都是在他的鼓励下完成的。[7]

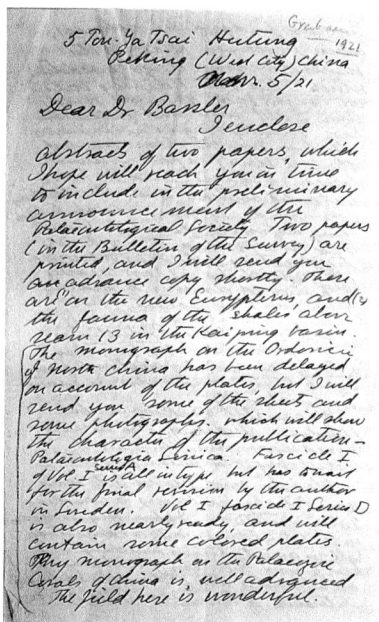

图 3　葛利普致 Ray Bassler 书信（史密森学会档案馆藏）

因此丁文江 1930 年这样评价葛利普：

> 作为一个教师，他的成功是显而易见的。尽管他讲课采用英语，学
> 生则基本都未留过洋，但他的课堂还是吸引了很多优秀的学生。他善于
> 启发式的教学、勤奋不辍的研究和清晰的讲授深深地赢得了学生的尊敬
> 和爱戴。在为《中国古生物志》撰写专论的 25 位中国学者中，有 19 位直
> 接受教于他，这雄辩地说明了他为中国地质学教育所作的卓越贡献。[6]

葛利普还认为要尽可能地选派优秀的年轻学子到欧美学习，但亦强调这
些学生需要在国内打下坚实的基础，否则难以达到理想的效果。[23]如 1923 年
他推荐杨钟健（北大地质学系当年毕业）前往德国慕尼黑大学，师从布罗里
（Ferdinand Broili，1874—1946）和施洛塞（Max Schlosser，1854—1932）；[34]1926
年，又推荐北大的年轻教师孙云铸赴德国哈勒大学，师从他的老朋友瓦尔特
（Johannes Walther，1860—1937）教授，[35]二人均于 1927 年获博士学位。也
许是葛利普认为欧洲的地质学实际训练要比美国的扎实（尽管美国的地质学
教育有更好的理论视野）[23]，他都将学生推荐到欧洲留学。

同为客居他国的科学家，葛利普与路易·阿加西的经历有着惊人的相似。
阿加西 1807 年生于瑞士，1829 年获慕尼黑大学和埃尔朗根大学哲学博士学位，
1830 年获慕尼黑大学医学博士学位。1832 年任新成立的瑞士纳沙泰尔
（Neuchâtel）学院教授。1846 年应邀前往美国，先在波士顿罗威尔学院（Lowell
Institute）任教，次年担任哈佛大学劳伦斯科学学院（Lawrence Scientific School）
教授，直至 1873 年逝世，在美国生活了 27 年。阿加西自认为是一名博物学家，
以现代科学的眼光来看，他主要在鱼类学、地质学、古生物学作出卓越贡献。
处于进化论和自然史研究的转折点，他属于传统描述式的博物学家，并极力
反对达尔文进化论。他在美国的教学和研究取得了巨大的成功：采取"阿加
西法"（事实上属于欧洲传统的方法）培养了美国第一、第二代博物学家，尤
为重要者如赖康忒（Joseph Le Conte，1823—1901）、斯廷普森（William
Stimpson，1832—1872）、斯卡德（Samuel Hubbard Scudder，1837—1911）、
海厄特（Alpheus Hyatt，1838—1902）、艾伦（Joel Asaph Allen，1838—1921）、
摩尔斯（Edward Sylvester Morse，1838—1925）、帕卡德（Alpheus Spring
Packard，1839—1905）、普特南（Frederick Ward Putnam，1839—1915）、维里
尔（Addison Emery Verrill，1839—1926）、谢勒（Nathaniel Southgate Shaler，
1841—1906）、韦尔德（Burt Green Wilder，1841—1925）、布鲁克斯（William

Keith Brooks，1848—1908）、古德（George Brown Goode，1851—1896）、约旦（David Starr Jordan，1851—1931）等，其子亚历山大·阿加西（Alexander Agassiz，1835—1910）亦成为著名的古生物学家。阿加西于 1859 年在哈佛大学建立比较动物学博物馆（Museum of Comparative Zoology），并使之成为集教育、研究、野外考察、出版为一体的研究机构；在生命的最后一年仍坚持自己的方式在彭尼基斯（Penikese）岛建立安德森博物学院，作为学生的暑期学校和海洋生物学研究站。（[32]，34-42 页；[36]）尽管他对理论的排斥以及对达尔文进化论的反对遭到不少后人的诟病，但这并不影响他对欧洲尤其美国科学的重要影响。[1]

从学术谱系而言，葛利普是在阿加西学生辈的影响下成长起来的，可以说是阿加西的继承者。葛利普对阿加西极为钦慕，常宣扬其事迹，[33]来华后信手拈来地以"阿加西法"训练中国学生。更为重要的是，以葛利普的生平和科学活动来看，他不仅在美国有重要影响，在华 26 年亦培养了中国第一、第二代的古生物学家，称其为"中国的阿加西"实当之无愧。当然，我们也应看到，与阿加西不同，葛利普对科学理论十分重视，[37]对进化论极力支持。或者说，阿加西尚只是传统的博物学家，而葛利普已是现代意义上的科学家了。这是科学发展的结果。

1.3 "研究教授"

1931 年，蒋梦麟被任命为北大校长，但他先因北大处境之艰难而婉拒。后在胡适、傅斯年、顾临（Roger S. Greene，1881—1947）等的筹措与帮助下，尤其是获得了中华教育文化基金会董事会（以下简称中基会）[2]的资助，蒋梦麟才重返北大，进行了卓有成效的改革，实现了北大的"中兴"。[20, 38]

胡适等的倡议在中基会 1931 年 1 月 9 日的第五次常会上获得通过，即同意中基会和北大 1931—1935 年每年提供国币 20 万元作为合作研究特款（即

---

1 参见 Christoph Irmscher, *Louis Agassiz: Creator of American Science*, Boston/New York: Houghton Mifflin Harcourt, 2013.

2 中华教育文化基金董事会（亦称中华教育文化基金会，China Foundation for the Promotion of Education and Culture）乃是利用美国第二次退还的庚款余数（本息 1200 万余美金，但因战争等因素，中基会所收款项，远低此数）设立的一个文教机构，1924 年 9 月于北京成立，宗旨为"促进中国教育及文化事业"。自成立起，中基会通过设立科学教授、研究教授、奖助学金，资助大学、科研机构，建立京师图书馆、静生生物调查所等，为中国的教育、文化事业做出了很大的贡献。1949 年，中基会从香港迁往纽约，1972 年迁至台北。参见杨翠华：《中基会对科学的赞助》，台北："中央研究院"近代史研究所，1991 年；赵慧芝："中基会和中国近现代科学"，《中国科技史料》1993 年第 3 期，68—82 页；左玉河："二三十年代'中基会'对中国学术研究之资助"，《扬州大学学报》（人文社会科学版）2012 年第 3 期，81—87 页。

5 年中基会和北大各 100 万，一共 200 万元）[1]，用于：（一）设立北京大学研究教授；（二）扩大北大图书仪器及其他相关设备；（三）设立北大助学金及奖学金。[39]

此一合作特款直至 1937 年方告终止，设立北京大学研究教授便是其中的重要举措。研究教授最多限 35 名，"以对于所治学术有所贡献见于著述为标准，经顾问委员会审定，由北大校长聘任。"年薪自 4800 元至 7200 元不等，此外每一教授每年应有 1500 元以内之设备费。研究教授每周授课至少六小时，并需担任学术研究及指导学生的工作。[39]此外，值得一提的是，北大的图书馆和地质馆亦是在这个合作特款的支持下于 1934 年建成的。[40]

1931—1932 年度首次聘任了 16 席研究教授：汤用彤（哲学）、陈受颐（史学）、周作人（文学）、刘复（文学）、徐志摩（西洋文学）、冯祖荀（数学）、王守竞（物理）、刘树杞（化学）、曾昭抡（化学）、许骧（植物）、汪敬熙（心理）、丁文江（地质）、李四光（地质）、赵迺抟（经济）、刘志敿（法律）、葛利普（古生物）。[41][2]

此后直至 1937 年共聘任了 6 次研究教授，[3]每次的聘任均会有所变化，最多的为 1932—1933 年度的 22 人。葛利普是极少数一直获得聘任的，这表明了他的学术成就及其对北大的贡献。

事实上，稍早于北京大学的研究教授制度，中基会自 1930 年起即特设"科学研究教授席"，在全国范围内，聘任著名学者。此举主要是配合中基会对各研究机构与大学的补助而设置的。由于经费有限，且中基会秉持宁缺毋滥的原则，故 1930 年以来获任该研究教席的仅有 8 人：[4]

翁文灏：1930—1933，地质学，地质调查所；

李济：1930—1948，考古学，中央研究院历史语言研究所；

秉志：1932—1948，动物学，静生生物调查所与中国科学社生物研究所；

庄长恭：1935—1946，化学，中央研究院化学研究所；

---

1 中基会 100 万元最后于 1936 年 8 月如数拨清，见《中华教育文化基金董事会第十二次报告（1936 年 7 月至 1937 年 6 月）》，1937 年 12 月，页 15b—16b；杨翠华：《中基会对科学的赞助》，台北："中央研究院"近代史研究所，1991 年，97 页。

2 胡适在《丁文江的传记》中记录了 15 名研究教授，其中独缺唯一的外国人葛利普，有误。参见文献[20]，113 页。

3 北京大学 1933 年和 1934 年分别出版第一次和第二次研究教授工作报告。感谢郭金海研究员提示此材料。

4 中华教育文化基金董事会历次报告。但一般误认为仅有 7 人，缺秦大钧，如杨翠华：《中基会对科学的赞助》，180—184 页；胡宗刚："关于中基会——档案中的历史"，《东方文化》2003 年 6 期，78—85 页。

陈焕镛：1935—1940，植物学，中山大学农林研究所；

葛利普：1938—1946，古生物学，地质调查所；

秦大钧：1939—1940，航空动力学，航空研究所；

胡先骕：1946—1948，植物学，静生生物调查所。

我们可注意到，北京大学的研究教授制度 1937 年结束，葛利普旋于 1938 年以地质调查所研究人员的身份被聘为为数甚少的中基会"科学研究教授"，直至 1946 年逝世。1937 年后北京大学南迁昆明，在此期间，因腿疾滞留在北京的葛利普拒绝与伪北大合作，薪水便全由中基会支付（虽然 1941 年以后可能因战争也无法兑现了）。抗战结束，中基会迅速接济葛利普的生活，葛氏 1946 年逝世后，中基会则协助处理其善后事宜。[42]

葛利普 1920 年来华即任北京大学地质学系教授，1931—1937 年一直被聘为北京大学的研究教授，1938 年后又被聘为中基会的科学研究教授。这种特有的情形，再以他一个外国人的身份，就更显罕见了。回想 1919 年他被哥伦比亚大学解聘，后又申请美国自然史博物馆的职位未果，[1]以此巨大的反差，或许可以理解他对中国所怀有的特殊情感。

1931 年被聘为北京大学研究教授，尤其是 1933 年自美国第 16 届国际地质大会返回中国后，葛利普除发表可观的古生物学著作之外，便将工作重心放在了更受其重视的脉动和极控理论，终于 1940 年以《年代的节律：从脉动理论和极控理论看地球的历史》（*The Rhythm of the Ages: Earth History in the Light of the Pulsation and Polar Control Theories*）一书进行系统的总结。[37]太平洋战争爆发后，他被囚禁于东交民巷集中营，期间不顾身体每况愈下，犹笔耕不辍，完成最后的著作《我们居住的世界：地球历史新论》（*The World We Live In: A New Interpretation of Earth History*），进一步申论其脉动和极控理论。[2]因此，他担任北大和中基会研究教授期间，最重要的研究工作便是其脉动和极控理论，或者可以说，他在生命的最后 15 年间，以北大和中基会研究教授的名义，完成了他一生的理论总结。

2　传播进化论和遗传学

2.1　"持续"一年的演讲

民国以后，因获取新知需求之高涨，西方学者来华演说络绎不绝，其中

1　此需另文讨论。

2　此书后由阮维周整理于 1961 年在台湾出版。

尤以杜威、罗素、泰戈尔最为引人瞩目，并产生了深远的影响。而从科学的角度而言，影响最大的当属葛利普来华不久开设的一个关于"地球与其生物之进化"的系列演讲。其所讲内容之丰富和前沿，引起反响之巨大，以致《申报》在 1921 年元旦曾有这样记载："在北京之外国名人讲演，以罗素、葛利普为最著。"[43]

罗素于 1920 年 10 月 12 日抵达上海，开始了他在中国近一年的巡回访问与演讲，至 1921 年 7 月 11 日返回英国，期间的访问与演说由赵元任担任翻译。[44-45]葛利普于 1920 年 10 月底抵达北京，随即于 1920 年 12 月至 1921 年 12 月举行了长达 16 次的系列演讲。当时的《北京大学日刊》等报刊上经常能同时看到两人的演讲公告或演讲录，可见到他们的演讲在那时风行之情形。

1920 年 11 月 17 日，《北京大学日刊》刊登葛利普将于 12 月起举行的关于地球及生物之进化的演讲，分作 12 次，并列出每次演讲的题目和纲要。[46]演讲始于 12 月 5 日，中间因 1921 年 3—10 月的教潮[1]和 10—11 月译者龚安庆出差，至 1921 年 12 月 19 日方告结束，正好一年稍余，演讲内容亦从 12 次扩为 16 次（表 1）。

表 1　葛利普演讲（1920—1921）

讲次	日期	地点	讲题	翻译	笔记
1	1920-12-05（星期日 14:00）	北大第三院第二教室	地球及生物之原始	王烈	赵国宾
2	1920-12-12（星期日 14:00）	北大第二院大讲堂	生物记载之造作及保存	王烈	赵国宾
3	暂未考	暂未考	地史上古生代主要之生物	王烈	赵国宾
4	1921-01-06（星期四 14:00）	北大第二院第一教室	陆上生物的初期	王烈	赵国宾
5	1921-01-23（星期日 14:00）	北大第二院第一教室	古生物界之大革命（古生代末期）[2]	王烈	赵国宾
6	1921-02-06（星期日 14:00）	北大第二院第一教室	中生代的异点和菊石的进化	王烈	赵国宾

1　1921 年 3 月，因北京国立八校（北大、高师、女高师、法政专门学校、医学专门学校、工业专门学校、农业专门学校、美术专门学校）索薪罢教而导致的教潮事件，演化成教职员被总统府守卫军殴打的"六三事件"，经多方长期交涉，至是年 10 月方才获得一定的解决。参见任伟："异心协力：索薪运动中之民国教员群像——以 1921 年国立八校索薪运动为中心"，《史林》2012 年第 3 期，149—160 页。葛利普演讲至第十次，即因教潮而告停止，至 1921 年 10 月恢复。

2　演讲录载《北京大学日刊》，1921-01-28、1921-02-01、1921-02-02、1921-02-03，并改名为"地史上的世界大革命"。

续表

讲次	日期	地点	讲题	翻译	笔记
7	1921-02-19 (星期六 14:00) [1]	北大第二院 第一教室	中生界之伟大动物	谭熙鸿	赵国宾[2]
8	暂未考	暂未考	中生界动物的衰灭和其紧要的关系	李四光	赵国宾
9	1921-03-06 (星期日 14:00)	北大第二院 大讲堂	近生代	李四光	赵国宾
10	1921-03-13 (星期日 14:00)	北大第二院 大讲堂	人类自然史	龚安庆[3]	田奇瑰[4]
11	1921-10-16 (星期日 14:00)	北大第二院 大讲堂	达尔文的学说	龚安庆	赵亚曾[5]
12	1921-10-23 (星期日 14:00)	北大第二院 大讲堂	天然的变异	龚安庆	赵国宾
13	1921-11-27 (星期日 14:00)	北大第二院 大讲堂	天然的变异	龚安庆	赵国宾
14	1921-12-04 (星期日 14:00)	北大第二院 大讲堂	环境之关系	龚安庆	赵国宾
15	1921-12-11 (星期日 14:00)	北大第二院 大讲堂	适应环境	龚安庆	赵国宾
16	1921-12-19 (星期一 16:00)	北大第二院 大讲堂	遗传性及结论	龚安庆	赵国宾

注:此表据《北京大学日刊》《晨报副镌》刊载关于葛利普演讲的信息整理。每一次演讲,往往会在《北京大学日刊》上作多次通告。除极少数因故调整外,演讲大部分都安排在星期日的下午。《北京大学日刊》在 1921 年 10 月多期上发表通告,表示葛氏之演讲将继续,且为使进化论的理论和事实讲解得更清楚,将余下的三讲扩为六讲,这样,整个演讲即为 16 次。

通过表 1,我们可想见当时演讲之情形。葛利普的这个系列演讲乃是由众多学者合作完成的结果,翻译、记录者中既有王烈、李四光、谭熙鸿、龚安庆等资深学者,亦有赵国宾、赵亚曾、田奇瑰这样的青年才俊。大部分演讲录曾在《北京大学日刊》《晨报副镌》上刊登过,最后经赵国宾、杨钟健整理成《地球与其生物之进化》一书。上文所述的葛利普在北大所开"进化论"课程很可能指的就是这个系列演讲。

1 《北京大学日刊》,1921-02-18,第一版,所载日期为星期日,疑误。
2 翻译者和笔记者见《北京大学日刊》,1921-03-01,第三版。
3 龚安庆,字展虞,安徽合肥人,英国剑桥大学及美国俄勒冈大学毕业,获文艺硕士及理科硕士学位。1920 年在北京大学地质学系任教授,讲授古生物学及实习、动物学、高等动物学、植物学、动植物实验等课程。后任外交官。参见《创立·建设·发展:北京大学地质学系百年历程(1909—2009)》,11 页。
4 笔记者为季瑜,即田奇瑰(字季瑜),见《晨报副镌》,1921-10-24 至 1921-11-05,第一版。
5 笔记者为予仁,即赵亚曾(字予仁),见《晨报副镌》,1921-10-21,第二版。

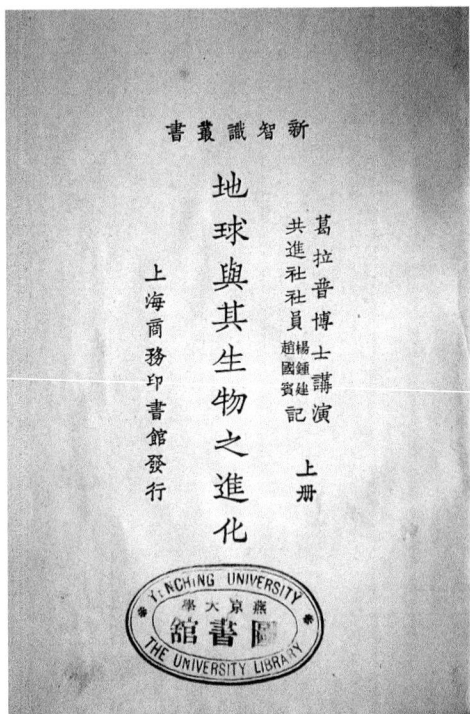

图 4 《地球与其生物之进化》扉页（1924）

关于《地球与其生物之进化》一书，上海泰东图书局先有前七讲的稿子，因 1921 年教潮未获新稿，误以为已讲完，遂于 1921 年出版前七讲的书稿。后北大新知书社刊行了全稿，并加入一些图像。但据赵国宾所说，由于他当时未在北京，疏于校阅，错误较多。有鉴于此，他后对全稿作了重新审阅，并配以葛利普《地质学教科书》（*Text Book of Geology*）和《历史地质学》（*Historical Geology*）两书中图片 40 余幅，1924 年由上海商务印书馆刊行（图 4），是为该讲稿的定本。[47]

此外，葛利普还多次在北京大学之外作过进化论的演讲。如 1922 年 2 月 12 日，适值达尔文诞辰 113 周年暨"南北统一纪念日"，北京高等师范学校博物学会召开达尔文诞辰 113 纪念大会，下午即有葛利普关于《达尔文的天然淘汰说》的演讲。[48]1 1924 年 7 月 1—5 日，中国科学社在南京举行第九次年会。葛利普赴南京参会，并于 3 日晚演讲进化论，历时三小时，引起与会者极大兴趣。4 日下午翁文灏推荐葛利普为中国科学社特社员，全场一致通过。[49]

1　另有北京高等师范学校博物学会主任黄以仁讲"徐变与疾病"，以及林长民讲"恋爱与婚姻"。

2.2 进化论和遗传学的传播

自严复所译《天演论》1898 年正式出版后[1]，进化论便迅速风行于世，成为近代国人的坚实信仰。然而，严复的译介主要着眼于进化论对社会的影响，而关于进化论的理论本身并未太多涉及。

较早介绍生物进化论的是 1903 年国民丛书社的《动物进化论》，虽内容已比较系统，[2]但似乎影响不大。1907 年，鲁迅译撰《人间之历史——德国黑格尔氏人类起源及系统即种族发生学之一元絜究诠解》[3]一文，介绍了海克尔（Ernst Haeckel, 1834—1919）的进化论学说，[50]但或是因在日本刊行，影响较为有限。作为"中国赫胥黎"的丁文江在其 1914 年编纂的《动物学教科书》中对进化论和魏斯曼的遗传学亦有介绍，但极为简略。1919 年，马君武翻译出版《达尔文物种原始》是进化论在中国传播的重要里程碑。[51]此后尤其是 1930 年后，关于进化论的译述层出不穷。关于遗传学，最早系统介绍现代遗传学及染色体知识的当属李积新编著的《遗传学》（商务印书馆，1923）和陈桢编译的《普通生物学》（商务印书馆，1924）。[52-54]此外，《科学》《民铎》等杂志对生物进化论和遗传学亦陆续有介绍。[51]

葛利普的系列演讲在生物进化论和遗传学的传播中起到了承前启后的作用。从接受者的角度而言，1920 年前后，对生物进化论和遗传学进行系统阐述的当属这个演讲。葛利普对进化论的历史和原理极为熟稔，而且还结合了古生物学、遗传学理论进行讲论，因此他的演讲既是以进化论的观点来解释古生物发展史和遗传学原理，亦可说是以古生物学和遗传学例子来理解进化论，是三者的有机结合。此外，葛利普虽以古生物学见长，但其着眼点却是整个地球的历史，终其一生均以此为念。[37]这在当时进化论思想的传播中是别具一格的。

通观葛利普的 16 次演讲，前 10 次为地球和各地质时代生物演化的历史，即古生物学部分，最后 6 讲集中阐述进化论和遗传学，分别为：生物进化的问题和原则、天然的变异（上下）、环境与适宜性（上下）、遗传性。[55]

在"生物进化的问题和原则"中，葛利普以一些浅显的例子，阐述了"天择"（Natural Selection）、"人择"（Artificial Selection）和"性择"（Sexual

---

1　关于进化论在此之前的传播，参见马自毅："进化论在中国的早期传播与影响——19 世纪 70 年代至 1898年"，《中国文化研究集刊》第 5 辑，上海：复旦大学出版社，1987 年。

2　[英]达尔文创义，[美]摩尔斯口述，[日]石川千代松笔记，国民丛书社重译：《动物进化论》，上海：国民丛书社，1903 年。

3　此文 1926 年被收入《坟》的第一篇，题名改为《人之历史——德国黑格尔氏种族发生学之一元研究诠解》。

Selection）三个概念及其相互关系。

在"天然的变异"中，他指出变异有两种：先天的和后天的，前者由前代遗传继承，后者由环境变化引起。"不平等是所有生物特有的一个性质"，有不平等才有竞争，有竞争才有进化；变化的地方越多，竞争愈盛，进化愈快。他以腹足类动物化石为例说明变异是有方向的，即正系的变异（orthogenetic variation），并指出达尔文派认为变异的偶然性是错误的，因为他们只注意了现在的生物，并未留心古代的生物；但变异的动因仍不清楚。

在"环境与适宜性"中，他主要讲了动植物所依赖的有机环境和无机环境，以赫胥黎实验和澳洲兔子增长为例，说明动植物与环境形成相互影响的食物链的关系，最后指出适者生存是铁的自然律，聪明的人类则能通过认识自然律以谋生存，主动面对环境的局限。

最后一讲综合魏斯曼（August Weismann，1834—1914）学说，孟德尔（Gregor Mendel，1822—1884）定律，荷兰德弗里斯（Hugo de Vries，1848—1935）、德国柯伦斯（Carl Correns，1864—1933）、奥地利柴马克（Erich von Tschermark，1871—1962）等的实验，以及摩尔根（Thomas Hunt Morgan，1866—1945）理论，集中讲述遗传学的历史和理论；并通过古生物学如头足类不同时期化石的研究，指出获得性遗传的正确性。

葛利普的演讲旁征博引，对各派学说如数家珍，娓娓道来，尤其与摩尔根有着相似的学科背景，又同在哥伦比亚大学同事多年，对他的研究应该十分熟悉。为照顾中国的听众，他常举中国的例子加以说明。他特别强调进化学说并非达尔文的发明，达尔文只不过是对进化论的方法和理论加了事实上的解释和理性的概括。[1]在其演讲中，他花了不少篇幅讲解遗传学，既有历史的叙述，更注重当时刚获得发展的新遗传学知识，多次提到了染色体（Chromosome）[2]的重要性。

---

[1] 葛利普曾在 1920 年的另一个演讲中指出："天择不是一种天然的力量，天择不是一种实体（Entity），天择不能做什么，天择不过是一种方法或是历程（Process），在天然选择的方法或历程中，最适宜于生存竞争者，就可以保存，而可以逃出一种破坏的力量，而他们不适于抵挡此种破坏的力量者，就不能生存了！这就叫做天择律。"见葛拉包著，斯行健译："生物进化的误解"，《现代青年》（广州），1920 年，第 31—35 期连载；又见《自然科学》，1928 年，1 卷 2 期，157—173 页。

[2] 关于染色体的概念，章炳麟早在 1900 年左右的《菌说》一文中即以"染色物"之名进行引介，后在《訄书》中有过进一步的讨论。参见蒋功成："章炳麟与西方遗传学说在近代中国的传播"，《自然辩证法通讯》2009 年第 8 期，86—90 页。此后，直至 1919 年陈寿凡在其编译的《人种改良学》中方有系统的介绍。1923 年冯肇传在《遗传学名词之商榷》一文中对当时传入的遗传学名词作了校订，其中 Chromosome 被译为"染色质体"。"染色体"之名后获得采用。葛利普 1921 年演讲时，此名的翻译尚处于不确定时期，因此在其演讲录中，未具中文译名。

民国以来关于进化论的著述首以此书最为全面。正如赵国宾所说，此书"在自然科学上，在中国的自然科学界，自然是占有很重要的位置。"他还借此机会对其同乡陕西人民提出了严厉的批判，认为生活在 20 世纪的陕西人，不应再沉迷于孔孟之道，而都应该看看此书，以获得正确的科学观念，"不然便非堕落到十八层地狱不可！"[47] 从演讲的角度看，因有临场感和互动的效果，葛利普的系列演讲从科学的意义上对进化论的澄清和普及无疑都起到不可忽视的作用。然而，从成书的角度而言，因有些部分翻译起来较有难度，且口语化明显，这对其传播和影响应该会有一些折扣。

3　结语

章鸿钊 1930 年在葛利普六十寿辰时赞誉其"东西地史因君重，灿烂勋名奕叶期"，表明了葛氏在中西地质学史上的卓越地位；同时亦以"他乡桃李发新枝"表达了葛氏对中国地质学教育的独特贡献。自 1920 年来华，葛利普在北大地质学系所开设的系统而全面的古生物学课程，与当时欧美地质学系相当；他因材施教对中国学生采取"阿加西法"训练，使学生深受影响；学生有相当基础后，他又积极推荐到国外留学。1937 年，北京大学南迁，葛利普因腿疾，未能前往，滞留北京。自 1920 年来华至 1937 年，除在地质调查进行研究外，他在北京大学的讲台上耕耘了 17 年。[1] 若以孙云铸、杨钟健 1928 年留学归国任北京大学地质学系教授算起，葛利普以其渊博的学识和崇高的精神在北大培养了中国第一代（大致以 1917—1927 年在学为界，如孙云铸、杨钟健、赵亚曾、尹赞勋、田奇㻪、张席禔、乐森璕、俞建章、陈旭、许杰、徐光熙、斯行健、黄汲清、朱森、计荣森等）和第二代（大致以 1928—1937 年在学为界，如高振西、赵金科、王钰、崔克信、阮维周、卢衍豪、王鸿祯等）古生物学家，奠定了中国的古生物学基础，堪称"中国古生物学之父"。他所扮演的角色和所作的独特贡献与路易·阿加西之于美国科学十分相似，因此亦可称为"中国的阿加西"。自 1931 年起，他相继连续被聘为北京大学和中基会研究教授，直至 1946 年逝世，这在当时中国科学界绝无仅有的情形，反映了其崇高的学术贡献和地位。

除了讲授系统的古生物学课程，葛利普还于 1920—1921 年在北大开设了"地球与其生物之进化"的系列演讲，不仅从地球历史的角度提供了古生物学

1　据北大地质学系 1935 级学生王鸿祯回忆，葛利普 1937 年犹在地史课上，讲其脉动和极控理论，"滔滔雄辩，令人心折"，"妙绪泉涌，引人入胜"。参见文献[13]。

的一个概貌，而且以古生物发展演化为背景，系统介绍了当时最新的生物进化论和遗传学理论，是关于古生物学、进化论和遗传学在中国的最新综合传播。这对当时古生物学尚属启蒙、进化论和遗传学亦只有浅显介绍的中国来说，无异于一场知识盛宴，在新文化运动中代表了科学的一面，其影响与罗素在中国的文化之旅相比肩。

致　谢　2014 年笔者赴史密森学会（Smithsonian Institution）查阅档案期间，承 Pamala Henson 博士给予热情帮助，搜集到葛利普的一些书信；韩琦研究员、张九辰研究员、郭金海研究员审阅本文初稿并提出宝贵建议；本文初稿曾在第 5 届"北京大学与中国现代科学"学术研讨会（北京大学，2015 年12 月 13 日）报告。谨此一并致谢。

参 考 文 献

[1] 致教育部送本校外国教员一览表[A]. 北京大学档案：BD1924006.

[2] 民国十四年六月国立北京大学职员表[A]. 北京大学档案：MC192503.

[3] 外国教员调查表[A]. 北京大学档案：BD1926002.

[4] 汪向荣. 日本教习[M]. 北京：商务印书馆，2013：77-78.

[5] 巴斯蒂. 京师大学堂的科学教育[J]. 历史研究，1998，（5）：47-55.

[6] Ting V K. Biographical Note [J]. *Bulletin of the Geological Society of China*, 1931, 10: iii-xviii.

[7] Sun Y C. Professor Amadeus William Grabau: Biographical Note[J]. *Bulletin of the Geological Society of China*, 1947, 27: 1-26.

[8] Shimer H W. Memorial to Amadeus William Grabau[J]. *Proceedings Volume of the Geological Society of America Annual Report for 1946*, 1947: 155-166.

[9] Kay M. Grabau, Amadeus William[C]//Gillispie C C. *Dictionary of Scientific Biography*, Vol. 5. New York: Charles Scribner's Sons, 1981: 486-488.

[10] 潘云唐. 葛利普——中国地质科学工作者的良师益友[J]. 中国科技史料，1982，（3）：22-30.

[11] 杨静一. 葛利普传略[J]. 自然科学史研究，1984，3（1）：83-89.

[12] 杨翠华. 历史地质学在中国的发展（1912—1937）[J]. "中央研究院"近代史研究所集刊，1986，（15）上：319-334.

[13] 王鸿祯. 葛利普教授——中国地质学界的良师益友[C]//王鸿祯. 中国地质事业早期史. 北京：北京大学出版社，1990：81-93.

[14] Mazur A. *A Romance in Natural History: The Lives and Works of Amadeus Grabau and Mary Antin*[M]. Garret: Syracuse, 2004.

[15] 孙承晟. 葛利普与北京博物学会[J]. 自然科学史研究，2015，34（2）：182-200.

[16] Chang H T. Chinese Poem[J]. *Bulletin of the Geological Society of China*, 1931, 10: 卷首.

[17] 翁文灏. 悼地质学大师葛利普先生[N]. 重庆大公报，1946-03-28：2.

[18] 李学通. 农商部地质研究所始末考[J]. 中国科技史料，2001，（2）：139-144.

[19] 李学通. 地质调查所沿革诸问题考[J]. 中国科技史料, 2003, （4）: 351-358.

[20] 胡适. 丁文江的传记[M]. 北京: 北京师范大学出版社, 2014: 22-26.

[21] 章鸿钊. 我对于丁在君先生的回忆[J]. 地质论评, 1936, （3）: 227-236.

[22] 佚名. 第二院注册通告[N]. 北京大学日刊, 1920-11-03: 2.

[23] Grabau to Bassler[A]. Smithsonian Institution Archives, Capital Gallery, 007234-Box 3.

[24] 黄汲清. 我的回忆: 黄汲清回忆录摘编[M]. 北京: 地质出版社, 2004.

[25] 胡伯素. 北京大学之地质系[J]. 国立北京大学地质学会会刊, 1930, （4）: 165-172.

[26] 孙云铸. 葛利普教授[J]. 科学, 1948, 30（3）: 70-72.

[27] Chang H T, et al. Letter from the Council[J]. *Bulletin of the Geological Society of China*, 1931, 10: I.

[28] 丁文江. 苏俄旅行记（1934—1935）[C]//欧阳哲生. 丁文江文集（第七卷）. 长沙: 湖南教育出版社, 2008: 108.

[29] 章鸿钊. 中国地质学发展小史[M]. 上海: 商务印书馆, 1937: 40-41.

[30] 张九辰. 科学史事的时代解读: 对中国地质学史的案例分析[J]. 自然科学史研究, 2015, 34（1）: 74-87.

[31] Bell I F A. Divine Patterns: Louis Agassiz and American Men of Letters. Some Preliminary Explorations [J]. *Journal of American Studies*, 1976, 10 (3): 349-381.

[32] Winsor M P. *Reading the Shape of Nature: Comparative Zoology at the Agassiz Museum*[M]. Chicago: The University of Chicago Press, 1991.

[33] 葛利普. 中国科学的前途[J]. 任鸿隽, 译. 科学, 1930, （6）: 759-777.

[34] 杨钟健. 杨钟健回忆录[M]. 北京: 地质出版社, 1983: 30-39.

[35] 王鸿祯. 中国地层古生物学奠基人孙云铸教授[J]. 中国地质教育, 1995, （3）: 9-14.

[36] Lurie E. Agassiz, Jean Louis Rodolphe[C]//Gillispie C C. *Dictionary of Scientific Biography*, Vol. 1. New York: Charles Scribner's Sons, 1981: 72-74.

[37] 孙承晟. 海进海退和大陆漂移之地球"沧桑"史——葛利普的脉动和极控理论[J]. 自然科学史研究, 2015, 34（4）: 470-486.

[38] 杨翠华. 蒋梦麟与北京大学（1930—1937）[J]. "中央研究院"近代史研究所集刊, 1988, （17）下: 261-305.

[39] 北京大学与中华教育文化基金董事会合作研究特款办法[C]//中华教育文化基金董事会第六次报告, 1931 年 12 月: 51a-52b.

[40] 中华教育文化基金董事会第十次报告[C]. 1935 年 12 月: 18b-19b.

[41] 中华教育文化基金董事会第七次报告[C]. 1932 年 11 月: 42a-43a.

[42] 中华教育文化基金董事会第十六次报告（1940 年 7 月至 1946 年 12 月）[C]. 1947 年 12 月: 5a-b.

[43] 北京通信[N]. 申报. 1921-01-01: 6.

[44] 陈满华. 1920—1921: 作为罗素译员的赵元任[J]. 中华读书报. 2013-11-06: 7.

[45] 裴毅然. 罗素首次访华细节: 由梁启超邀请并筹措经费[J]. 新民晚报. 2012-05-16.

[46] 本校新闻[N]. 北京大学日刊. 1920-11-17: 2.

[47] 赵国宾. 笔记者言[C]//葛利普演讲, 赵国宾, 杨钟健笔记. 地球与其生物之进化（上）. 上海: 商务印书馆, 1924: 1-2.

[48] 野云. 纪北京高师达尔文百十三周纪念会[N]. 申报. 1922-02-15: 7.

[49] 中国科学社年会纪[N]. 申报. 1924-07-07: 10.

[50] 陈福康. 《人之历史》的再认识——兼述评日本中岛长文先生对鲁迅此文的研究[J]. 东北师大学报, 1984, （4）: 54-60.

[51] 李楠. 生物进化论在中国的传播（1873—1937）[D]. 西安: 西北大学博士学位论文, 2012.

[52] 付雷. 现代遗传学知识在近代中国的传播——中学生物教科书的视角[J]. 中国科技史杂志，2014，（2）：147-157.

[53] 曹育. 孟德尔遗传学是怎么传入我国的[J]. 中国科技史料，1988，（1）：89-91.

[54] 冯永康. 20世纪上半叶中国遗传学发展大事记[J]. 中国科技史料，2000，（2）：175-185.

[55] 葛利普演讲，赵国宾、杨钟健笔记. 地球与其生物之进化（下）[M]. 上海：商务印书馆，1924：75-164.

The Father of China's Paleontology: Amadeus W. Grabau and the Department of Geology of Peking University

SUN Chengsheng

Abstract   Invited by V. K. Ting, Amadeus W. Grabau (1870–1946) came to China from America in 1920, and acted concurrently as professor at Peking University (hereafter PKU) and chief paleontologist of the Geological Survey of China. Based on Chinese and Western archives and journals of the Republican period, this essay investigates Grabau's teaching activities and lectures at PKU, and his appointments as "Research Professor" of PKU and China Foundation for the Promotion of Education and Culture from 1931–1946. With his outstanding contributions to paleontological education and research in China, Grabau can be called the "Father of China's Paleontology". His role was very similar to that of Louis Agassiz in American science in the 19th century, so he can also be regarded as the "Chinese Agassiz". In addition, through his lecture series entitled "The Earth and the Evolution of Its Creatures" in 1920–1921, Grabau systematically introduced the most up-to-date paleontology, including the Theory of Evolution and genetics, which represented a scientific side of the New Cultural Movement at the time.

Keywords   Amadeus W. Grabau; Peking University; China Foundation for the Promotion of Education and Culture; geology; paleontology; Theory of Evolution; genetics

——原载《自然科学史研究》2016年第3期